K.S. Masonde

Principles of
AUTOMATIC CONTROL

To Vincent and Bill

Principles
of
AUTOMATIC
CONTROL

MARTIN HEALEY, Ph.D., M.Sc., C.Eng., M.I.E.E.
Professor of Microprocessor Engineering
Department of Electrical and Electronic Engineering
University College, Cardiff

HODDER AND STOUGHTON
LONDON SYDNEY AUCKLAND TORONTO

ISBN 0 340 17671 7

First printed 1967
Second edition 1970   Reprinted 1971
Third edition 1975   Reprinted 1977, 1979, 1980

Printed in Great Britain for
Hodder and Stoughton Educational,
a division of Hodder and Stoughton Ltd,
Mill Road, Dunton Green, Sevenoaks, Kent, by
Biddles Ltd, Guildford, Surrey

General Editor's Foreword

PROFESSOR J. H. CALDERWOOD

M.Eng., Ph.D., C.Eng., F.I.E.E., F.Inst.P.

Chairman of Electrical Engineering Department
University of Salford

Control Engineering has a strong claim to be recognised as a primary technology, equal in status to Electrical Engineering, Mechanical Engineering and Civil Engineering.

It is built on a substantial, though recently developed, theoretical framework, and its future development promises to be the most exciting of all the technologies. Although it will inevitably continue to be closely linked with Electrical Engineering because of a common interest in frequency analysis and because Electrical Engineering will nearly always offer the most suitable means of putting control into practice, its field of application is extremely wide, overlapping practically every other technology. It is, however, important not to allow this wide field of application to have too much influence in determining the discipline and body of knowledge to be incorporated in an undergraduate course in the new technology. Control Engineering is not an amalgam of portions of Electrical, Mechanical, Hydraulic, Production, and Chemical Engineering; its raw material is not sometimes an electrical property, sometimes a mechanical one. The raw material of Control Engineering is information, and Control Engineering is Information Engineering.

Most technologies have developed from the application of a branch of physics or chemistry. This is not so with Control Engineering. Physics for once is a tool and not the basis of the technology; here lies a possible danger to its healthy growth. Without a fundamental science as a parent subject there is a danger that control engineering will develop in a fragmented form, each fragment tailored to suit the technology to which it is applied. Recognition of Control Engineering as a primary technology, and the consequent introduction of undergraduate courses of study will do a lot to meet this danger, provided the fragmentation is not built into the courses. It is hoped that this book will help to indicate a coherent approach to the subject.

"The beauty of mathematics lies in the fact that it furnishes a universal language which characterizes in a uniform way processes of quite different physical appearance."

RICHARD BELLMAN

Preface

The textbook presents the reader with the fundamentals for the analysis and design of automatic, or closed loop, control systems. It is a valid addition to the long list of control textbooks as it is specifically written for an undergraduate course in control engineering. This is not an original piece of work, but a deliberate, careful, organisation of the standard technology into such a form as to be of maximum use to a student. Brevity is a key word; pure mathematics is minimised; the references are carefully selected and not just a list of all available books; terminology, symbols and units are standard and up to date.

An introduction to the modern techniques of analysis can be found here, including frequency response, s-plane and state variable methods and it is intended that this will serve as a bridge between elementary and highly specialised works. In particular it is hoped that this book will provide a common foundation to post-graduate courses now in existence. However, while the scope is wide, the format is such that, by eliminating Part Three, students following T.E.C. courses containing control engineering will find this text acceptable.

To be able to use this book, the student must understand the complex operator j. In addition, before starting Part Two, he should acquire a working knowledge of differential equations and their solution using Laplace Transforms, as well as an acquaintance with the Fourier series, integral and transform. To this end a mathematical appendix is included.

The presentation is in four parts. In Part One, the basic ideas of closed loop control are presented, and some applications explained. Nomenclature and the methods of solution of typical problems are also discussed.

Part Two covers the methods of analysis and design, and great care has been taken to provide a more logical presentation than has previously been produced. Chapters 3, 4 and 5 define and explain the differential equation, transfer function and frequency response techniques of linear systems analysis. In the next three chapters, particular problems of closed loop analysis—stability, specifications and compensation—are solved using the general techniques of chapters 4 and 5. This section covers the bulk of any undergraduate course.

Part Three comprises a less detailed discussion of the more advanced topics: nonlinear systems, statistical analysis, sampled data systems and the more recent methods of multivariable system analysis.

The practical aspects of servomechanisms are described in Part Four.

Hydraulic, d.c. and a.c. servos are covered and, in addition, there are chapters on measuring, analogue computing, component rating and a synopsis of the problems of instrumentation.

Control engineering is recognised as a discipline in its own right rather than as a section of either the electrical or mechanical branches of engineering. However, this recognition does imply additional priming for both groups, as students in a mechanical engineering department will not have as great an understanding of transients and transform techniques, frequency response, or electronics as their equivalents on the electrical side. On the other hand their knowledge of dynamics will be wider. This, however, affects only the starting point, the subject being wide open to study at more advanced levels and with a rapidly expanding field of application.

My thanks are due to my colleagues J. G. Thomas and T. D. R. Hughes for their suggestions, and to the latter in particular for his help in writing chapter 12. I am also indebted to Professor Calderwood for his comments on the manuscript.

MARTIN HEALEY

Preface to Third Edition

In this new edition of *Principles of Automatic Control* I have continued with the original philosophy of aiming the book at the student on T.E.C. and B.Sc. courses. I have resisted the temptation to include chapters on optimal control, estimation theory and multivariable frequency response methods, since I feel such material warrants a special book. Nevertheless the chapter on multivariable systems has been revised and extended, including observability, controllability and discrete state equations. The chapter on digital control has been completely rewritten, stressing the computer control (D.D.C.) concept.

The introductory chapters have been revised, starting at an earlier point by introducing a first-order velocity control system. Short sections have also been added on signal flow graphs and sensitivity functions.

I must comment that I have experimented with intermingling the teaching of classical and modern control theory, with very poor results. It is my current opinion that it is necessary to teach the single input–single output linear techniques first, before embarking on a study of nonlinear systems and state–space concepts. The broader concept is most attractive in principle but is too confusing to undergraduate students in practice.

I would like to thank all the people who have spoken or written to me about the presentation and contents of the previous editions. I would welcome any constructive comments from teachers or students which may help me with a future edition.

M. H.

Contents

Preface   vii

Part One   CLOSED LOOP CONTROL

1. The Fundamentals and Applications of Closed Loop Control

    1.1   A brief history of automatic control   3
    1.2   The principle of closed loop control   4
    1.3   Modes of automatic control   8
    1.4   Terminology   10
    1.5   Some examples of the application of closed loop control   12

2. The Analytic Viewpoint

    2.1   The need for analysis   14
    2.2   Mathematical models   14
    2.3   The feedback amplifier   16
    2.4   System performance   17
    2.5   Methods of analysis   19

Part Two   ANALYSIS AND DESIGN OF LINEAR SYSTEMS

3. Differential Equation Analysis

    3.1   Application to linear systems   23
    3.2   Notation used for velocity and position control systems   24
    3.3   First order systems   24
    3.4   The remote position control servomechanism   29
    3.5   General form and solution of the second order equation   33
    3.6   Errors   38
    3.7   The effect of additional time constants   38
    3.8   Summary   40

4. Transfer Function Analysis

    4.1   Definition of transfer functions   42
    4.2   Validity of transfer functions   43
    4.3   Block diagrams   43

4.4 Manipulation of transfer functions in closed loops   44
4.5 Major and minor loops   46
4.6 The open loop equivalent of a closed loop   47
4.7 Direct feedback equivalent of a general system   47
4.8 Application of transfer function analysis to the example
        of a position control servomechanism   48
4.9 Systems with multiple inputs   51
4.10 Systems with multiple inputs and outputs   52
4.11 The general form of a transfer function   54
4.12 Signal flow diagrams or graphs   57
4.13 Summary   63

5. Frequency Response Analysis

5.1 Frequency response functions   64
5.2 Graphical representation of $G(j\omega)$ 67
5.3 Some features of $G(j\omega)$ 69
5.4 Minimum and non-minimum phase systems   72
5.5 Logarithmic plotting   74
5.6 Asymptotic approximation on Bode diagrams   77
5.7 Use of asymptotic approximation to construct Bode dia-
        grams   83
5.8 Use of measured frequency response to find $G(j\omega)$ 85
5.9 Summary   87

6. Stability Analysis

6.1 Stable and unstable systems   88
6.2 Stability analysis by pole–zero location   89
6.3 The Hurwitz–Routh stability criterion   92
6.4 Root loci   94
6.5 The Nyquist stability criterion   102
6.6 Bode's theorems   109
6.7 Methods of determining the closed loop from the open
        loop frequency response function   111
6.8 Summary   117

7. Design Procedure and Specifications

7.1 Introduction to design   118
7.2 Specifications   118
7.3 Sensitivity functions   127
7.4 Design procedure   129

8. Compensating Techniques

8.1 Methods of compensation   131

8.2 The uncompensated proportional error system 132
8.3 Proportional error system with derivative of output feedback 134
8.4 Series compensation 135
8.5 Parallel compensation 149
8.6 Input compensation 154
8.7 Compensation for two inputs 155
8.8 Design techniques using the *s*-plane 156

Part Three AN INTRODUCTION TO SOME FURTHER TECHNIQUES

9. Nonlinear Systems

9.1 Introduction 167
9.2 The common types of nonlinearities 168
9.3 Some effects of nonlinearities in closed loop control systems 170
9.4 Describing functions 171
9.5 Stability analysis using describing functions 179
9.6 Closed loop response of a nonlinear system 184
9.7 Frequency dependent describing functions 186
9.8 An introduction to phase-plane analysis 186

10. The Application of Statistics to Closed Loop Control Systems

10.1 The field of statistical application 194
10.2 Some important statistical terms 195
10.3 The weighting function and the convolution integral 200
10.4 Error criteria 202
10.5 Analysis in the time domain 204
10.6 Analysis in the frequency domain 210
10.7 Conclusions 215

11. Digital Control Systems

11.1 Sampled data systems 217
11.2 Mathematical representation of the sampled signal 219
11.3 Analysis of $f^*(t)$ 221
11.4 The Z transform 224
11.5 Manipulation of transfer functions in sampled systems 225
11.6 Stability analysis of sampled systems 227
11.7 Direct digital control 229

12. Multivariable Systems—State Variables and Matrices

12.1 Introduction 231

12.2 Matrix algebra   232
12.3 State space analysis   234
12.4 Solution of the state vector differential equation   238
12.5 The transition matrix   239
12.6 State variable diagrams   241
12.7 Representation of input functions by state variables 244
12.8 Transformation of the state vector   246
12.9 Observability and controllability   247
12.10 Transfer function matrices   249
12.11 State variable equations for digital systems   249
12.12 Conclusions   251

Part Four   PRACTICAL ASPECTS

13. Methods of Measuring, Computing and Simulating Systems

13.1 Measurement of transfer functions   255
13.2 Computation and simulation   261

14. System Components

14.1 General considerations   265
14.2 Comparison of d.c., a.c. and hydraulic servomechanisms 266
14.3 Servomotors and drives   267
14.4 Referred inertia and friction   268
14.5 Gearing   269
14.6 Motor rating   270
14.7 Structural components   274
14.8 Coarse and fine systems   275
14.9 Assisted braking   277
14.10 Transducers   278
14.11 Synchros and resolvers   279

15. D.C. Servomechanisms

15.1 Introduction   283
15.2 D.C. servomotors   283
15.3 Field control   284
15.4 Armature control   285
15.5 Power amplifiers   285
15.6 Other forms of continuous control   290
15.7 Relay servomechanisms   293
15.8 Transfer functions of continuous d.c. motor systems 295
15.9 Motor torque–speed characteristics   298
15.10 D.C. tachometers   299

16. A.C. Servomechanisms

16.1	Introduction	300
16.2	Modulators and demodulators	302
16.3	A.C. servomotors	302
16.4	Modulator–demodulator systems	304
16.5	All-a.c. systems	304
16.6	A.C. tachometers	306
16.7	Thyristor speed control of a.c. motors	307

17. Hydraulic Servomechanisms

17.1	Introduction	308
17.2	Pumps and motors	308
17.3	Methods of control	309
17.4	Flow control valves	312
17.5	System transfer functions	317
17.6	Oil	323

Appendix. Laplace and Fourier transforms   324

References   332

Problems   334

Index   347

Conversion Table

1 m	$= 3\cdot28$ ft
1 kg	$= 2\cdot2$ lb
1 N	$= 0\cdot224$ lbf
1 N m	$= 0\cdot738$ lbf-ft
1 kg m^2	$= 0\cdot738$ lbf-ft-s^2
1 N/m^2	$= 1\cdot45 \times 10^{-4}$ lbf/in^2
(1 MN/m^2	$= 145$ lbf/in^2)
1 cm^3/s	$= 6\cdot1 \times 10^{-2}$ in^3/s
	$= 1\cdot32 \times 10^{-2}$ gallons/min
g	$= 9\cdot81$ m/s$^2 = 32\cdot17$ ft/s^2

Part One

CLOSED LOOP CONTROL

1
The Fundamentals and Applications of Closed Loop Control

1.1 A Brief History of Automatic Control

It was as long ago as 1788 that James Watt decided that a man controlling the opening and closing of steam valves was not the best way of keeping the speed of his steam engines constant. So the Watt governor, which used the 'lift' of rotating balls as a speed monitor, automatically shutting off the steam as the speed tended to increase and vice versa, was the first automatic control system brought to prominence, although undoubtedly not the first to be applied.

Instability, the unfortunate by-product of closed loop control, was recognised early in, for example, hunting of steam engines and ship steering, which led to an early attempt at analysis by Maxwell (1868). More lasting contributions, on problems of stability in general, were made by Hurwitz (1875), Routh (1884) and Liapounov (1892) along with the essential work of such mathematicians as Laplace (1749–1827), Fourier (1758–1830) and Cauchy (1789–1857) on which basis the modern engineering methods of analysis are founded.

The increase in the use of automatic control was understandably slow, since engineering itself was in its infancy. However, with the development of electronics the application and theoretical understanding of closed loop control increased rapidly, since the feedback amplifier soon became a necessity. A most significant step was the work of Nyquist in 1932 in respect of stability in terms of the open loop frequency response. Also of significance were the contributions of the telecommunications pioneers, particularly Heaviside in the 1920s, leading to a better understanding of Laplace and Fourier transforms and the logarithmic unit, the decibel.

Thus in 1934 Hazen presented the first precise analytical approach to the design of closed loop control systems, drawing on the work of the above-mentioned originators. From then on, accelerated by the Second World War, the need for faster and more accurate systems led to rapid development in the field, with the engineer gradually taking over the development from the mathematician. Prominent in this up-surge were Bode and Nichols for the frequency response techniques; Guillemin for the network synthesis approach; Evans for the root locus method; Weiner and Phillips for the statistical approach; and Tustin, Kochenburger, Lur'e, Ragazzini, Zadeh, Shannon, Bellman, Kalman, etc., for work on nonlinear, discrete and multivariable systems. Most of the development may appear to have been done in the U.S.A., but much has been done in parallel in the U.S.S.R.; relatively little has stemmed from Europe.

Modern progress stems mainly from flight and process control and, rather unfortunately, purely academic exercises, with particular accent on nonlinearity, optimisation and self-organising systems. The state–space approach is also receiving a lot of attention as an alternative method of analysis to the *s*-plane and frequency response methods used in this book.

The much-bandied word 'Automation' must also be mentioned, mainly to dispel any ideas that it has any great relation to automatic control. Automation is essentially sequence-controlled mechanisation.

One final classification not covered in this context is the oscillator, a device which also uses feedback, but in a positive sense not in the negative sense essential to closed loop control systems.

1.2 The Principle of Closed Loop Control

1.2.1 The System
Systems are encountered, of varying degrees of complexity, in which one or more parameters require careful control. Typical examples are the control of temperature or pressure in a chemical process; of position or velocity in mechanical systems; and of output voltage from an electrical generator. Fig. 1.1 represents a general categorisation. The examples

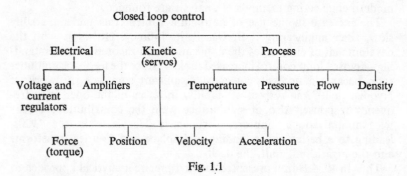

Fig. 1.1

in the preceding list are diverse in physical nature but are similar in more obscure ways, i.e. they are all prone to vary due to external or internal 'disturbances', and it is likely that the parameter being controlled will need its value continuously updated to meet some new requirement.

The parameter of the system which it is required to control, e.g. the temperature, velocity, etc., is called the *controlled variable, c*. The basic elements of the system, e.g. the heater and liquid bath, motor and gearing, etc., are called the *plant*. The term 'output' has commonly been used for c, a hangover from the early work on feedback amplifiers, but this is misleading, since for a simple hot water supply the output is the water drawn off; the controlled variable is the temperature of the water.

If it is to be possible to alter, and eventually to control, the controlled variable, then the plant must be capable of variation, and some form of *correcting unit* is required to manipulate the plant, e.g. an amplifier to supply the field current of an electrical machine, or a throttle valve in a flow control system (possibly indirectly controlled by an electric or pneumatic motor).

It is most important that the power required to feed the correcting unit is small. The low power input signal can then *control* the flow of power to the controlled variable, but not *supply* it. Essentially the plant and its correcting unit form a high power amplifier. Thus in the electric generator the prime mover supplies the power, the generator field current controls it.

The signal fed to the correcting unit is processed by a *controller*, which is 'told' the required value of the controlled variable, the *desired value*, and converts this into an appropriate form to suit the correcting unit. The controller output is termed the *manipulated variable, m*. The controller could be no more complex than a calibrated dial on a valve, but for automatic systems it will be required to perform more complex functions. It is normally an electronic device, although pneumatic controllers have played a major role in process control systems. Broadly speaking, all systems of which a variable quantity is to be controlled can be put into one of two categories: (*a*) open loop and (*b*) closed loop, as shown schematically in Fig. 1.2.

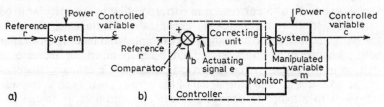

Fig. 1.2 (*a*) an open loop, and (*b*) a closed loop system.

1.2.2 Open Loop System

The *reference, r*, is fed direct to the controller at a value calculated to give the desired value of c. As an idealised example, consider a room heating system with a radiator fed with constant temperature water. The flow control valve setting, r, must be adjusted until the heat input is sufficient to supply the natural heat losses and to maintain the desired temperature. Such a system is inaccurate, since any change in the status quo will affect the temperature, e.g. changes in the outside temperature, or the opening and closing of doors.

1.2.3 Closed Loop System

In a closed loop system, the manipulated variable, m, is a function of the *actuating signal*, e, which is the difference between the reference and the monitored or measured value, b, of the controlled variable. In simple controllers m is proportional to e; more sophisticated controllers add additional terms such as the derivative and integrand of e. The controller must also perform the differencing of r and b. Physically the monitor will normally comprise a sensor mounted on the plant and an associated measuring element in the controller.

Thus, if c is at the value dictated by r, then $b = r$ and e is zero. No further action takes place. Whenever b differs from r, indicating that c is incorrect, a finite actuating signal is developed, of such a polarity as to produce action from the correcting unit and plant to cause e to reduce. The action should continue until e reaches zero. This is termed *negative feedback*; positive feedback would mean that, with $b \neq r$, the finite value of e causes e to change in such a way as to cause the difference between r and b to increase, thus losing all control. Positive feedback is used for oscillators, and has no application here. Closed loop systems or negative feedback systems can also be descriptively termed 'error actuated'. The use of the term 'error signal' for the variable e is, however, discouraged, since while e exists only when the system is in error, the true error is the difference between the desired value of c and the actual value, and not $r - b$.

The closed loop system has two distinctly desirable features:

(i) Regulation

Consider c to be at a correct value corresponding to a particular fixed value of r, so that e is zero. Some uncontrolled 'disturbance' on or in the system now causes c to change, for example wind loading on a radar aerial, change in viscosity of a fluid being pumped, or increase in friction in a constant velocity drive. Immediately c changes, the controller produces an actuating signal of the correct polarity so as to cause the plant to attempt to change c in the opposite polarity to the disturbance, thus counteracting the effect. In a well designed system, e can be kept very small to produce the necessary reaction from the plant so that the resulting error is also small. The system is said to have good *regulation* against external disturbances and/or undesirable changes in the plant itself.

(ii) Controllability

The correcting unit in practice may control the flow of large powers in the plant, but is actuated by low power signals e and r. Thus, since e only *controls* and does not provide the work effort, the controlled variable c can be dictated by low power references. Thus, for instance a computer can be used to generate voltage levels proportional to desired values, the resulting high power effect being produced by the control system.

An early application of these principles came about in the steering of ships. In early ships, a mechanical linkage connected the helmsman's wheel to the rudder, with sufficient gearing that the man's effort produced adequate forces at the rudder. High seas caused severe trouble in the rudder 'steering' the helmsman! Further reduction in the gearing meant more problems in excessive wheel turning and backlash, etc. This situation became intolerable with the introduction of steam ships. However, the steam engine also meant that a source of motive power became available so that closed loop control of the rudder displacement (c) could be achieved from the wheel position (r) as desired, without disturbances on the rudder affecting the wheel.

Further (unimportant but illustrative) examples can be taken from a man driving a car:

(a) *Speed control.* In order to maintain speed on differing inclines, the driver monitors the speed via the speedometer and compares it to the desired speed in his mind. When the speed changes, the *detected* error induces the driver to adjust the accelerator pedal position to compensate.

(b) *Steering action.* The angular rotation of the steering wheel and the direction of the car are not directly related; in fact the steering wheel deflection controls the rate of turn. The driver uses his eyes as a monitor and the road ahead as the desired position. The brain acts as error detector, and the steering wheel is used as a correcting unit to direct the car correctly.

Many of the essentials and problems of closed loop control are illustrated by this example:

(i) An open loop system, i.e. a predetermined wheel position pattern for a given road, is impossible. This is clear from the impossibility of steering down a straight road without adjusting the steering wheel to counteract wind and surface irregularities.

(ii) Nonlinearities such as backlash and saturation (maximum lock) increase the steering difficulties and reduce accuracy.

(iii) The effects of inertia on speed of response are clear in the comparison of large and small cars.

(iv) System gain in the form of steering gear ratio affects the response so that high-geared sports car steering is sharp in response, but liable to oscillation in inexperienced hands, while the steering of a large car, requiring a lot of turns on the wheel, is 'woolly' but free from overshoot.

(v) Damping is introduced mainly by the driver's own reflexes, so that a learner-driver is 'under-damped' and can steer wavy 'straight' paths.

In fact the above example is one of extreme sophistication, in that the human brain is a very capable computer and adds refinement to the control in the form of predictions (by viewing the road well ahead) and adaption to the environment. Such refinements would be nearly impossible to build into an automatic system.

1.3 Modes of Automatic Control

A closed loop system in which the control, monitoring and differencing are performed by mechanical, electrical or pneumatic devices is said to be *automatically controlled*. If the correcting action is performed by a human, the system is said to be *manually controlled*.

Consider the historical example of the speed control of steam engines. Prior to 1788 the engine speed control was open loop, set to a fixed steam

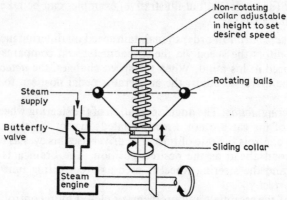

Fig. 1.3 A steam-engine governor

flow valve setting to give nearly the desired speed and then left, relying on high inertia to give some degree of speed holding. An 'improvement' was to manually control the engine by a man reading a speed indicator and increasing or decreasing the steam flow accordingly to correct for fall or rise in speed respectively. Here the speed indicator was the monitor, the desired speed the reference, the man's eye and brain the controller, and the man's hands and the steam control valve the correcting unit; the engine itself, of course, being the plant. Then, in 1788, James Watt introduced the now famous governor, which is an automatic system, shown in Fig. 1.3. The engine speed is monitored by the 'lift' of rotating balls, working against a spring, the initial tension of which is the reference, determined by the desired speed. As the rotating balls rise or fall, they close or open the steam control valve via a mechanical linkage in such a sense as to compensate for the speed change.

Two other commonplace examples of automatic control systems are room temperature control and water level control in cisterns.

The temperature control comprises a heating element which is switched on when a thermostat senses that the temperature is below the 'set' value and switched off when the desired temperature is reached. Water level control is achieved by sensing the water height with a floating ball which controls the flow of inlet water to an extent dependent upon the fall in level: for a small fall in level the valve will be only 'cracked' open, giving a small inflow; for a large fall in level the valve will be fully open, giving maximum refill rate.

The preceding three examples demonstrate the variety in the mode of operation of control systems, in addition to the obvious choice of manual or automatic.

The first choice is between *continuous* or *discontinuous* action. The simplest example of continuous control is the proportional control, where the controller action is directly proportional to the actuating signal. The speed control is an example of this; the level control is similar except that the action of a ball-valve does not give flow directly proportional to opening, and therefore introduces *nonlinearity*. The temperature control, on the other hand, is discontinuous in action, having only two possible states, on or off. Such a system is thus termed *on–off* or *two-step* action.

A second consideration is whether or not the system is capable of absorbing as well as supplying power. Both the temperature and level controls are incapable of reverse action, e.g. if some other source, say sunshine, causes the temperature to rise, then the thermostat will switch the heater off, but there will be no correcting (cooling) action. To correct for the temperature rise, a power absorption unit (cooler) must be switched on. The resultant system would still be discontinuous but would be a *two-step action with dead-zone*, i.e. low temperature, heater on; high temperature, refrigeration on; temperature within a small band either side of set value, both off.

The water level control is also capable of supplying but not absorbing power, i.e. if some external source overfills the tank there is no corrective action to bring the level down. An overflow, however, will give an on–off absorption action resulting in a system with continuous filling control and discontinuous over-filling control!

The speed control goes some way towards full control, since any attempt of the load to over-run, requiring the engine to brake, will shut the steam supply off, leaving the inherent friction loading. More comprehensive systems must provide full driving/braking effort, either drawing power from or regenerating back to the supply.

While many control systems are designed so that the reference signal is readily varied, other systems are designed to maintain a condition set by a fixed reference signal. Such systems could be termed *regulators*, and the value of the reference signal which specifies the value of the controlled variable to be held is called the *set point*.

1.4 Terminology

Fig. 1.4 is a more detailed schematic diagram of the closed loop system introduced in Fig. 1.2 (*b*).

The following definitions have been extracted from the A.I.E.E. committee report, October 1951, p. 905, *Elec. Eng.*, Vol. 70:

1. *A feedback control system* is a control system which tends to maintain a prescribed relationship of one system variable to another by comparing functions of these variables and using the difference as a means of control.

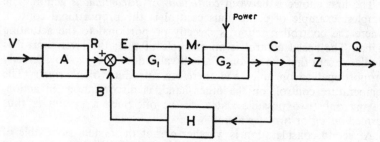

Fig. 1.4 Schematic diagram of a closed loop system.

2. *The controlled variable, C*, is that quantity or condition of the controlled system which is directly measured and controlled.

3. *The indirectly controlled variable, Q*,* is that quantity or condition which is controlled by virtue of its relation to the controlled variable and which is not directly measured for control.

4. *The command, V*,* is the input which is established or varied by some means external to, or independent of, the feedback control system under consideration.

5. *The reference input, R*, is a signal established as a standard of comparison for a feedback control system by virtue of its relation to the command, *V*. A constant value of *R* is also termed a *set point*.

6. *The primary feedback signal, B*, is a function of the controlled variable, *C*, which is compared with the reference input, *R*, to obtain the actuating signal.

* The symbols θ_i and θ_o which are used in many, more elementary, textbooks are here used for the particular case of angular position only. F_i and F_o are used to denote general input and output respectively.

7. *The actuating signal*, E, is the reference input, R, minus the primary feedback signal, B.

8. *The manipulated variable*, M, is the quantity or condition which the controller applies to the controlled system.

9. *A disturbance*, U, is a signal other than R which tends to affect the value of the controlled variable, C.

10. *A parametric variation* is a change in the system properties which may affect the performance of the feedback control system.

11. *The system error*, Y_e, is the ideal or desired output minus the value of the ultimately controlled variable, i.e. Q, or C in the absence of an indirectly controlled system.

12. *The desired ouput*, Q_d or C_d, is the output which would result from an idealised system operating from the same command as the actual system.

13. *The idealised system*, G_d, is the theoretical system relating the desired output to the command.

14. *The controlled system*, G_2, is the body, process or plant, a particular quantity or condition of which is to be controlled.

15. *The indirectly controlled system*, Z, is the body, process or plant which determines the relation between the indirectly controlled variable, Q, and the controlled variable, C.

16. *The feedback controller* is a mechanism which measures the value of the controlled variable, C, accepts the value of the command, V, and as the result of a comparison manipulates the controlled system in order to maintain an established relation between the controlled variable and the command. The feedback controller comprises the feedback elements, control elements, reference input elements and summing point.

17. *The feedback elements*, H, comprise the portion of the feedback control system which establishes the relation between the primary feedback, B, and the controlled variable, C.

18. *The control elements*, G_1, comprise the portion of the feedback control system which is required to produce the manipulated variable from the actuating signal.

19. *The reference input elements*, A, comprise the portion of the feedback control system which establishes the relation between the reference input, R, and the command, V.

20. *The summing point* is a descriptive symbol used in block diagrams to denote the algebraic summation of two or more signals. The direction of information flow is indicated by arrows and the algebraic nature of the summation by plus or minus signs.

All the terms V, Q, R, C, E, etc., can be expressed as functions of time, t, the Laplace operator, s (or p), complex frequency, $j\omega$, or real frequency, ω. In general, small letters will be used exclusively for time functions; thus, for example, $r(t)$, $R(s)$, $c(t)$, $E(j\omega)$, etc. The significance of these functions will be explained in Part Two.

In this text there is a bias towards kinetic control or SERVOMECHANISMS. Unfortunately there are various definitions of a servomechanism, or more simply a SERVO, the following being used here:

> "A servomechanism is a closed loop control system with a kinetic output involving a power amplifier in the forward path which is actuated by the difference between the input and the monitored output."

Other definitions (the above is basically Hazen's original definition) exist, e.g. the A.I.E.E. definition states that the output must be mechanical position, but as long as each individual has the overall picture of Fig. 1.1 clear in his mind, then no significant trouble can arise from this lack of consistency.

In addition, a system may be continuous or discontinuous. Most systems fall into the continuous class, but on–off control is fairly common and will be considered later.

1.5 Some Examples of the Application of Closed Loop Control

1. *Aircraft Control Systems.* The response of an aeroplane to the pilot's commands can be poor, making it difficult to fly. Control systems utilising gyro's and accelerometers as sensors, and the elevators as actuators are fitted to improve the handling, such systems being termed 'stability augmentation systems'. Autopilots are further applications of closed loop control whereby constant velocity is maintained by control of the engine thrust and constant altitude by replacing the pilots 'stick' signals by an actuating signal derived from a given reference altitude and an altimeter. These are complex control problems, due to the wide range of variation of system parameters due to changes in flight conditions, e.g. velocity, altitude and mass.

2. *Automatic Landing of Aircraft.* The desired altitude and velocity of the aircraft are computed from the ground and transmitted as control signals to the aircraft.

3. *Missile Control.* This combines many of the problems of 1 and 2 and, in addition, includes complex telecommunications links.

4. *Radar and Gun Control.* Position control systems with particular problems of fast positioning and good regulation against wind disturbances. They may also have to track targets, so that velocity errors as well as position errors must be minimised. The feedback signal is derived from the transmitted–received beam of the radar, the output position being moved to keep the received signal a maximum.

5. *Ship Steering and Roll Stabilisation.* Very similar techniques to

flight control but with far higher inertias, lower speeds of response and higher powers.

6. *Machine-tool Control.* The use of position control systems for profile copying, a low-force contact being maintained with a template, controlling a high cutting force at the output.

More complex is the numerical control of the position or velocity of a cutter, drill, etc., from numerical input information, i.e. the co-ordinates of a hole are processed on to a magnetic or paper tape and the table positioned accordingly.

7. *Remote Position Control.* In radar systems, computers, nuclear reactors, etc., remote position control systems of low power ratings are used extensively to position potentiometers and indicator dials.

8. *Nuclear Power Control.* The rate of feeding fuel into a reactor to give a required output is another use of feedback control which presents many particular problems and is leading to modern developments in nonlinear control.

9. *Speed Control.* Speed control is used in many industrial examples, e.g. grinding-wheel speeds for precision grinding, instrument tape-recorders, strip-rolling, wire-drawing, etc. In most cases, in addition to the advantages of speed regulation, the ease of changing the reference input over a wide range to control the actual speed is most advantageous.

10. *High-speed Mechanical Systems.* High-speed interconnected mechanical systems, such as used in textiles and some machining processes, are limited by the wear, stressing and elasticity of the linkages. Separate drives to individual parts, synchronised by using closed loops can allow much higher speeds.

11. *Voltage and Current Stabilisers.* The output voltage is compared with a stable reference voltage and the amplified difference fed to a transistor in series with the output, the resistance of the transistor varying proportional to the actuating signal. Thus if the output load current increases and the voltage tends to fall, the actuating signal causes the effective resistance of the transistor to fall, reducing the internal volt drop and maintaining the output voltage.

12. *Process Controls.* In the chemical industry, e.g. synthetic yarn production, oil refining, chemical plants, etc., such quantities as flow and temperature must be critically controlled. There is also great scope for computer-controlled inputs to such processes.

In addition to the particular problems of closed loop control associated with these examples, they all have their own practical problems, e.g. temperature range and weight in flight control, rigidity of mechanical structures in radar sets and machine tools, suitable selection of actuators, etc. The application of electric, hydraulic or pneumatic control is roughly equal. Electric control is usually the cheapest, hydraulic the most responsive and pneumatic the most suitable for the long time constants of process controls.

2
The Analytic Viewpoint

2.1 The Need for Analysis

In the previous chapter the basic ideas of closed loop control were expounded. While it was not stressed, the very definition of a closed loop control as being error actuated implied that control systems are not free from error. It is the aim of the design engineer then to produce a system which has the minimum error, although it is far better to start with a figure for the maximum allowable error and to attain this specification at the minimum cost, i.e. the best design being a compromise between accuracy and cost and not necessarily the most accurate.

Trial and error attempts to design a closed loop control system show that it can very easily be made into an oscillator! This condition, known as 'instability', is caused by the fact that in transmitting a signal from the input around the loop back to the summing point, time delays are invariably encountered due to energy storage in inertia, capacitance, etc. The feedback signal can thus arrive at such an instant, and with such an amplitude, as to appear as positive feedback, causing the actuating signal to increase rather than decrease. To solve this problem, and to attempt any accurate design, therefore, requires an accurate analysis of the system with appropriate measuring techniques. It is clear that if a system is unstable no useful measurements can be made so that techniques have been developed for measuring the open loop system performance and from these results predicting the performance of the closed loop.

2.2 Mathematical Models

The preceding chapter was used to introduce the philosophy and application of automatic closed loop control on a purely descriptive basis. The later chapters, 6, 7 and 8, are aimed at putting design techniques on a firmer basis, which simply and unavoidably means a mathematical basis. Pure mathematics has never been popular with engineers, so the aim of this work is to present these methods in as simple and painless a fashion as possible, yet without throwing away those techniques which simplify the problem. In short, the aim is to encourage an interest in mathematical techniques which are directly applicable to real engineering problems.

The ultimate objective of control system engineering is to produce a working device which will perform a specified control task to an accepted degree of accuracy, as reliably and cheaply as possible. Thus a great deal of attention must be paid to the hardware, as discussed in Part Four of this book. The next requirement is to produce a *design* which will determine the size of motor, etc., and the value of time constants and gains, etc., to meet the specification. In most cases it is impossible to design a satisfactory system by inspired trial and error methods; this is particularly so in closed loop systems. Thus the design stage must be preceded by a closer understanding of the system: it must be *analysed*.

The basis of any analytical method is a mathematical model of the system, the differential equations of which can be written and solved, thus indicating the performance of the given system. The design comes in changing the system equations to give a better performance and then interpreting the mathematical changes with actual hardware.

It will be shown in Chapter 3 that direct solution of the differential equations is an unwieldy tool, and superior, more practical, methods have been developed and are explained in Chapters 4 and 5. These three chapters present a short course on 'linear systems analysis'. Two important points arise here:

(i) No practical system can be perfectly linear, due to such obvious causes as saturation, backlash, hysteresis, etc. The analysis of non-linear systems, discussed in Chapter 9, is a much more complex problem than that of linear systems, so that wherever possible the assumptions are made that the system is 'nearly linear' and that linear analytical techniques will give an acceptably accurate answer.

(ii) The analysis is equally applicable to open or closed loop systems, so that Chapters 3, 4 and 5 quite generally apply to *any* linear dynamic system. Chapters 6, 7 and 8 extend those techniques to the particular problems of closed loops.

Mathematical modelling involves the representation of physical devices by terms such as inertia, damping force, capacitance, resistance, etc., and writing the differential equations relating the variables of the system, e.g. velocity, current, temperature, etc., to these constants. The reader must be very clear about the difference between the constants of the system and the variables! The use of the word 'constant' to describe inertia, resistance, etc., gives the definition of linearity: such commonly used 'constants' aren't; some more so than others! As an example, consider a simple rotating mass, the inertia of which is near enough constant, but the viscous friction coefficient is not. Consider also a gearbox for which the output displacement is considered to be a constant ratio of the input displacement, which is not true due to back-lash and twisting of shafts under load.

One advantage of mathematical modelling is that electrical and mechanical systems lead to similar equations, so that combined systems such as an electric motor can be directly analysed; there is no mathematical difference between an inertia–friction time constant and a resistance–inductance time constant.

2.3 The Feedback Amplifier

Fig. 2.1 shows a schematic diagram of a feedback amplifier. It has a forward gain of G, i.e. the output–input ratio of the amplifier stages, and the feedback signal is H times the output voltage. H is in practice a fraction. This example is simple since all variables, f_i, f_o and e, are voltages. It should be admitted that this model is tailored to serve as an introduction to general control system analysis; in specialised

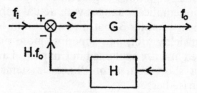

Fig. 2.1 A feedback amplifier

electronics texts the notation will be different, and in practice the *sub-traction* of Hf_o from f_i is achieved by ensuring that the amplifier has an odd number of stages, thereby making f_o of opposite polarity to e, and adding the monitored output to f_i.

From Fig. 2.1 it can be deduced that:

$$f_o = Ge$$

and

$$e = f_i - Hf_o$$

Hence

$$f_o = G(f_i - Hf_o)$$

$$\therefore \frac{f_o}{f_i} = \frac{G}{1 + GH}$$

If now $GH \gg 1$

$$\frac{f_o}{f_i} \simeq \frac{G}{GH} = \frac{1}{H}$$

Thus if the product GH, the 'open loop gain' as it is called, is high compared to 1, the resulting 'closed loop gain', $1/H$, is independent of G. Now G will be adversely affected by change in temperature, ageing of transistors, etc., and is also prone to distortion. On the other hand,

H would be made up from a resistance attenuator chain, with a stable, distortion-free value. Thus the resulting closed loop amplifier will be virtually distortion-free. The price paid for this marked improvement in performance over the open loop amplifier is the extra gain required to make $GH \gg 1$. Thus an amplifier with an overall gain of 100 must have $H = 10^{-2}$, so that G will need to be of the order of 10^5 or more.

The block diagram of Fig. 2.1 and the resulting closed loop gain formula is, however, misleadingly simple, since it totally neglects any dynamic effects in the amplifier. The equation $f_o = Ge$, implied by the diagram, cannot possibly be correct, since it assumes that a step change in e will cause an instantaneous, proportional response in f_o. This cannot possibly happen, due to the effects of stray capacitance. In fact f_o will be related to e by a differential equation, so the diagram is in error in depicting G as a constant multiplier. In order to use a diagram such as Fig. 2.1, the Laplace transform is introduced so that the differential equation relating f_o to e is replaced by an equivalent algebraic one; this is discussed in Chapter 4.

Thus, while the steady state performance of the closed loop system can be improved by making the open loop gain higher, the dynamic performance deteriorates and will lead to instability.

This example, then, highlights the need for introducing more sophisticated mathematical models into the analysis and design of systems.

2.4 System Performance

The system error is the desired output minus C, where the desired output is directly related to R.

The sources of error then are:

(i) Errors in H, e.g. inaccuracies in the output monitor.

(ii) Changes and nonlinearity in G, which have been minimised, but not completely eliminated since GH cannot be infinite. It also includes the effects of imperfect practical systems, e.g. static friction, dead zones, backlash and hysteresis.

(iii) Change of C due to the disturbance U. U is, for example, wind gusts on an aerial, cutting forces on a machine tool, changes in load current in a voltage regulator, etc.

The resulting error can be split into two parts: a steady state error and a transient error.

Fig. 2.2 shows the output voltage of a voltage regulated supply in which the steady state and transient errors can easily be isolated.

The transient error can reverse sign, i.e. the output can overshoot, due to energy storage in inertia, etc., as is shown in Fig. 2.3, which is the response of a position control system to a step change in desired position.

The steady state errors of this sort of system are mainly due to static friction as the output ceases to move when the output force falls below

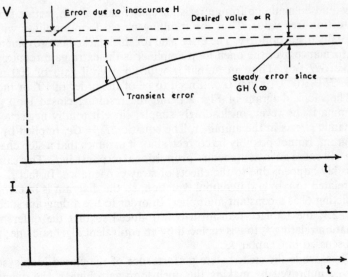

Fig. 2.2 Transient response of a voltage regulator.

the static friction force. Thus an error will exist such as to produce an output force less than or equal to the static friction force. The steady state error is therefore likely to lie within a band about the desired value. Changes in the load on the output will also require the control system to produce a reaction force, therefore introducing a further error.

The 'goodness' of a system cannot be clearly defined, but it is related to the response of the system and the form of the input. A strict measure of the goodness is difficult to define.

Consider, then, the two position control systems of Fig. 2.3; case (a)

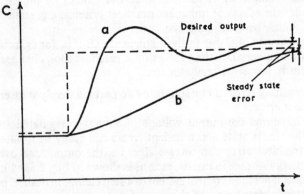

Fig. 2.3 Output response of a control system to a step input: (a) under-damped, and (b) over-damped.

has the faster response and therefore a smaller, or rather shorter duration, transient error, but has the larger steady state error. Which is better depends upon the nature of the input, the step inputs only being introduced as a means of comparison. If the desired output varies rapidly, then (*a*) will be the best system, but if the variation is much slower than the response of either system, (*b*) will be more accurate since the steady state error is smaller. This is a simple example and the design of optimum control systems is a complex subject, but it should make clear the necessity for producing a specification for any system, based on a knowledge of the desired output.

It must be stressed that to achieve higher steady state accuracy and quicker response the open loop gain GH must be increased, which is also more likely to produce an oscillatory system and eventually an unstable one, therefore:

> "**High accuracy and good stability are incompatible and a good design is a compromise between the two.**"

2.5 Methods of Analysis

The following chapters deal with the common methods of system analysis in some detail, the presentation of which can be better appreciated if some brief knowledge of the contents is first absorbed. The following, then, is a summary of Parts Two and Three.

The most obvious method of analysis is to consider the system performance in the time domain by measuring the output response for given inputs. Common forms of input, known as singularity functions, are step, impulse and ramp functions, and sinusoidal functions. The transient and steady state response to these functions will give some measure of the dynamics of the system so as to enable the differential equations representing the system to be calculated. Accurate interpretation of these responses is, however, difficult and can give relatively little information as to the nature of any design modifications required.

The most common method of analysis is in the frequency domain. Here the output response to a sinusoidal input is considered in the steady state only, any transients being allowed to die away before measurement is made. The results, however, must be taken over a complete range of frequencies, in theory over all frequencies up to infinity, but in practice up to such a frequency at which the response is negligible. These results can be more easily related to the differential equations of the system so that mathematical techniques can be introduced to solve and design systems, notably the Laplace transform and the closely associated transfer function.

It is very common in practice to work solely in the frequency domain, even to the extent of writing the specification, which is basically in the time domain, as an equivalent frequency response; suitable methods are clearly defined for determining system stability and for designing any compensation which may be needed to meet the specification. Design is possible in terms of transfer functions, commonly referred to as design in the 'complex frequency domain', e.g. the *s*-plane, although it is not considered in this text. However, an introduction to the *s*-plane and root loci is made so that the design procedure can be followed up in Truxal (ref. 1).

Nonlinearity, i.e. saturation, etc., and noise, i.e. components in the output unrelated to the input, cause great problems in analysis in the frequency or complex frequency domains, to such an extent that there is no completely satisfactory solution available. Some indication of possible methods of tackling nonlinear problems is given in Chapter 9 and the application of statistics is outlined in Chapter 10. Chapter 11 gives an introduction to sampled data control systems and Chapter 12 presents the newer time domain methods, including multivariable systems.

It must once again be stressed that most measurements are made with the loop open, the ensuing results being used to determine the closed loop response with any possible compensation.

Part Two

ANALYSIS AND DESIGN OF LINEAR SYSTEMS

3
Differential Equation Analysis

3.1 Application to Linear Systems

In the preceding chapters the basic rudiments of closed loop control have been discussed and some inference made as to methods of analysis. Chapters 3, 4 and 5 cover methods of representing mathematically and graphically the characteristics, and particularly the dynamic characteristics, of any closed loop control problem from process controls to feedback amplifiers. The particular example of the position control servomechanism is used frequently so that easy comparison of different techniques can be made. It must be stressed, however, that any problem, the dynamics of which can be formulated, can be treated by the following methods.

No practical system can be perfectly linear, due to such obvious causes as saturation, backlash or hysteresis, although the bulk of the theory presented here assumes that the system can be represented by differential equations with constant coefficients. While one of the advantages of closed loop control is to minimise the effects of nonlinearities it must be the designer's aim to first reduce as far as possible, all nonlinear effects.* The inference then, is that the amplitude of the input signal has no effect on the basic equations of motion, so that the linear theory will suffice to solve many design problems and also to give the necessary background for more rigorous solutions of nonlinear problems as outlined in Chapter 9.

In deriving the system differential equations the effects of energy storage, dissipation and conversion must be considered whether the energy is electrical, mechanical or thermal, etc. Considered from a circuit viewpoint there is a strong case for the use of all-electric (i.e. $R-L-C$) equivalent networks with associated e.m.f.s to represent **any** system and it is recommended that this approach be followed up (see Cheng, ref. 7), particularly where more complex mechanical and process systems are concerned. For this text the straightforward derivation of the differential equation is used since the examples considered are simple.

* This is an oversimplification of the problem since in practice certain nonlinearities are deliberately introduced, such as current limiting in motors to limit acceleration forces on structures which could otherwise be damaged; in fact, certain nonlinear effects are introduced to improve performance (see Section 9.4.5).

3.2 Notation used for Velocity and Position Control Systems

e actuating signal, volts
i motor field current, amperes
I_a motor armature current, amperes
θ_i position of reference (input) potentiometer, rad
θ_o output position, rad
ω output velocity ($= \dot{\theta}_o$), rad/s
A amplifier voltage gain
R field resistance, ohms
L field inductance, henry
G amplifier transconductance ($= A/R$), amperes/volt
n gear ratio
J_m motor inertia, kg m^2
J_L load inertia, kg m^2
J total inertia referred to output, kg m^2
F_m motor viscous friction coefficient, N m s
F_L load viscous friction coefficient, N m s
F viscous friction coefficient referred to output, N m s
T_f field time constant ($= L/R$), seconds
T_m motor and load time constant ($= J/F$), seconds
K_1 potentiometer bridge constant, volts/rad difference in θ_i and θ_o
K_2 motor torque constant (I_a constant), N m/ampere of field current
K_3 tachogenerator constant, volts/(output rad/s)

In the following examples the motor is used with a constant armature current, so that the developed torque is directly proportional to the field current, independent of speed. For this reason it is sometimes referred to as a 'torque-controlled' servo. Other modes of connection of the motor and further comment on this method are included in Chapters 14 and 15. Nonlinear effects such as field saturation, backlash, etc. are ignored. The problem is further simplified in the earlier analysis by assuming perfect mechanical components so that, for example, twisting of shafts is neglected. It should be stressed that the development of later chapters is aimed at methods of allowing for more detailed and exact analysis.

3.3 First Order Systems

3.3.1 A Resistance–Inductance Circuit

Consider the simple electrical network of Fig. 3.1. Capital letters are used to denote the system constants, in this case R and L, and small letters to denote variables as functions of time, i.e. v and i. A constant voltage would be treated as a special case of v.

The differential equation governing the response of the current, i, to the applied voltage, v, is

$$v = iR + L\frac{di}{dt}$$

This is termed a first-order differential equation, since it has only one derivative. The system it represents is called a first-order system. The solution, the value of i, will depend primarily on the applied voltage, v,

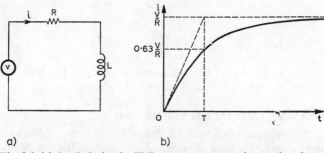

Fig. 3.1 (*a*) An *R–L* circuit. (*b*) Response to a step changes in v from 0 to V applied at time $t = 0$.

but will also depend upon the 'initial condition' of the system. In fact an initial condition must be specified for each order of the system. In this example only one initial condition is required, that of the current.

As an example, let v be a constant voltage, V, applied by closing a switch at time $t = 0$. Since prior to closing the switch no current can flow, and since the current through an inductance cannot change instantaneously, the initial value of i, termed $i(0)$, is zero. Thus it is required to solve, for $t > 0$,

$$L\frac{di}{dt} + Ri = V$$

given that $i(0) = 0$. The form of writing the equation has been changed only to conform to a more usual general notation.

Rearranging

$$dt = \frac{L}{V - iR}\,di$$

Integrating both sides,

$$t = -\frac{L}{R}\log_e(V - iR) + C$$

where C is the constant of integration.

Substituting $i = 0$ when $t = 0$ gives

$$C = \frac{L}{R} \log_e V$$

so that

$$t = -\frac{L}{R}[\log_e(V - iR) - \log_e V]$$

$$\therefore \quad -\frac{R}{L}t = \log_e\left[\frac{V - iR}{V}\right]$$

$$\therefore \quad \frac{V - iR}{V} = e^{-Rt/L}$$

$$\therefore \quad i = \frac{V}{R}(1 - e^{-t/T})$$

where $T = L/R$ is called the time constant, and is the time taken for the current to build up to 63% of its full value.

3.3.2 An Open Loop Velocity Control System

Using the notation of section 3.2, consider a d.c. motor (with constant I_a) directly driving a load of inertia J and viscous friction coefficient F (viscous friction is discussed more fully in section 3.4.2). Then the developed motor torque $= K_2 . i$

$$= GK_2 . v$$

where v is the voltage applied to the amplifier feeding the field coil (the inductance has been neglected).

The opposing torque $= J\dot{\omega} + F\omega$
since $\dot{\omega} = d\omega/dt$ is the acceleration.

Thus, equating torques,

$$J\dot{\omega} + F\omega = GK_2v$$

This is again a first-order system, and the differential equation can be compared to that for the R–L circuit. In each case the 'forcing function' is the voltage, v. The dependent variables, however, are physically dissimilar, in one case velocity and in the other current. Nevertheless, taking full advantage of mathematical similarity, and specifying the initial condition of the velocity control system as $\omega(0) = 0$, the solution of ω in response to a step change in v is

$$\omega = \frac{GK_2}{F} V(1 - e^{-t/T_1})$$

where $T_1 = J/F$. This time constant is the time taken for the speed to build up to 63% of the full speed. The full speed is given by $(GK_2/F)V$, and is therefore proportional to V.

Note that both systems achieve a steady state only after an interval of time (in theory infinity but in practice after say ten time constants). The period before steady state is reached is called the 'transient period'.

3.3.3 A Simple Closed Loop Velocity Control System

Now consider the application of feedback control to the velocity control system by monitoring the speed with a tachogenerator and subtracting this voltage from the input (or reference) voltage.

Thus the tacho voltage $= K_3\omega$

and v is replaced by the actuating signal

$$e = v - K_3\omega$$

Hence

$$J\dot{\omega} + F\omega = GK_2(v - K_3\omega)$$

$$\therefore \quad J\dot{\omega} + (F + GK_2K_3)\omega = GK_2v$$

Solving, with $\omega(0) = 0$ and $v = V$,

$$\omega = \frac{GK_2}{F + GK_2K_3} V(1 - e^{-t/T_2})$$

where $T_2 = J/(F + GK_2K_3)$.

If now GK_2K_3 is chosen to be much bigger than F, then

$$\omega \simeq \frac{1}{K_3} V (1 - e^{-t/T_2})$$

Comparing the open and closed loop systems it can be seen that the time constant must be reduced by the action of the closed loop and that the speed–voltage ratio is independent of G, K_2 and F, all of which are prone to variation in practice. The cost of the improvement in performance offered by the closed loop is the increase in the open loop gain required to make $GK_2K_3 \gg F$, and of course the necessity for the tachogenerator.

The difference in 'open loop gain' can be achieved by adjusting G only. If for the open loop system G is chosen so that $GK_2/F = 1/K_3$ and for the closed loop system a larger value of G is used so that, say, $GK_2K_3 = 100F$, then, by examining the two equations for the response in speed to a step change in V, it can be established that:

    (i) the steady state speeds are similar,

    (ii) the time constant for the closed loop system is approximately 100 times smaller than for the open loop.

3.3.4 The Effect of External Load Torque on the Simple Velocity Control System

If an external load torque,* T_L, is applied to the system of Section 3.3.3,

---

  * Strictly a small letter should be used to denote a function which can vary with time. The rule is broken here because of the obvious association of t with time!

then the motor must produce an opposing torque. Thus the developed motor torque must be equated to

$$J\dot{\omega} + F\omega + T_L$$

Following the previous closed loop example, the dynamic equations become

$$J\dot{\omega} + (F + GK_2K_3)\omega = GK_2v - T_L$$

In effect the system has two inputs: the reference, v, and the 'disturbance', T_L.

Consider the system to be running in the steady state with $v = V$ and no load. The steady speed, from the previous example (just as easily deduced by noting that for a steady speed $\dot{\omega} = 0$), is

$$\omega = \frac{GK_2}{F + GK_2K_3} \cdot V$$

Now let a step load, T_L, be applied at the instant of time $t = 0$. Separating the variables,

$$dt = \frac{J}{-(F + GK_2K_3)\omega + GK_2V - T_L} \, d\omega$$

Integrating both sides,

$$t = \frac{-J}{F + GK_2K_3} \log_e \left[-(F + GK_2K_3)\omega + GK_2V - T_L\right] + C$$

Substituting $\omega = GK_2V/(F + GK_2K_3)$ when $t = 0$ to find C, and then simplifying, yields

$$\omega = \frac{GK_2}{F + GK_2K_3} \, V - \frac{1 - e^{-t/T_2}}{F + GK_2K_3} \, T_L$$

where $T_2 = J/(F + GK_2K_3)$.

Thus the applied load torque slows the system down exponentially to a new steady speed of the initial speed minus $T_L \cdot 1/(F + GK_2K_3)$.

Now define stiffness as the torque required to produce a unit change in speed, then

$$\text{stiffness} = F + GK_2K_3$$

The effect of the closed loop on the stiffness can be inferred by noting that for the open loop system $K_3 = 0$ and the open loop stiffness is F. Thus, with the figure of $GK_2K_3 = 100F$ used in the previous section, the closed loop stiffness is approximately 100 times higher than the open loop stiffness.

It has been possible here to simply demonstrate mathematically the advantages of closed loop control which were descriptively introduced in Chapters 1 and 2. The disadvantage, instability, is however not shown up, due purely to the oversimplification of the problem. It is thus necessary to study higher order systems than the first-order one considered so far.

3.4 The Remote Position Control Servomechanism

3.4.1 The Basic Second-Order System

In this example the output position is monitored by a slide-wire potentiometer, while the desired output position is simulated by an identical input potentiometer. The two potentiometers connected in bridge form act as their own error detector, the amplified output feeding a d.c. motor field coil. The motor is fed with constant armature current so that the output torque is proportional to the field current, and the motor output position is geared down to the desired output. Field saturation is avoided in order to maintain linear working conditions.

The feedback must be negative so that unbalance of the bridge circuit produces an error signal driving the output in a direction such as to reduce the error to zero.

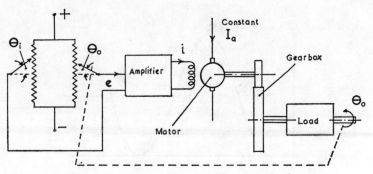

Fig. 3.2 A remote position control servomechanism.

Consider the circuit of Fig. 3.2.

The amplifier input = Actuating signal, e
$$= K_1(\theta_i - \theta_o).$$
The field current, $i = Ge = GK_1(\theta_i - \theta_o).$
The developed motor torque $= K_2 i$
$$= GK_1K_2(\theta_i - \theta_o)$$
and $\therefore$ the torque at the output $= nGK_1K_2(\theta_i - \theta_o)$ \hfill (3.1)

Output torque = Acceleration + Viscous + Static and load torques.
For the moment neglect all except the acceleration torque
$$= J\ddot{\theta}_o \, \text{N m}.$$

If the motor position $= \theta_m$

then $n\theta_o = \theta_m$

and $n\ddot{\theta}_o = \ddot{\theta}_m$

Acceleration torque required for the motor $= J_m\ddot{\theta}_m$

which referred to the output $= nJ_m\ddot{\theta}_m$

Acceleration torque required for the load $= J_L\ddot{\theta}_o$

Total acceleration torque referred to the output $= nJ_m\ddot{\theta}_m + J_L\ddot{\theta}_o$

$$= (n^2J_m + J_L)\ddot{\theta}_o$$

$$\therefore J = n^2J_m + J_L$$

Similarly, as will be required later, $F = n^2F_m + F_L$. Note that the inertia and friction of the gearing itself must be included.

Thus $nGK_1K_2(\theta_i - \theta_o) = J\ddot{\theta}_o$

$$\therefore \frac{J}{nGK_1K_2}\ddot{\theta}_o + \theta_o = \theta_i \qquad (3.2)$$

This equation relates the motion of the output, θ_o, at any instant of time for any given input, θ_i. It is the equation of simple harmonic motion and Fig. 3.3 shows the 'response' of the system to a step change in θ_i.

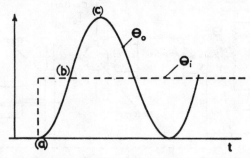

Fig. 3.3 Response of an undamped system to a step input.

The output will oscillate with an amplitude equal to the applied step at a frequency $\omega_n = \sqrt{\dfrac{nK_1K_2G}{J}}$ and is called 'the undamped natural frequency of oscillation'.

This phenomenon can be explained thus: At the instant of applying θ_i the 'error', $\theta_i - \theta_o$, is a maximum, and therefore maximum torque is applied to the load, resulting in maximum acceleration. As θ_o approaches the value set by θ_i, the error, the torque and therefore the acceleration fall until they are zero when $\theta_o = \theta_i$. However, the output velocity is at its maximum value at this instant of coincidence so that the kinetic energy stored by the inertia causes the output to swing past the desired position. In so doing a decelerating torque is produced which halts the output when it has overshot an amount equal to the initial displacement. At this point the error is again a maximum, producing maximum acceleration so that the process repeats itself *ad infinitum*.

3.4.2 The Effect of Viscous Friction

Friction is a nonlinear effect as shown in Fig. 3.4. The true diagram, (*a*), can be approximated as in (*b*) which can be split into two components: (i) the static or coulomb friction, and (ii) the viscous friction.

The effect of static friction is to produce steady state errors since when the output force is less than or equal to the static force the output

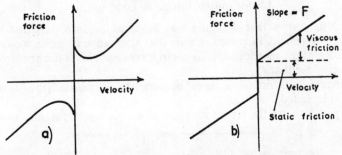

Fig. 3.4 Friction characteristics: (*a*) actual, and (*b*) linearised.

will cease to move (inertia will affect this somewhat), leaving the resulting error. For the sake of linearity this is neglected here.

Viscous friction forces, or torques, are assumed proportional in amplitude to velocity and therefore produce the desirable effect of damping. In practice there are better ways of producing damping, but this must first be considered.

Referring to the system of Fig. 3.2, let the viscous friction constant

$$= F \, \text{N m/(rad/s)} \, (\equiv \text{N m s})$$

so that viscous torque at the output

$$= F\dot{\theta}_o \, \text{N m}$$

Substituting in equation 3.1

$$\text{developed output torque} = nGK_1K_2(\theta_i - \theta_o)$$
$$\text{Load torque} = J\ddot{\theta}_o + F\dot{\theta}_o$$

Hence by equating:

$$J\ddot{\theta}_o + F\dot{\theta}_o + nGK_1K_2\theta_o = nGK_1K_2\theta_i$$

and

$$\frac{J}{nGK_1K_2}\ddot{\theta}_o + \frac{F}{nGK_1K_2}\dot{\theta}_o + \theta_o = \theta_i \qquad (3.3)$$

This is a second-order differential equation, the solution of which will be dealt with after considering velocity feedback.

Note, however, that presence of the $\dot{\theta}_o$ term in equation 3.3 means that the output response to an applied step will not continue to oscillate as in Fig. 3.3, due to the absorption of the stored kinetic energy by the viscous friction, i.e. the system is said to be DAMPED.

3.4.3 Velocity or Output Derivative Feedback

Damping is introduced into a system by viscous friction. This is clearly not a good form of damping since the idea of a friction force $F\dot{\theta}_o$ is in the first place an approximation, i.e. F is not truly a constant, and further it entails utilisation of some of the useful motor output torque.

Consider the balance equation:

Motor output torque = Load torque

It was shown that damping was achieved by increasing the load torque as velocity increased; an obvious alternative to which would be to decrease the motor output as speed increases. This is the principle of 'velocity' feedback.

Since viscous friction can never be completely eliminated, the $F\dot{\theta}_o$ term is still included.

Fig. 3.5 shows the circuit diagram of a r.p.c. system similar to that

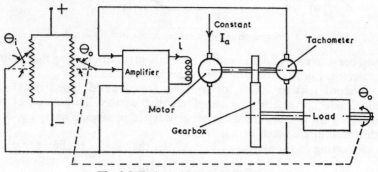

Fig. 3.5 Tachometric velocity feedback.

of Fig. 3.2, but with the output velocity monitored by a tachogenerator. The tacho produces a voltage proportional to velocity and is usually mounted on the motor shaft rather than the output shaft in order to minimise brush ripple.

If the tacho generates K_3' volts/(rad/s), then the velocity feedback constant referred to the output

$$= K_3 = nK_3' \text{ volts/(output rad/s)}$$

The amplifier input is now $K_1(\theta_i - \theta_o) - K_3\dot{\theta}_o$

so that the output torque $= nGK_2(K_1(\theta_i - \theta_o) - K_3\dot{\theta}_o)$

Load torque $= J\ddot{\theta}_o + F\dot{\theta}_o$

Hence $\quad J\ddot{\theta}_o + (F + nGK_2K_3)\dot{\theta}_o + nGK_1K_2\theta_o = nGK_1K_2\theta_i$

$$\therefore \frac{J}{nGK_1K_2}\ddot{\theta}_o + \frac{F + nGK_2K_3}{nGK_1K_2}\dot{\theta}_o + \theta_o = \theta_i \qquad (3.4)$$

It can be seen by comparison of this equation with equation 3.3 that the system is again damped and that if F is minimised the same degree of damping can be achieved by increasing K_3.

Velocity feedback damping, then, is superior to viscous friction damping since:

(i) it is much more linear and predictable;
(ii) no output torque is required, thus reducing the work load on the motor.

3.5 General Form and Solution of the Second-Order Equation

It will be shown later that for a practical system the equation of motion will be of a higher order than second. Many important principles, however, can be introduced and explained very adequately by considering a simple second-order system.

In solving the differential equations of a system the Laplace transform is of the utmost importance and any reader unfamiliar with the transform approach should first read the Appendix. Here the general equation will be solved by the classical approach, but the same problem is used as an example in the Appendix.

First, however, note that the second-order differential equation just derived for a *closed loop* system is the same equation as could be derived for some open loop system, i.e. a closed loop system is only a particular example of a dynamic problem.

Thus for the spring-mass-damper system of Fig. 3.6, where

$$\text{Spring stiffness} = \lambda \text{ N/m}$$
$$\text{Mass} = M \text{ kg}$$
$$\text{Damping coefficient} = F \text{ N/(m/s)}$$
$$\text{Displacement} = x \text{ m}$$
$$\text{Initial displacement} = x_i \text{ m}$$

Then
$$(x_i - x)\lambda = \frac{M}{g}\ddot{x} + F\dot{x}$$

or
$$\frac{M}{g\lambda}\ddot{x} + \frac{F}{\lambda}\dot{x} + x = x_i$$

Similarly for the R–L–C circuit of Fig. 3.7:

$$v = iR + L\frac{di}{dt} + \frac{1}{C}\int i\,dt$$

Substituting
$$v_c = \frac{1}{C}\int i\,dt$$

$$CL\frac{d^2v_c}{dt^2} + CR\frac{dv_c}{dt} + v_c = v$$

Thus three examples have been derived from totally dissimilar physical systems with similar defining differential equations. Thus a general form of the mathematical equation will be most helpful. The simplest general form is

$$a\ddot{x} + b\dot{x} + cx = y$$

where x is the dependent variable and y is the forcing term. This equation is better normalised to

$$\frac{a}{c}\ddot{x} + \frac{b}{c}\dot{x} + x = \frac{1}{c}y$$

For reasons not clear until the solution of this equation is considered, there is a preferable general form which, while more awkward in

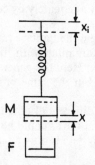

Fig. 3.6 A spring, mass and damper.

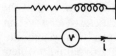

Fig. 3.7 *R–L–C* circuit.

appearance, expresses the nature of the transient response more explicitly. Thus if $b = 0$, the response is simple harmonic, of frequency

$$\omega_n = \sqrt{\frac{c}{a}}\,\text{rad/s}$$

This is called the undamped or natural frequency, and the equations can be rewritten as

$$\frac{1}{\omega_n{}^2}\ddot{x} + \frac{b}{c}\ddot{x} + x = \frac{1}{c}y$$

The solution of this equation has two parts:

    (i) a complementary function or transient term;
    (ii) a particular integral or steady state term.

To find the complementary function, an auxiliary equation is written, with $T = b/c$, as

$$\frac{1}{\omega_n{}^2}m^2 + Tm + 1 = 0$$

the roots of which are

$$m = \frac{-T \pm \sqrt{T^2 - 4/\omega_n{}^2}}{2/\omega_n{}^2}$$

Define critical damping as the condition for which both roots are equal, and let T_c be the value of T corresponding to this condition, then

$$T_c{}^2 = \frac{4}{\omega_n{}^2}$$

or $\qquad T_c = \frac{2}{\omega_n}$ (the possible negative value can be neglected)

Now define the term 'damping ratio' as

damping ratio $\zeta = \dfrac{\text{actual damping coefficient}}{\text{critical damping coefficient}} = \dfrac{T}{T_c} = \dfrac{T}{2/\omega_n}$

Hence

$$T = \frac{2\zeta}{\omega_n}$$

The general form of the second-order differential equation is thus

$$\frac{1}{\omega_n{}^2}\ddot{x} + \frac{2\zeta}{\omega_n}\dot{x} + x = \text{forcing terms}$$

The constants ω_n and ζ are more meaningful physically than the simple a and b.

Thus the position control servo is described by the equation

$$\frac{1}{\omega_n{}^2}\ddot{\theta}_o + \frac{2\zeta}{\omega_n}\dot{\theta}_o + \theta_o = \theta_i$$

The auxiliary equation roots become:

$$m = \frac{-\dfrac{2\zeta}{\omega_n} \pm \sqrt{\dfrac{4\zeta^2}{\omega_n{}^2} - \dfrac{4}{\omega_n{}^2}}}{2/\omega_n{}^2} = -\zeta\omega_n \pm \omega_n\sqrt{\zeta^2 - 1}$$

If $\zeta < 1$, then

$$m = -\sigma \pm j\omega \quad \text{where} \quad \sigma = \zeta\omega_n \quad \text{and} \quad \omega = \omega_n\sqrt{1 - \zeta^2}$$

The complementary function then has three possible forms:

(i) Under-damped, $\zeta < 1$

The roots are imaginary, $m = -\sigma \pm j\omega$

$$\text{C.F.} = e^{-\sigma t}(A \sin \omega t + B \cos \omega t)$$

σ is called the DECAY RATE (s^{-1}) and ω the FREQUENCY OF OSCILLATION (rad/s).

(ii) Critically damped, $\zeta = 1$

The roots are equal, $m = -\sigma$

$$\text{C.F.} = (A + Bt)e^{-\sigma t}$$

(iii) Over-damped, $\zeta > 1$

The roots are real, $m = -\sigma_1$ and $-\sigma_2$

$$\text{C.F.} = Ae^{-\sigma_1 t} + Be^{-\sigma_2 t}$$

The coefficients A and B need to be determined from given initial conditions, but it should be noted that the transient performance has been investigated *without reference to the input* θ_i. The auxiliary equation is of particular importance in determining system stability, when in operational form it is called the 'characteristic equation' (see Chapter 6).

At this stage two important input functions must be considered, the step and the ramp. The appropriate particular integrals must be calculated and added to the complementary function to find the complete solution for θ_o.

(*a*) Step function input:

$$\theta_i = P \quad \text{for} \quad t \geqslant 0$$
$$= 0 \quad \text{for} \quad t < 0$$

By inspection the particular integral is $\theta_o = P$, since all derivatives are zero in the steady state.

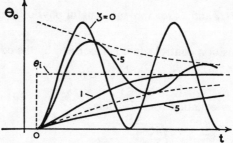

Fig. 3.8 Response of a second-order system to a step input.

Putting the initial conditions that θ_o and $\dot{\theta}_o$ are zero when $t = 0$, the constants A and B can be calculated giving the following results, which are shown graphically in Fig. 3.8.

(i) Under-damped $\zeta < 1$

$$\theta_o = P\left[1 - e^{-\zeta\omega_n t}\left(\cos \omega t + \frac{\zeta}{\sqrt{1 - \zeta^2}} \sin \omega t\right)\right]$$

(ii) Critically damped $\zeta = 1$

$$\theta_o = P[1 - e^{-\omega_n t}(1 + \omega_n t)]$$

(iii) Over-damped $\zeta > 1$

$$\theta_o = P\left[1 - e^{-\zeta\omega_n t}\left(\cosh \omega t + \frac{\zeta}{\sqrt{\zeta^2 - 1}} \sinh \omega t\right)\right]$$

$$= P\left[1 + \frac{\sigma_2}{\sigma_1 - \sigma_2}\,e^{-\sigma_1 t} - \frac{\sigma_1}{\sigma_1 - \sigma_2}\,e^{-\sigma_2 t}\right]$$

(*b*) Ramp function input:

$$\theta_i = Qt \quad \text{for} \quad t \geqslant 0$$
$$= 0 \quad \text{for} \quad t < 0$$

Let the particular integral be of the form $\theta_o = a + bt$

then $\qquad\qquad \dot{\theta}_o = b \quad \text{and} \quad \ddot{\theta}_o = 0$

Substituting in the basic equation:

$$0 + \frac{2\zeta}{\omega_n}\cdot b + a + bt = Qt$$

Hence $\qquad\qquad\qquad b = Q$

and $\qquad \dfrac{2\zeta}{\omega_n}b + a = 0 \quad \text{or} \quad a = -\dfrac{2\zeta}{\omega_n}Q$

The particular integral then is

$$\theta_o = Qt - Q\frac{2\zeta}{\omega_n}$$

Note straight away that $\theta_o \neq \theta_i$ in the steady state.

Adding the particular integral and the complementary function with the initial conditions that $\theta_o = \dot{\theta}_o = 0$, when $t = 0$ gives the following results as plotted in Fig. 3.9.

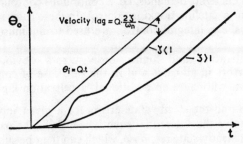

Fig. 3.9 Response of a second-order system to a ramp input.

(i) Under-damped $\zeta < 1$

$$\theta_o = Q\left[t - \frac{2\zeta}{\omega_n} + \frac{e^{-\zeta\omega_n t}}{\omega_n}\left(2\zeta\cos\omega t + \frac{2\zeta^2 - 1}{\sqrt{1 - \zeta^2}}\sin\omega t\right)\right]$$

Desired output — Steady state error — Transient error

(ii) Critically damped $\zeta = 1$

$$\theta_o = Q\left[t - \frac{2}{\omega_n} + e^{-\omega_n t}\left(\frac{2}{\omega_n} + t\right)\right]$$

(iii) Over-damped $\zeta > 1$

$$\theta_o = Q\left[t - \frac{2\zeta}{\omega_n} + \frac{e^{-\zeta\omega_n t}}{\omega_n}\left(2\zeta \cosh \omega t + \frac{2\zeta^2 - 1}{\sqrt{\zeta^2 - 1}} \sinh \omega t \right) \right]$$

The steady state error, the output lagging the input, is misleadingly called the 'VELOCITY LAG ERROR', misleading since it is a position lag *due* to a velocity or rate input.

3.6 Errors

A closed loop control system is actuated by the 'error' existing between the input and the output, an error which theoretically reduces to zero when steady state is reached. In practical systems, however, the following steady state errors do exist:

(i) Static position error. Forces applied to the system by static friction and steady loads, due for example to cutting forces on a machine tool or wind loads on a radar aerial, will require the control system to produce a steady output force at the final position which is achieved only by a steady state position error. The magnitude of this error will depend upon the system gain so that a high gain system will need only a small error to give the required output force.

(ii) Velocity error. Fig. 3.9 shows that if θ_o is changing at a constant rate, then an additional error to that caused by loading exists. This velocity error is particularly important in systems subjected to continually changing demands, i.e. a continuous contouring machine-tool control or a satellite tracking control. In these cases special design considerations, e.g. integral control, are used to minimise the velocity error.

(iii) Acceleration and higher order errors obviously exist as transient errors in all cases and in the rare case of constant acceleration input, will cause an additional 'acceleration lag' of position.

In order to find the steady state errors for a given input, it is not necessary to solve the differential equations in full. The particular integral gives the steady state response, which can then be subtracted from the desired response to give the error. Other methods are discussed in Chapter 7, and examples are included in Sections 8.2 and 8.4. Note that the steady state response may not be constant, e.g. with sinusoidal θ_i the steady state response of θ_o will also be sinusoidal; steady state is here defined as that condition obtaining when the transient terms have decayed away.

3.7 The Effect of Additional Time Constants

In deriving the second-order differential equation so far considered, a number of simplifications, or rather omissions, were made. Such effects as time constants in the amplifier, compliance of the load, twisting of structures, gearing and lead-screws will all affect the differential equation. Some of these factors are enlarged upon in Part Four. Here the simplest example of the motor-field time constant will be analysed in order to show the necessity for a better method of solution than the 'long-hand' method so far used.

The example of Fig. 3.5 will now be reworked, allowing for the field inductance L.

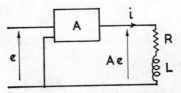

Fig. 3.10 Equivalent circuit of motor field and amplifier.

The amplifier was defined to have a gain or 'transconductance' of G amperes out per volt in. The effective voltage gain is A and the field circuit resistance is R.

Thus
$$A = G \cdot R$$

The equivalent circuit of the amplifier is shown in Fig. 3.10, from which

$$Ae = iR + L\frac{di}{dt} = R\left(i + \frac{L}{R}\frac{di}{dt}\right)$$

$$\therefore \quad Ge = i + T_f\frac{di}{dt}$$

This expression is in itself limiting and clumsy since i cannot be expressed as a simple function of e. From the previous example

$$e = K_1(\theta_i - \theta_o) - K_3\dot{\theta}_o$$

and
$$J\ddot{\theta}_o + F\dot{\theta}_o = nK_2 i$$

Thus
$$A[K_1(\theta_i - \theta_o) - K_3\dot{\theta}_o] = iR + L\frac{di}{dt}$$

$$= R\frac{J\ddot{\theta}_o + F\dot{\theta}_o}{nK_2} + L\frac{d}{dt}\frac{J\ddot{\theta}_o + F\dot{\theta}_o}{nK_2}$$

Hence
$$nK_2A[K_1(\theta_i - \theta_o) - K_3\dot{\theta}_o] = LJ\dddot{\theta}_o + (LF + RJ)\ddot{\theta}_o + RF\dot{\theta}_o$$

$$\therefore \quad \frac{LJ}{nK_1K_2A}\dddot{\theta}_o + \frac{LF + RJ}{nK_1K_2A}\ddot{\theta}_o + \frac{RF + nK_2K_3A}{nK_1K_2A}\dot{\theta}_o + \theta_o = \theta_i$$

This is a third-order equation which is appreciably more difficult to solve (factorising the third-order auxiliary equation), so that the Laplace transform should be put to use. More important, however, is that the simultaneous equations used to derive this expression could have been written in operational form, thereby simplifying the derivation of the system equation. This approach is expanded in the following chapter under the general title of 'Transfer Functions'. The point is to *derive* the equation in operational form (Laplace transform form) rather than simpy to *solve* the equations:

For example, the amplifier just considered gave the following defining equation:

$$Ae = iR + L\frac{di}{dt}$$

Transforming, with *all* initial conditions at zero, e.g. i, θ_0 and $\dot{\theta}_o = 0$ at $t = 0$

$$AE(s) = I(s)R + LsI(s)$$

$$\therefore I(s) = \frac{A}{R + sL} E(s) = \frac{A}{R} \cdot \frac{1}{1 + s\dfrac{L}{R}} \cdot E(s)$$

$$= G \cdot \frac{1}{1 + sT_f} E(s)$$

The transform equation enables i to be expressed as a simple function of e, which is not possible from the original equation.

Similarly

$$nK_2I(s) = Js^2\theta_o(s) + Fs\,\theta_o(s)$$

$$= s(F + sJ)\theta_o(s) = sF\left(1 + s\frac{J}{F}\right)\theta_o(s)$$

$$= sF(1 + sT_m)\,\theta_o(s)$$

$T_m = \dfrac{J}{F}$ has the dimensions of time and is therefore a time constant.

Finally $$E(s) = K_1(\theta_i(s) - \theta_o(s)) - K_3s\theta_o(s)$$

So that

$$nK_2G \cdot \frac{1}{1 + sT_f} [K_1(\theta_i(s) - \theta_o(s)) - K_3s\,\theta_o(s)] = sF(1 + sT_m)\theta_o(s)$$

Hence

$$\left(\frac{FT_mT_f}{nK_1K_2G} s^3 + \frac{F(T_f + T_m)}{nK_1K_2G} s^2 + \frac{F + nK_2K_3G}{nK_1K_2G} s + 1\right)\theta_o(s) = \theta_i(s)$$

This solution is still clumsy, so that the following chapters will enlarge upon the ideas of transfer functions and introduce frequency response functions as means of simplifying the design by avoiding the full solution.

3.8 Summary

This chapter constitues a precis of linear system analysis, with particular attention to servomechanisms. Cheng (ref. 7) or Aseltine (ref. 8) are good texts on this topic.

The simple differential equations encountered here were solved by classical methods, and it is suggested that the reader should be quite certain of repeating the solution using the Laplace transform. It has been shown that the major impact of the Laplace transform lies in the simplification of the derivation of the equations, rather than the solution, since the equations are transformed from differential to algebraic equations. Thus the next chapter aims to show how the system can be analysed by interpreting the transform equation directly, without inverse transformation.

4
Transfer Function Analysis

4.1 Definition of Transfer Functions

The transfer function of an element is defined as the ratio of the Laplace transform of the output to the Laplace transform of the input, provided the initial conditions are zero; its use is to simplify the derivation of the operational form of the system differential equation. Since it defines an output over an input the transfer function of any element can be found whether it is an open or a closed loop. In addition, the input and output can be in any dimension, e.g. a position output in radians and an electrical input in volts; or a torque output and a spool valve-displacement input. As it is a ratio of Laplace transforms the transfer function is an algebraic function of s.

Thus the transfer function is

$$\frac{F_o(s)}{F_i(s)} = G(s)$$

The symbol $G(s)$ is normally used in closed loop control theory to denote transfer functions of forward paths, and $H(s)$ to denote the transfer function of a feedback path.

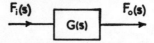

Fig. 4.1 Single transfer function block.

N.B. While $F_i(s)$ and $F_o(s)$ have inverse transforms $f_i(t)$ and $f_o(t)$, the inverse transform of $G(s)$ is not an equivalent time domain function.* (It is, in fact, called the weighting function and is of no significance until Section 10.3.) In the time domain the equivalent of the transfer function is a differential equation; *this is the real advantage of transfer functions that no equivalent simple algebraic input–output relationship can be established in the time domain.*

* Some authors have differentiated between direct transforms by using a bar notation and transfer functions by using capital letters, thus:

$$f_o(s) = G(s) f_i(s)$$

As long as the difference is appreciated this is accepted to be an unnecessary complication.

42

4.2 Validity of Transfer Functions

The conditions for validity of a transfer function stem from the definition as a ratio of Laplace transforms, so that to be meaningful it must be an algebraic function with constant parameters. This eliminates all systems with non-constant coefficients such as nonlinear systems with coefficients dependent upon the magnitude of f_i, and time-variant systems where the system parameters vary with time, i.e. aircraft elevator characteristics which change with altitude and therefore with time. However, since the major use of closed loop control is to eliminate the effects of change in open loop parameters, i.e. $G(s)$, it is assumed in general that the rate and magnitude of changes with time are small enough to allow a sufficiently accurate transfer function to be used.

4.3 Block Diagrams

The use of transfer functions enables a complete linear system to be represented by a series of interconnected blocks, each with its own

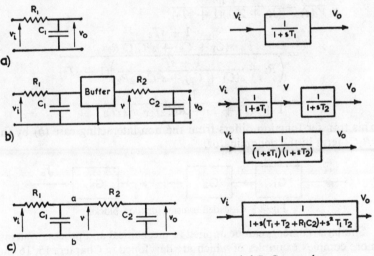

Fig. 4.2 Effect of buffer stages on cascaded R–C networks.

transfer function. It is essential that blocks are non-interacting, otherwise the transfer function of a block can be modified by the loading effect of a following block, as in the examples of Fig. 4.2.

In case (a):

$$\frac{V_o(s)}{V_i(s)} = \frac{\frac{1}{sC_1}}{R_1 + \frac{1}{sC_1}} = \frac{1}{1 + sC_1R_1} = \frac{1}{1 + sT_1}$$

where $\quad T_1 = C_1R_1$

In case (*b*) the elements are separated by a buffer stage, so as to be non-interacting:

$$\frac{V(s)}{V_i(s)} = \frac{1}{1 + sT_1} \quad \text{and} \quad \frac{V_o(s)}{V(s)} = \frac{1}{1 + sT_2}, \text{ where } T_2 = C_2 R_2$$

$$\therefore \frac{V_o(s)}{V_i(s)} = \frac{V_o(s)}{V(s)} \cdot \frac{V(s)}{V_i(s)} = \frac{1}{(1 + sT_1)(1 + sT_2)}$$

Since block diagrams must be noninteracting the block diagram of Fig. 4.2 (*b*) does not need to show the buffer.

In case (*c*), the effective transform impedance between points *a* and *b* is given by C_1 in parallel with R_2 and C_2.

$$\therefore Z_{ab} = \frac{\dfrac{1}{sC_1}\left(R_2 + \dfrac{1}{sC_2}\right)}{\dfrac{1}{sC_1} + R_2 + \dfrac{1}{sC_2}} = \frac{1 + sC_2 R_2}{sC_1 + sC_2 + s^2 C_1 C_2 R_2}$$

$$\frac{V(s)}{V_i(s)} = \frac{Z_{ab}}{R_1 + Z_{ab}} \quad \text{and} \quad \frac{V_o(s)}{V(s)} = \frac{1}{1 + sT_2}$$

$$\therefore \frac{V_o(s)}{V_i(s)} = \frac{Z_{ab}}{(R_1 + Z_{ab})(1 + sT_2)}$$

$$= \frac{\dfrac{1 + sT_2}{s(C_1 + C_2) + s^2 C_1 C_2 R_2}}{\left(R_1 + \dfrac{1 + sT_2}{s(C_1 + C_2) + s^2 C_1 C_2 R_2}\right)(1 + sT_2)}$$

$$= \frac{1}{1 + s(T_1 + T_2 + R_1 C_2) + s^2 T_1 T_2}$$

This transfer function differs from the non-interacting case (*b*) by the $sR_1 C_2$ term in the denominator.

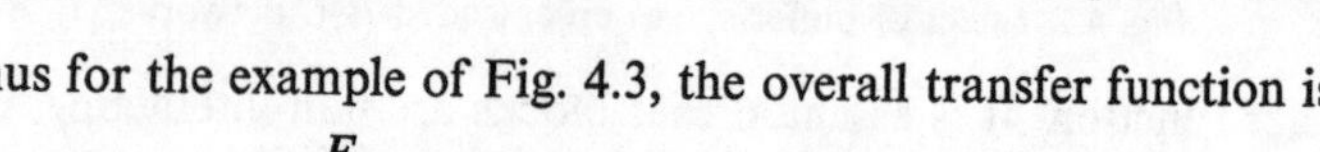

Fig. 4.3 Cascaded transfer function blocks.

Similar arguments can be applied to mechanical and process systems, more complex examples of which are developed in Chapters 15, 16 and 17.

Thus for the example of Fig. 4.3, the overall transfer function is

$$\frac{F_o}{F_i} = G = G_1 \cdot G_2 \cdot \ldots \cdot G_n$$

4.4 Manipulation of Transfer Functions in Closed Loops

So far the symbols f_i and f_o have been used to denote the general case of any input and any output. For use in specific closed loop problems a

new set of symbols will be used, as were listed in detail in Section 1.4
Thus

Reference (input) $r(t)$ or $R(s)$ or $R(j\omega)$
Controlled variable (output) $c(t)$ or $C(s)$ or $C(j\omega)$
Actuating signal $e(t)$ or $E(s)$ or $E(j\omega)$
Transfer function of forward paths $G(s)$ or $G(j\omega)$
Transfer function of feedback paths $H(s)$ or $H(j\omega)$

The input, output and actuating signal are functions of t, s or $j\omega$ to
represent time functions, Laplace transforms and frequency response

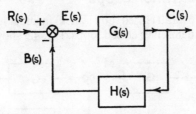

Fig. 4.4 Basic block diagram of a closed loop system.

functions (see Chapter 5); remember that $G(s)$ does not have an equivalent $g(t)$. $G(j\omega)$ is called a frequency response function. Fig. 4.4 shows
the block diagram of a general closed loop system.

The following points must be observed:

(i) It is essential to indicate a direction of signal 'flow' on all
block diagrams.

(ii) The summing point is assumed to be non-interacting and of
polarity as indicated. Positive feedback would be indicated by two
plus signs.

It is advisable now to refer back to the completely general case of
Section 1.4.

The overall closed loop transfer function is given by $\dfrac{C(s)}{R(s)}$.

$$C(s) = G(s) \cdot E(s)$$
$$B(s) = H(s) \cdot C(s)$$

and
$$E(s) = R(s) - B(s)$$
$$= R(s) - H(s) \cdot C(s) \qquad (4.1)$$
$$\therefore \quad C(s) = G(s)(R(s) - H(s)C(s))$$

Hence
$$\frac{C(s)}{R(s)} = \frac{G(s)}{1 + G(s)H(s)} \qquad (4.2)$$

Also
$$\frac{E(s)}{R(s)} = \frac{1}{1 + G(s)H(s)} \qquad (4.3)$$

The open loop transfer function, independent of where the loop is
opened, is $G(s)H(s)$. Obviously for a given open loop transfer function

there can only be one resulting closed loop transfer function as given by equation 4.2. This fact is of major significance since it means that a complete knowledge of the open loop system gives a complete knowledge of the closed loop. The problem of system stability will be dealt with later, but it is clear that if a closed loop system is unstable, no measurements can possibly be made upon it, while open loop measurements are quite practical, enabling a designer to see what compensation is necessary to make the closed loop stable.

4.5 Major and Minor Loops

In many cases the overall closed loop system includes in the loop certain blocks which are themselves closed loops. Fig. 4.5 shows an example of

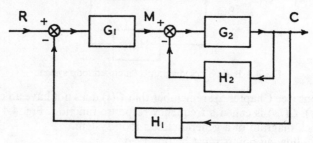

Fig. 4.5 A multiple loop system.

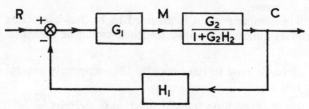

Fig. 4.6 Reduction of Fig. 4.5.

this which is similar to the position control system using tachometric velocity feedback. (H_2 would be the tachometer constant.)

The 'minor' loop G_2 and H_2 can be reduced by applying equation 4.2, leaving the circuit of Fig. 4.6, which is called the major loop.

From Fig. 4.6

$$\frac{C(s)}{R(s)} = \frac{G_1(s) \cdot \dfrac{G_2(s)}{1 + G_2(s)H_2(s)}}{1 + H_1(s)G_1(s) \cdot \dfrac{G_2(s)}{1 + G_2(s)H_2(s)}}$$

$$= \frac{G_1(s)G_2(s)}{1 + G_2(s)H_2(s) + G_1(s)G_2(s)H_1(s)}$$

4.6 The Open Loop Equivalent of a Closed Loop

The two networks of Fig. 4.7 are equivalent, i.e. have the same overall transfer function if:

$$G(s)G'(s) = \frac{G(s)}{1 + G(s)H(s)}$$

so that

$$G'(s) = \frac{1}{1 + G(s)H(s)}$$

From this it can be seen that the same response for a system can be achieved by either open or closed loop control. It must be remembered,

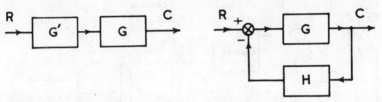

Fig. 4.7 Equivalent open and closed loop systems.

therefore, that the importance of closed loop control is not primarily to achieve good dynamic response but to compensate for changes in $G(s)$ and to regulate against disturbances.

4.7 Direct Feedback Equivalent of a General System

A direct feedback system is one for which $H = 1$.
    In general

$$\frac{C(s)}{R(s)} = \frac{G(s)}{1 + G(s)H(s)} = \frac{1}{H(s)} \cdot \frac{G(s)H(s)}{1 + G(s)H(s)}$$

The term $\dfrac{G(s)H(s)}{1 + G(s)H(s)}$ would be the transfer function of a direct feedback system with a forward transfer function of $G(s)H(s)$ as shown in the equivalent circuit of Fig. 4.8.

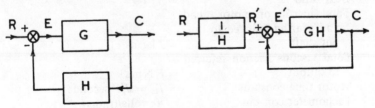

Fig. 4.8 Equivalent direct feedback system.

The equivalent direct feedback circuit is of importance, for instance, when using a Nichols Chart (Chapter 6) which applies only to direct feedback loops. The chart can be used to find $\dfrac{C(s)}{R'(s)}$ and multiplying by $\dfrac{1}{H(s)}$ gives $\dfrac{C(s)}{R(s)}$.

4.8 Application of Transfer Function Analysis to the Example of a Position Control Servomechanism

The example used in Section 3.7 can now be repeated, using transfer function analysis. The circuit is redrawn in Fig. 4.9.

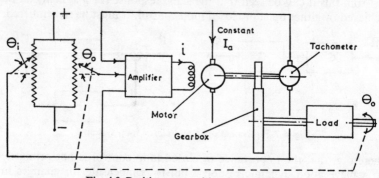

Fig. 4.9 Position servo with velocity feedback.

The constants of this system are:

Potentiometer bridge constant $= K_1$ volts/rad of $\theta_i - \theta_o$
Amplifier gain $= A$ volts/volt
                              $= G$ amperes/volt (N.B. do not confuse G and $G(s)$)
Field circuit resistance $= R$ ohms
Field circuit inductance $= L$ henry
Field circuit time constant $= T_f = L/R$ seconds
Field current $= i$ amperes
Motor torque constant (I_a constant) $= K_2$ Nm/ampere of field current
Gear ratio $= n : 1$
Total inertia of motor, load and gearing referred to output shaft $= J$ kgm^2
Total viscous friction referred to output $= F$ Nms
Motor time constant $= T_m = J/F$ seconds
Tachometer constant $= K_3$ volts/(rad/s)
Output velocity $= \omega_o = \dot{\theta}_o$ rad/s

Certain component transfer function blocks will be first considered separately:

(i) Velocity and position. Since velocity is the derivative of position:

$$\Omega_o(s) = s\Theta_o(s) \quad \text{or} \quad \frac{\Theta_o(s)}{\Omega_o(s)} = \frac{1}{s}$$

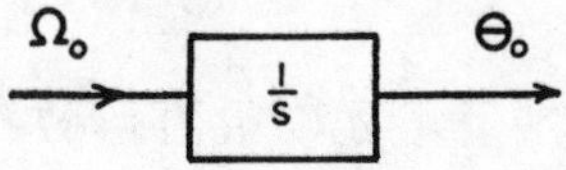

$$\Omega_o \longrightarrow \boxed{\frac{1}{s}} \longrightarrow \Theta_o$$

Fig. 4.10

(ii) Torque and velocity:

Total torque $\qquad T(t) = J\ddot{\theta}_o + F\dot{\theta}_o \qquad$ (4.4)

therefore $\qquad T(s) = (Js + F)\Omega_o(s)$

therefore $\qquad \dfrac{\Omega_o(s)}{T(s)} = \dfrac{1}{sJ + F}$

A relation between torque and position could be written:

$$\frac{\Theta_o(s)}{T(s)} = \frac{1}{s(sJ + F)}$$

It is, however, not desirable to eliminate $\dot{\theta}_o$ initially since, for example, the tachometer feedback is derived from this variable.

Further $\qquad \dfrac{\Omega_o(s)}{T(s)} = \dfrac{1}{F\left(1 + s\dfrac{J}{F}\right)} = \dfrac{1}{F(1 + sT_m)}$

where $T_m = \dfrac{J}{F}$ is the motor time constant.

Alternatively viscous friction can be expressed as a feedback in the following way:

From equation 4.4:

$$\text{Acceleration torque} = J\ddot{\theta}_o = T(t) - F\dot{\theta}_o$$

and $\qquad \therefore \ddot{\theta}_o(s) = \dfrac{\text{Accelerating torque}}{J}$

$$\therefore \Omega_o(s) = \frac{1}{s} \cdot \frac{1}{J} \cdot (T(s) - F\Omega_o(s))$$

This expression is the solution of the closed loop of Fig. 4.11.

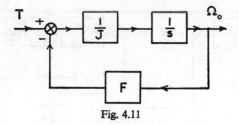

Fig. 4.11

(iii) Field current and amplifier input voltage:

$$\frac{V_o(s)}{V_i(s)} = A \quad \text{and} \quad \frac{I(s)}{V_o(s)} = \frac{1}{R + sL}$$

$$\therefore \frac{I(s)}{V_i(s)} = \frac{A}{R + sL} = \frac{A}{R} \cdot \frac{1}{1 + sT_f} = G \cdot \frac{1}{1 + sT_f}$$

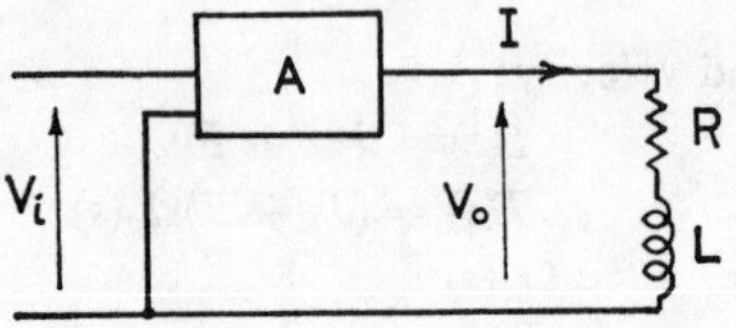

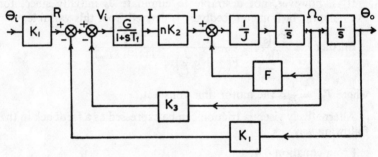

Fig. 4.12

The complete circuit of Fig. 4.9 can now be drawn in block diagram form thus:

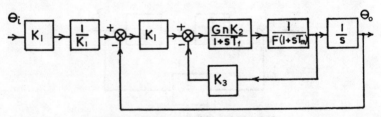

Fig. 4.13 Block diagram of the position control system of Fig. 4.9.

This diagram can now be reduced in stages as follows, by using the $\dfrac{G}{1 + GH}$ relationship, the technique being to reduce all minor loops first. Fig. 4.14 also includes the direct feedback modification of Fig. 4.8.

Fig. 4.14 First reduction of Fig. 4.13.

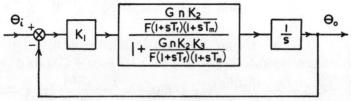

Fig. 4.15 Second reduction of Fig. 4.13.

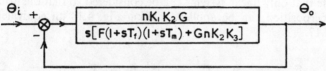

Fig. 4.16 Simplification of Fig. 4.15.

Finally, simplifying Fig. 4.16:

$$\frac{\theta_o(s)}{\theta_i(s)} = \frac{\dfrac{nK_1K_2G}{s[F(1 + sT_f)(1 + sT_m) + GnK_2K_3]}}{1 + \dfrac{nK_1K_2G}{s[F(1 + sT_f)(1 + sT_m) + GnK_2K_3]}}$$

$$= \frac{nK_1K_2G}{nK_1K_2G + s[F(1 + sT_f)(1 + sT_m) + GnK_2K_3]}$$

$$= \frac{1}{1 + \dfrac{nK_2K_3G + F}{nK_1K_2G}s + \dfrac{F(T_f + T_m)}{nK_1K_2G}s^2 + \dfrac{FT_fT_m}{nK_1K_2G}s^3}$$

On inspection this expression is the same as derived in Section 3.7.*

4.9 Systems with Multiple Inputs

Fig. 4.17 shows a simple system which has two inputs, e.g. the position control system previously analysed with the additional consideration

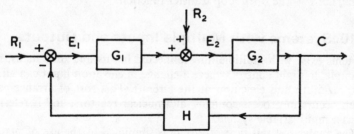

Fig. 4.17 A simple system with two inputs.

of applied load torques, in which case R_1 would be the reference input in radians and R_2 the load torque in N m. This is analysed in more detail in Chapters 15 and 17.

* The reader is now recommended to glance through Sections 14.7, 15.8 and 17.5.

$$C(s) = G_2(s)E_2(s) = G_2(s)[G_1(s)E_1(s) + R_2(s)]$$

and $\qquad E_1(s) = R_1(s) - H(s)C(s)$

$$\therefore \ C(s) = G_2(s)[G_1(s)(R_1(s) - H(s)C(s)) + R_2(s)]$$

$$\therefore \ C(s)(1 + G_1(s)G_2(s)H(s)) = G_1(s)G_2(s)R_1(s) + G_2(s)R_2(s)$$

$$\therefore \ C(s) = \frac{G_1(s)G_2(s)}{1 + G_1(s)G_2(s)H(s)} \cdot R_1(s) + \frac{G_2(s)}{1 + G_1(s)G_2(s)H(s)} \cdot R_2(s)$$

Note that only linear systems can be considered so that the principle of superposition can be applied, making Fig. 4.18 an equivalent of Fig. 4.17.

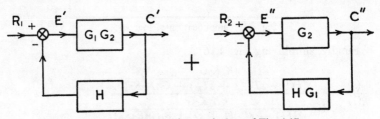

Fig. 4.18 Superposition equivalent of Fig. 4.17.

Thus $\qquad C'(s) = \dfrac{G_1(s)G_2(s)}{1 + G_1(s)G_2(s)H(s)} \cdot R_1(s)$

and $\qquad C''(s) = \dfrac{G_2(s)}{1 + G_1(s)G_2(s)H(s)} \cdot R_2(s)$

so that by superposition $C(s) = C'(s) + C''(s)$

giving the answer derived above.

It should be noted that while the response of the system output to the two inputs is different, the denominator of the two expressions is the same, i.e. $1 +$ the open loop transfer function.

4.10 Systems with Multiple Inputs and Outputs

Systems exist with multiple interconnected inputs and outputs as, for example, in flight control, where a change in elevation input can affect both azimuth and elevation of the aircraft. This sort of arrangement often occurs in process controls, and nuclear reactors, and is referred to as a multivariable system.

The analysis of this type of system is simplified by the use of matrix notation particularly when there are more than two inputs and outputs. It is unnecessary that the number of inputs equals the number of outputs.

Fig. 4.19 shows a block diagram of a general system.

Consider the two input, two output problem of Fig. 4.20.

$$C_1 = G_1G_2E_1 + G_6G_2E_2$$

and $$C_2 = G_3G_5E_1 + G_4G_5E_2$$

i.e. $$\begin{bmatrix} C_1 \\ C_2 \end{bmatrix} = \begin{bmatrix} G_1G_2 & G_6G_2 \\ G_3G_5 & G_4G_5 \end{bmatrix} \begin{bmatrix} E_1 \\ E_2 \end{bmatrix}$$

or $$[C] = [G][E]$$

Also $$E_1 = R_1 - H_1C_1$$
$$E_2 = R_2 - H_2C_2$$

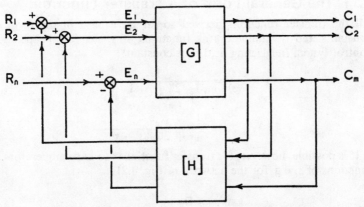

Fig. 4.19 Matrix representation of a multivariable system.

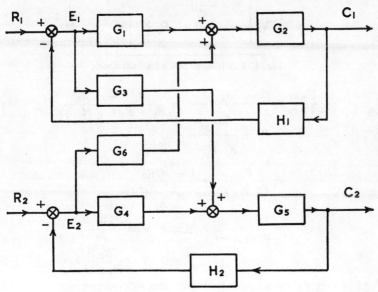

Fig. 4.20 A two-input, two-output interconnected system.

i.e.
$$\begin{bmatrix} E_1 \\ E_2 \end{bmatrix} = \begin{bmatrix} R_1 \\ R_2 \end{bmatrix} - \begin{bmatrix} H_1 & 0 \\ 0 & H_2 \end{bmatrix} \begin{bmatrix} C_1 \\ C_2 \end{bmatrix}$$

or
$$[E] = [R] - [H][C]$$

Manipulating
$$[C] = [I + [G][H]]^{-1} [G] [R]$$

This is the matrix equivalent of the conventional closed loop formula. Multivariable aspects are considered further in Chapter 12.

4.11 The General Form of a Transfer Function

All the transfer functions derived so far have been in the form of constants and denominators as functions of s, i.e. for the position control system (neglecting field time constant):

$$\frac{C(s)}{R(s)} = \frac{1}{1 + \dfrac{2\zeta}{\omega_n} s + \dfrac{1}{\omega_n{}^2} s^2}$$

$$= \frac{\omega_n{}^2}{\omega_n{}^2 + 2\zeta\omega_n s + s^2}$$

It is possible, however, for transfer functions* to have numerators as functions of s, e.g. for the network of Fig. 4.21:

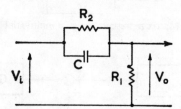

Fig. 4.21 A passive phase lead network.

$$\frac{V_o(s)}{V_i(s)} = \frac{R_1}{R_1 + \dfrac{R_2 \cdot \dfrac{1}{sC}}{R_2 + \dfrac{1}{sC}}} = \frac{R_1(1 + sCR_2)}{R_2 + R_1(1 + sCR_2)}$$

$$= \frac{R_1}{R_2 + R_1} \cdot \frac{1 + sCR_2}{1 + sC \dfrac{R_2 R_1}{R_2 + R_1}}$$

$$= \frac{1}{a} \cdot \frac{1 + saT}{1 + sT}, \quad \text{where} \quad a = \frac{R_2 + R_1}{R_1}$$

$$\text{and} \quad T = \frac{R_2 R_1}{R_2 + R_1} \cdot C$$

* Examples of networks which have numerators with complex roots are given in Section 16.5.

A transfer function can thus be manipulated into the general form of a rational polynomial:

$$G(s) = \frac{N(s)}{D(s)}$$

$$= \frac{b_o + b_1 s + b_2 s^2 + \ldots + b_m s^m}{a_o + a_1 s + a_2 s^2 + \ldots + a_n s^n}$$

where the a's and b's are real constants. $N(s)$ is the numerator and $D(s)$ the denominator polynomial.

This expression can be factorised thus:

$$G(s) = K' \cdot \frac{(s - z_1)(s - z_2) \ldots (s - z_m)}{(s - p_1)(s - p_2) \ldots (s - p_n)}$$

where $z_1, z_2, \ldots$, etc., are the roots of the equation $N(s) = 0$, and $p_1, p_2, \ldots$, etc., are the roots of the equation $D(s) = 0$. $K' = \dfrac{b_m}{a_n}$.

The roots z_1, z_2, etc., are called zeros, since if $s = z_1, z_2$, etc., the overall function will be zero.

The roots p_1, p_2, etc., are called poles, since if $s = p_1, p_2$, etc., the overall function will be infinite.

$(s - z_i)$ is called a zero factor and $(s - p_i)$ a pole factor. z and p may be real or complex.

The complex roots can now be expressed as complex numbers, e.g., $z_1 = \sigma_1 + j\omega_1$. From the fact that the a's and b's of the general equation must be real for any physical system all complex roots must exist in conjugate pairs:

i.e. if
$$z_1 = -(\sigma + j\omega) \quad \text{and} \quad z_2 = -(\sigma - j\omega)$$

then
$$(s - z_1)(s - z_2) = (s + \sigma + j\omega)(s + \sigma - j\omega)$$
$$= s^2 + 2\sigma s + (\sigma^2 + \omega^2) \tag{4.5}$$

while if
$$z_1 = -(\sigma_1 + j\omega_1) \quad \text{and} \quad z_2 = -(\sigma_2 + j\omega_2)$$

then
$$(s - z_1)(s - z_2) = (s + \sigma_1 + j\omega_1)(s + \sigma_2 + j\omega_2)$$
$$= s^2 + (\sigma_1 + \sigma_2 + j\omega_1 + j\omega_2)s$$
$$+ (\sigma_1 + j\omega_1)(\sigma_2 + j\omega_2) \tag{4.6}$$

Equation 4.5 has real coefficients, while 4.6 has possible imaginary coefficients and is therefore inadmissible. Since conjugate pairs of roots come from the solution of second-order equations $G(s)$ can be rewritten:

$$G(s) = K' \cdot \frac{(s - z_1)(s - z_2) \ldots (s^2 + \beta_1 s + \alpha_1)(s^2 + \beta_2 s + \alpha_2) \ldots}{(s - p_1)(s - p_2) \ldots (s^2 + \delta_1 s + \gamma_1)(s^2 + \delta_2 s + \gamma_2) \ldots}$$

where all coefficients z, p, α, β, γ and δ are real.

It is equally useful to normalise the transfer function by expressing the polynomial in s in the form $1 + as + \ldots$

As an example, the pole factor can be written as

$$s - p_1 = -p_1\left(1 - \frac{1}{p_1}s\right).$$

For such a simple factor the pole can be identified as minus the reciprocal of a time constant, so that the pole factor becomes $(1 + sT_1)/T_1$. The second-order factor (with complex roots) can be similarly normalised by writing it in the common ζ and ω_n form, with $\zeta < 1$.

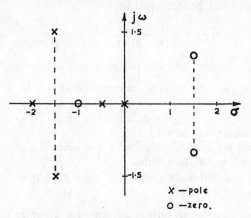

Fig. 4.22 s-Plane plot of
$$G(s) = \frac{K'(s - 1\cdot5 + j1)(s - 1\cdot5 - j1)(s + 1)}{s(s + 0\cdot5)(s + 2)(s + 1\cdot5 + j1\cdot5)(s + 1\cdot5 - j1\cdot5)}$$

The exception is a root or roots at the origin, which is left as a common s in the denominator raised to the power k; positive k indicating poles and negative k zeros at the origin.

Thus the transfer function can be rewritten as

$$G(s) = \frac{K(1 + sT_a)(1 + sT_b)\ldots\left(1 + \dfrac{2\zeta_a}{\omega_{na}}s + \dfrac{1}{\omega_{na}^2}s^2\right)\ldots}{s^k(1 + sT_1)(1 + sT_2)\ldots\left(1 + \dfrac{2\zeta_1}{\omega_{n1}}s + \dfrac{1}{\omega_{n1}^2}s^2\right)\ldots}$$

Thus $G(s)$ is made up of combinations of all or some of the simple set of types of factors indicated above.
Note that

$$K = \frac{(-z_1)(-z_2)\ldots\alpha_1\alpha_2\ldots K'}{(-p_1)(-p_2)\ldots\gamma_1\gamma_2\ldots}$$

The notation of the 'gain constant' K (without a prime) for $G(s)$ normalised to a $1+$ functions of s form and K' for the form normalised to $s^n + \ldots$ is used throughout this book.

The roots of $G(s)$ can be plotted on an Argand diagram, known as the '*s-plane*'. Root loci techniques, which are explained later use the s-plane to plot the loci of the closed loop poles from a known set of open loop roots as the gain constant is increased.

Systems having zeros in the R.H. half of the s-plane are called *non-minimum phase* systems, while those having zeros only in the L.H. half plane or on the $j\omega$ axis are called *minimum phase systems*. Systems with poles in the R.H. half plane are unstable as is shown in Chapter 6.

4.12 Signal Flow Diagrams or Graphs

4.12.1 Signal Flow Diagrams

Signal flow diagrams* are primarily an alternative pictorial representation to block diagrams. All signals (variables) are represented by dots,

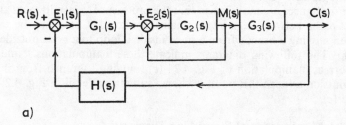

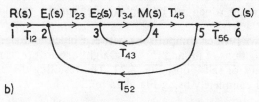

Fig. 4.23 Equivalent block diagram and signal flow graph representation.

called *nodes*, related variables being joined by lines called *directed branches*. Each branch has an associated transmittance, T_{jk}, which links node j to node k with zero transmittance from k to j, i.e. transmittance is unidirectional and is indicated by an arrow.

Thus Fig. 4.23 (*b*) shows the equivalent signal flow graph for the system represented by the block diagram of Fig. 4.23 (*a*)

Note the following points:

(i) The signal at a node is equal to the sum of all signals transmitted to the node, i.e. a node with more than one input is a summing

* See the articles by S. J. Mason, 'Feedback theory—some properties of signal flow graphs', *Proc. I.R.E.*, Sept. 1953 and 'Feedback theory—further properties of signal flow graphs', *Proc. I.R.E.*, July 1956.

point. Note that the transmittance may be negative. Thus at node 3

$$E_2(s) = T_{23}E_1(s) + T_{43}M(s)$$

(ii) The transmittances are simply related to the transfer functions, thus:

$$T_{23} = G_1(s); \ T_{43} = -1; \ T_{52} = H(s); \text{ etc.}$$

(iii) Nodes 1 and 6 are the input and output nodes respectively. Thus the transmittances T_{12} and T_{56} are both unity, and merely help make the diagram clearer.

(iv) The same rules of non-interaction as discussed in Section 4.3 apply to signal flow diagrams.

4.12.2 Manipulation of Signal Flow Diagrams

Branches can be moved from one node to another provided that the transmittances are modified so that the product of transmittances along any paths between *all* nodes remains unaltered. Care must be taken when moving a branch from a node inside a loop to a node outside the loop. The following diagrams indicate these manipulations. Note the incorrect manipulation of Fig. 4.24 (*e*), moving a branch from inside to outside a loop; this point is shown in greater detail in Fig. 4.26 (*b*) and (*c*).

4.12.3 Simplification of Signal Flow Diagrams

Fig. 4.25 shows a table of basic rules for simplification (reduction) of signal flow diagrams, which can be applied after suitable manipulation of the graph.

Fig. 4.26 shows further comparisons between block and signal flow graphs. The reductions of (*a*) and (*b*) are straightforward; (*c*) is reduced by creating a pair of self-loops, Fig. 4.27. The resulting diagram can then be reduced completely using Fig. 4.25 (*c*).

4.12.4 Mason's Signal Flow Formula

Input–output relationships, for both block and signal flow diagrams, have been derived by reducing the complexity of the diagram step by step. Mason gives a formula, stated here, to derive the input–output relation directly from the original signal flow diagram as an alternative to reduction.

An input variable is represented by a *source node* and an output (response) variable by a *sink node*. A path from a source to a sink node, without passing through any other node more than once, is called an *open* or a *forward path*, e.g. x_1–x_3–x_4–x_5–x_6–x_2 and x_1–x_3–x_5–x_6–x_2 in Fig. 4.26 (*a*), where x_1 is the source and x_2 the sink node.

A closed (feedback) path is called a *loop*, e.g. x_3–x_4–x_3 and x_4–x_5–x_4 in Fig. 4.26 (*c*). The product of the transmittances of the branches

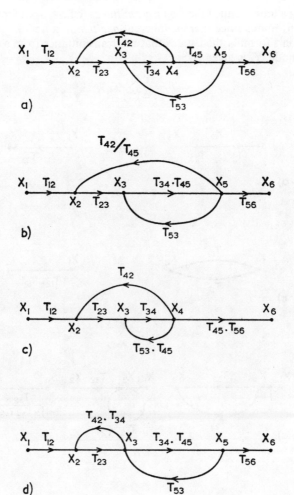

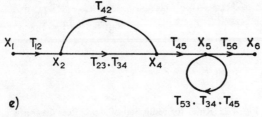

Fig. 4.24 Manipulation of branches ((e) shows a common error—see text).

forming a loop is called the *loop transmittance* (c.f. — open loop transfer function; minus since there is inherent *addition* at a node).

Mason's formula then states that net transmittance from a source to a sink node, *T*, is given by:

$$T = \frac{1}{\Delta} \sum_{k=1}^{l} T_k \Delta_k$$

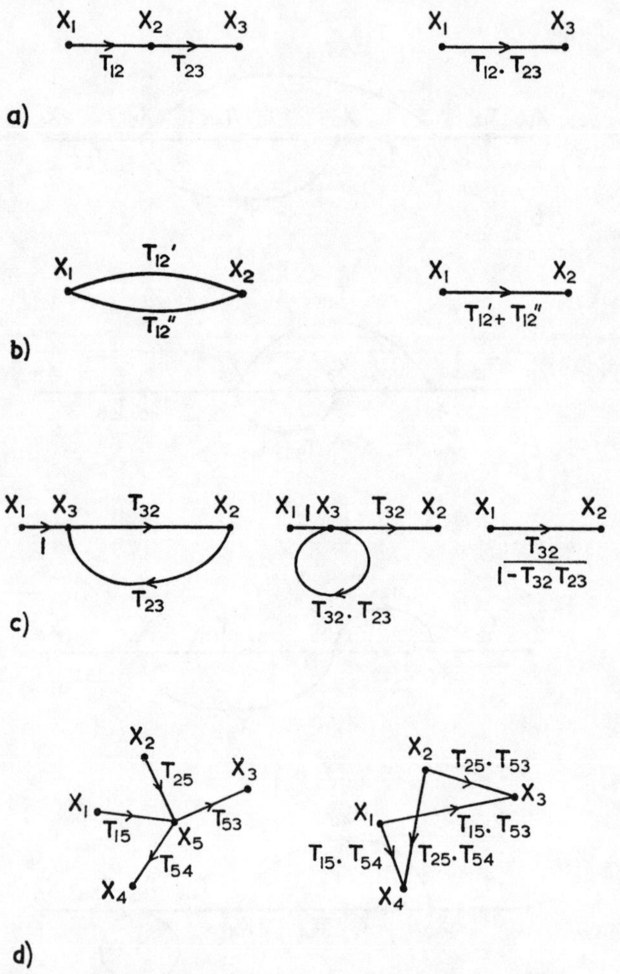

Fig. 4.25 Rules for reduction of signal flow diagrams

(*a*) Series paths. (*b*) Parallel paths. (*c*) Feedback loop. This is equivalent to $G/(1 + GH)$. Note that x_3, the signal *at* the node, is equivalent to *E*, the signal *after* the summing point on a block diagram (see Fig. 4.26). (*d*) Elimination of a node.

where there are l open paths between the source and sink nodes under consideration.

$\Delta = 1 -$ (sum of all loop transmittances)
  $+$ (sum of products of loop transmittances of all possible non-touching loops taken in pairs)
  $-$ (sum of similar products taken three at a time)
  $+$ etc.

$\Delta_k =$ value of Δ calculated for that part of the graph not touching the kth open path.

Thus for Fig. 4.26 (*a*):

two open paths (i) x_1–x_3–x_4–x_5–x_6–x_2, $T_1 = G_1 G_2 G_3$
         (ii) x_1–x_3–x_5–x_6–x_2, $T_2 = G_3$

one loop   (*a*) x_5–x_6–x_5, $T_a = -G_3 H$

There are no non-touching feedback loops and the loop touches both open paths, hence

$$\Delta = 1 - T_a = 1 + G_3 H$$
$$\Delta_1 = 1, \quad \Delta_2 = 1$$

$$\therefore \ T = \frac{1}{1 + G_3 H}\,[G_1 G_2 G_3 \,.\, 1 + G_3 \,.\, 1] = \frac{G_3(1 + G_1 G_2)}{1 + G_3 H}$$

For Fig. 4.26 (*b*):

one open path (i) x_1–x_3–x_4–x_5–x_6–x_2, $T_1 = G_1 G_2$

two loops   (*a*) x_3–x_4–x_3, $T_a = -G_1$
         (*b*) x_5–x_6–x_5, $T_b = -G_2$

The loops do not touch, but both touch the open path.

$$\therefore \ \Delta = 1 - (T_a + T_b) + T_a T_b = 1 + G_1 + G_2 + G_1 G_2$$
$$\Delta_1 = 1$$

$$\therefore \ T = \frac{1}{1 + G_1 + G_2 + G_1 G_2}\,.\,G_1 G_2 = \frac{G_1 G_2}{(1 + G_1)(1 + G_2)}$$

For Fig. 4.26 (*c*):

one open path (i) x_1–x_3–x_4–x_5–x_2, $T_1 = G_1 G_2$

two loops   (*a*) x_3–x_4–x_3, $T_a = -G_1$
        (*b*) x_4–x_5–x_4, $T_b = -G_2$

No non-touching feedback loops and both loops touch the open path

$$\therefore \ \Delta = 1 - (T_a + T_b) = 1 + G_1 + G_2$$
$$\Delta_1 = 1$$

$$\therefore \ T = \frac{1}{1 + G_1 + G_2}\,.\,G_1 G_2 = \frac{G_1 G_2}{1 + G_1 + G_2}$$

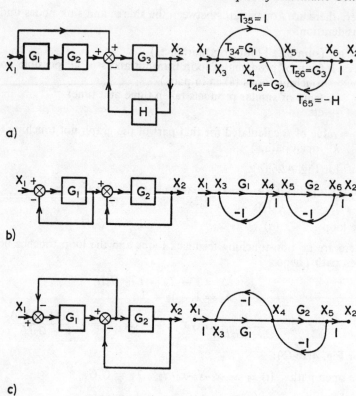

a)

b)

c)

Fig. 4.26 Further examples of signal flow diagrams.

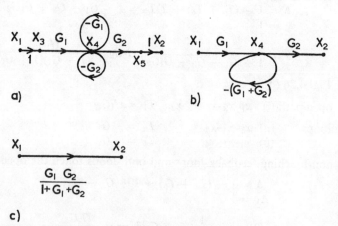

a)

b)

c)

Fig. 4.27 Reduction of Fig. 4.26 (c)

4.13 Summary

It has been shown in this chapter that it is possible to give significance to the Laplace transformed differential equation directly. The system dynamics are considered in the s-domain rather than the physical time-domain. The objective is to achieve an understanding of the significance of the system dynamics represented by the characteristics of the transfer function, e.g. the poles and zeros, without recourse to inverse transformation. The importance of such an approach is to retain the algebraic nature of functions in the s-domain compared to the differential nature of the time-domain. The disadvantage is the loss of physical significance; it is possible to display $f_0(t)$ on an oscilloscope, but what does $F_0(s)$ look like physically?

To take maximum advantage of the algebraic properties of transfer functions, two diagrammatic aids have been introduced: block diagrams and signal flow graphs.

Towill (ref. 5) gives a detailed treatment of transfer function techniques.

5

Frequency Response Analysis

5.1 Frequency Response Functions

It has been shown that the dynamic performance of a system can be analysed in terms of the system constants by the classical differential equation approach, or by Laplace transform techniques leading to the use of transfer functions. These methods, directly applicable to systems with known parameters, are limited in practice by the problem of measurement. Chapter 13 outlines a method of measuring the polynomial form of $G(s)$, and more conventional transient methods, i.e. measuring the response to say a step input, are also applicable, but both are limited in practice. The accepted practical technique is to measure the response in amplitude and phase to a sinusoidal input function, based on Fourier methods.

Let the system to be analysed, it may be any system either open or closed loop, mechanical or electrical, etc., have a general transfer function $G(s)$,

then
$$\frac{F_o(s)}{F_i(s)} = G(s) = \frac{b_o + b_1 s + \ldots + b_m s^m}{a_o + a_1 s + \ldots + a_n s^n}$$

Now let $f_i(t)$ be a sinusoid $E \sin \omega t$, so that

$$F_i(s) = E \frac{\omega}{s^2 + \omega^2}$$

and
$$F_o(s) = EG(s) \frac{\omega}{s^2 + \omega^2}$$

$$= E \left\{ \sum_{k=1}^{n} \frac{A_k}{s - p_k} + \frac{B_1}{s - j\omega} + \frac{B_2}{s + j\omega} \right\}$$

by partial fraction expansion, where $p_1, p_2, \ldots p_k, \ldots p_n$ are the n poles of $G(s)$. The poles may be real or imaginary, but are assumed distinct for simplicity (for a repeated pole, say $(s - p)^2$, two terms $A_1/(s - p)^2 + A_2/(s - p)$ are needed).

If any of the system poles, p, have positive real parts the system is unstable and need be considered no further.

Inverse transforming, the first term results in terms such as $EA_k e^{p_k t}$, which, since p_k has a negative real part, decay away to zero. They are, in fact, the transient terms.

The interest here centres on the steady state terms. By Heaviside's expansion theorem (see Cheng, ref. 7):

$$B_1 = (s - j\omega) \frac{G(s)\omega}{(s + j\omega)(s - j\omega)}\bigg]_{s=j\omega} = \frac{G(j\omega)}{2j}$$

and

$$B_2 = (s + j\omega) \frac{G(s)\omega}{(s + j\omega)(s - j\omega)}\bigg]_{s=-j\omega} = \frac{G(-j\omega)}{-2j}$$

Expressing in polar form $G(j\omega) = |G(j\omega)|e^{j\phi}$

and $G(-j\omega) = |G(j\omega)|e^{-j\phi}$

since $G(-j\omega)$ is the complex conjugate of $G(j\omega)$, also written $G^*(j\omega)$. It is always the case that complex conjugate poles, such as $s = \pm j\omega$ above, when split by partial fractions have complex conjugate constants, i.e. B_1 and B_2 above.

Thus the steady state output is

$$F_0(s)_{s.s.} = E \cdot |G(j\omega)|\left\{\frac{e^{j\phi}}{2j(s - j\omega)} - \frac{e^{-j\phi}}{2j(s + j\omega)}\right\}$$

$$= E \cdot |G(j\omega)|\left\{\frac{s(e^{j\phi} - e^{-j\phi}) + j\omega(e^{j\phi} + e^{-j\phi})}{2j(s - j\omega)(s + j\omega)}\right\}$$

$$= E \cdot |G(j\omega)|\left\{\frac{s[\cos\phi + j\sin\phi - (\cos\phi - j\sin\phi)]}{2j(s^2 + \omega^2)}\right\}$$

$$= E \cdot |G(j\omega)|\left\{\frac{\omega\cos\phi + s\sin\phi}{s^2 + \omega^2}\right\}$$

$\therefore$ inverse transforming (see table in the Appendix)

$$f_0(t)_{s.s.} = E \cdot |G(j\omega)| \sin(\omega t + \phi)$$

which means that the steady state output is the input multiplied in amplitude by $|G(j\omega)|$ and shifted in phase by ϕ.

Thus for a linear stable system subjected to a sinusoidal input of frequency ω, the steady state output and input are related by a frequency response function, $G(j\omega)$, found by replacing s by $j\omega$ in the transfer function, $G(s)$.

In polar form $G(j\omega) = |G(j\omega)|\underline{/\phi(\omega)}$

where $|G(j\omega)| = \dfrac{\text{Amplitude of the steady state output}}{\text{Amplitude of the input}}$

and $\phi(\omega) = $ Angle by which the output leads the input (lags for negative $\phi(\omega)$).

s was previously shown to be $\sigma + j\omega$, so that the frequency response function, $G(j\omega)$, is the limiting case of the transfer function as $\sigma \to 0$. An alternative approach based on this surmise is via the Fourier transform, which is discussed briefly in the Appendix.

As an example of the substitution of s by $j\omega$ take the R–L–C network of Fig. 5.1, which can be described by a transform and also, for sinusoidal signals only, by impedances,

Transform impedance　$Z(s) = R + sL + \dfrac{1}{sC}$

A.C. impedance　　　　　$Z = R + j\omega L + \dfrac{1}{j\omega C}$

While $G(j\omega)$ can be found from a given value of $G(s)$, it is more important practically that the converse applies and $G(s)$ can be found

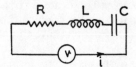

Fig. 5.1 A series R–L–C network.

from $G(j\omega)$, the point being that $G(j\omega)$ is relatively easy to measure. However, in order to bring out the salient features of $G(j\omega)$, the effect of substituting $j\omega$ for s in known transfer functions will first be considered.

The general form of the transfer function can be written as a frequency response function thus:

$$G(s) = \frac{b_0 + b_1 s + b_2 s^2 + \ldots + b_m s^m}{a_0 + a_1 s + a_2 s^2 + \ldots + a_n s^n}$$

and

$$G(j\omega) = \frac{b_0 + b_1(j\omega) + b_2(j\omega)^2 + \ldots + b_m(j\omega)^m}{a_0 + a_1(j\omega) + a_2(j\omega)^2 + \ldots + a_n(j\omega)^n}$$

It is quite common to leave $G(j\omega)$ in the form shown, involving $(j\omega)^n$, to show the connection with $G(s)$. However, the expression can be manipulated conventionally, i.e.

$$G(j\omega) = \frac{(b_0 - b_2\omega^2 + b_4\omega^4 - \ldots) + j(b_1\omega - b_3\omega^3 + b_5\omega^5 - \ldots)}{(a_0 - a_2\omega^2 + a_4\omega^4 - \ldots) + j(a_1\omega - a_3\omega^3 + a_5\omega^5 - \ldots)}$$

$$= X(\omega) + j\,Y(\omega)$$

where $X(\omega)$ and $Y(\omega)$ are real functions of ω.

For example, the general form of the second-order transfer function is:

$$G(s) = \frac{\omega_n^2}{s^2 + 2\zeta\omega_n s + \omega_n^2}$$

$$\therefore\ G(j\omega) = \frac{\omega_n^2}{(j\omega)^2 + 2\zeta\omega_n(j\omega) + \omega_n^2}$$

$$= \frac{\omega_n^2}{(\omega_n^2 - \omega^2) + j2\zeta\omega_n\omega} = \frac{\omega_n^2[(\omega_n^2 - \omega^2) - j2\zeta\omega_n\omega]}{(\omega_n^2 - \omega^2)^2 + (2\zeta\omega_n\omega)^2}$$

Hence

$$X(\omega) = \frac{\omega_n{}^2(\omega_n{}^2 - \omega^2)}{(\omega_n{}^2 - \omega^2)^2 + (2\zeta\omega_n\omega)^2}$$

and

$$Y(\omega) = -\frac{2\zeta\omega_n{}^3\omega}{(\omega_n{}^2 - \omega^2)^2 + (2\zeta\omega_n\omega)^2}$$

$G(j\omega)$ can be written in terms of a normalised frequency $u = \dfrac{\omega}{\omega_n}$ and it must be stressed that ω is an applied excitation frequency, while ω_n is the undamped natural frequency of the *system*, which is a constant.

Thus put $\quad u = \dfrac{\omega}{\omega_n},$ so that

$$G(ju) = \frac{1}{(1 - u^2) + j2\zeta u}$$

In general, if $G(j\omega)$ can be expressed as a product of several factors thus:

$$G(j\omega) = \frac{N_1(j\omega) \cdot N_2(j\omega) \ldots N_m(j\omega)}{D_1(j\omega) \cdot D_2(j\omega) \ldots D_n(j\omega)}$$

then the amplitude

$$G(\omega) = |G(j\omega)| = \frac{|N_1| \cdot |N_2| \ldots |N_m|}{|D_1| \cdot |D_2| \ldots |D_n|}*$$

and the phase

$$\phi(\omega) = \underline{/N_1} + \underline{/N_2} + \ldots + \underline{/N_m} - \underline{/D_1} - \underline{/D_2} - \ldots - \underline{/D_n}$$

It was explained previously in Section 4.11 that transfer functions, and therefore frequency response functions, can be factorised into the product of first- and second-order terms only, the second-order terms having complex conjugate roots. Second-order terms with real roots can be factorised. This form is expounded later in Section 5.3.

5.2 Graphical Representation of $G(j\omega)$

In the previous section it was shown that $G(j\omega)$ can be represented as a complex function of ω thus:

$$G(j\omega) = X(\omega) + jY(\omega) = G(\omega)e^{j\phi(\omega)}$$
$$= G(\omega)\underline{/\phi(\omega)}$$

where $G(\omega)$ is the amplitude of $G(j\omega)$ and $\phi(\omega)$ is the phase angle, i.e.

$$G(\omega) = \sqrt{X(\omega)^2 + Y(\omega)^2}$$

and

$$\phi(\omega) = \tan^{-1}\frac{Y(\omega)}{X(\omega)}$$

---

* Note the use of the symbol $G(\omega)$ to mean $|G(j\omega)|$. The significance is that a function of ω is real, a complex quantity being a function of $j\omega$.

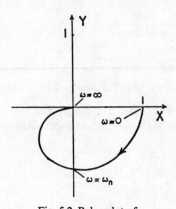

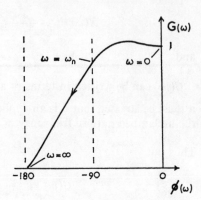

Fig. 5.2 Polar plot of

$$G(j\omega) = \frac{\omega_n^2}{\omega_n^2 + 2\zeta\omega_n(j\omega) + (j\omega)^2}$$

Fig. 5.3 Amplitude-phase plot of

$$G(j\omega) = \frac{\omega_n^2}{\omega_n^2 + 2\zeta\omega_n(j\omega) + (j\omega)^2}$$

(see also Fig. 5.10)

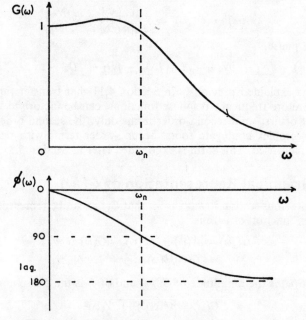

Fig. 5.4 Amplitude and phase versus frequency plots for

$$G(j\omega) = \frac{\omega_n^2}{\omega_n^2 + 2\zeta\omega_n(j\omega) + (j\omega)^2}$$

(see also Fig. 5.11)

In practice, most systems will involve a lagging phase angle as typified by the second-order example in the previous section.

$G(j\omega)$ can be represented graphically by plotting $G(\omega)$ and $\phi(\omega)$ as ω varies. The most common methods are:

(i) Polar diagrams, i.e. $Y(\omega)$ against $X(\omega)$ with ω as a parameter (Fig. 5.2).

(ii) Amplitude-phase diagrams, i.e. $G(\omega)$ against $\phi(\omega)$ with ω as a parameter (Fig. 5.3).

(iii) Amplitude and phase/frequency diagrams, i.e. $G(\omega)$ against ω *and* $\phi(\omega)$ against ω (Fig. 5.4).

It must again be stressed that any function, either open or closed loop, can be expressed in these forms. The plotting of open loop frequency response, however, has led to an association of names with these diagrams, which while having no particular significance until Chapter 6, may as well be introduced at this stage. They are:

(i) The polar plot is called a NYQUIST diagram.

(ii) The amplitude-phase diagram, with $G(\omega)$ in decibels, a logarithmic unit which is introduced in Section 5.5, is the basis of the Nichols chart, Section 6.7.3, and is thus called a Nichols diagram in this text.

(iii) The pair of graphs, $G(\omega)$ and $\phi(\omega)$ to a base of ω, used to interpret Bode's theorems, Section 6.6, are the basis of Bode diagrams. Again the decibel is used for $G(\omega)$ and the horizontal axis, ω, is also plotted on a logarithmic scale.

5.3 Some Features of $G(j\omega)$

Consider the general polynomial form of $G(j\omega)$:

$$G(j\omega) = \frac{b_0 + b_1(j\omega) + b_2(j\omega)^2 + \ldots + b_m(j\omega)^m}{a_0 + a_1(j\omega) + a_2(j\omega)^2 + \ldots + a_n(j\omega)^n}$$

For any physical system $n > m$, otherwise there would be a non-zero gain at infinite frequency, which is impossible. $n = m$ can be considered possible over a practical working frequency range. Factorising the general equation (see Section 4.11):

$$G(j\omega) = \frac{K(1 + j\omega T_a)(1 + j\omega T_b) \ldots \left(1 + \frac{2\zeta_a}{\omega_{na}}(j\omega) + \frac{1}{\omega_{na}^2}(j\omega)^2\right) \ldots}{(j\omega)^k(1 + j\omega T_1)(1 + j\omega T_2) \ldots \left(1 + \frac{2\zeta_1}{\omega_{n1}}(j\omega) + \frac{1}{\omega_{n1}^2}(j\omega)^2\right) \ldots}$$

Thus it can be seen that only four types of factors are necessary to make up the frequency response function of any system, viz.:

(i) A constant gain factor, K.

(ii) Integrating or differentiating factors, $j\omega$.

(iii) First-order terms, $1 + j\omega T$.

(iv) Second-order terms, with imaginary roots,

$$1 + \frac{2\zeta}{\omega_n}(j\omega) + \frac{1}{\omega_n^2}(j\omega)^2$$

At high frequency

$$G(j\omega) \xrightarrow[\omega \to \infty]{} \frac{b_m(j\omega)^m}{a_n(j\omega)^n} = \frac{b_m}{a_n} \cdot \frac{1}{(j\omega)^{n-m}}$$

If it is remembered that multiplying by j corresponds to a phase shift of 90° lead, and therefore $1/j$ to a phase shift of 90° lag, the ultimate high frequency phase shift will be $(n - m) \cdot 90°$ lag. Assuming that $n > m$ the high frequency amplitude tends to zero.

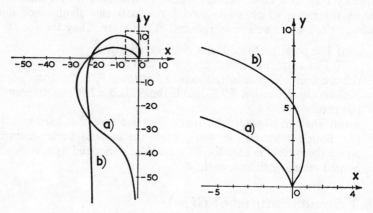

Fig. 5.5 Frequency response locus (Nyquist diagrams) for:

$$(a)\ G(j\omega) = \frac{10(1 + 2j\omega)}{j\omega(1 + j\omega)(1 + j\omega + 2(j\omega)^2)}$$

$$(b)\ G(j\omega) = \frac{10(1 + 0 \cdot 02j\omega)}{j\omega(1 + j\omega)(1 + j\omega + 2(j\omega)^2)}$$

Closed loop systems are sometimes typed from the form of the open loop frequency response function by the value of k. This is only applicable provided H does not involve a simple $j\omega$ term. Thus

$k = 0$ is a type 0 system, e.g. a velocity control

$k = +1$ is a type 1 system, e.g. a position control

$k = +2$ is a type 2 system, e.g. an integral compensated position control

Negative values of k are not considered, i.e. simple $j\omega$ numerator terms, although they can occur in frequency response functions for certain components of the system.

The phase at infinite frequency was shown to be lagging by 90° times the *order* of the denominator minus numerator: similarly, the limiting phase as $\omega \to 0$ will depend upon the *type* of the system.

Thus
$$G(j\omega) \xrightarrow[\omega \to 0]{} \frac{K}{(j\omega)^k}$$

The amplitude $G(0)$ tends to infinity except for $k = 0$, and the phase $\phi(0)$ tends to $k \cdot 90°$ lag.

Confusion between the type and order of a system must be avoided, the **order** of the system being equal to the order of the denominator, n.

Example

$$\begin{aligned}
G(s) &= \frac{10 + 21s + 2s^2}{s + 2 \cdot 1s^2 + 3 \cdot 2s^3 + 2 \cdot 3s^4 + 0 \cdot 2s^5} \\
&= \frac{10(1 + 2 \cdot 1s + 0 \cdot 2s^2)}{s(1 + 2 \cdot 1s + 3 \cdot 2s^2 + 2 \cdot 3s^3 + 0 \cdot 2s^4)} \\
&= \frac{10(1 + 2s)(1 + 0 \cdot 1s)}{s(1 + s)(1 + 0 \cdot 1s)(1 + s + 2s^2)} \\
&= \frac{10(1 + 2s)}{s(1 + s)(1 + s + 2s^2)}
\end{aligned}$$

Note the following:

(i) The major problem is factorising the polynomials, particularly the fourth order one in the numerator. The ideal answer would be to use a computer, but the various mathematical methods as described in Chestnut and Mayer (ref. 2) may have to be used.

(ii) A common term $(1 + 0 \cdot 1s)$ occurs in the numerator and denominator and can be cancelled. An unstable mode can be hidden by cancelling a $(1 - sT)$ pole and zero. Known as an 'uncontrollable mode' and occurring but rarely, this has no effect for zero initial conditions, but can otherwise be excited.

(iii) The term $1 + s + 2s^2$ is not factorised since it has complex roots, i.e.

$$-\tfrac{1}{4} \pm j\tfrac{1}{4}\sqrt{7}$$

(iv) Since the substitution $s = j\omega$ is to be made it can be made immediately and manipulation done in terms of $j\omega$. Conversely, and probably easier, manipulation can be done in terms of s, substituting $j\omega$ finally.

Hence
$$G(j\omega) = \frac{10(1 + 2j\omega)}{j\omega(1 + j\omega)(1 + j\omega + 2(j\omega)^2)}$$

This, then, is a type 1 system of fourth order with $n - m = 3$. Fig. 5.5 (*a*) shows a polar plot of this function.

Fig. 5.5 (*b*) shows that it is possible for the phase of a system to exceed $(n - m) \cdot 90°$ at an intermediate frequency. Fig. 5.6 shows some typical loci which should be studied carefully.

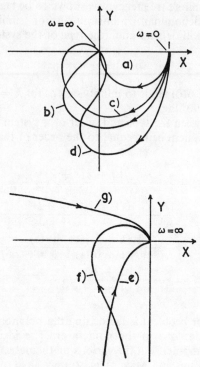

Fig. 5.6 Typical Nyquist diagrams for:

$(a)\ G(s) = \dfrac{1}{1 + sT};$ $(b)\ G(s) = \dfrac{1}{1 + bs + as^2};$

$(c)\ G(s) = \dfrac{1}{(1 + sT)(1 + bs + as^2)};$ $(d)\ G(s) = \dfrac{1 + sT}{1 + bs + as^2};$

$(e)\ G(s) = \dfrac{1}{s(1 + sT)};$ $(f)\ G(s) = \dfrac{1}{s(1 + bs + as^2)};$

$(g)\ G(s) = \dfrac{1}{s^2(1 + sT)}$

5.4 Minimum and Non-Minimum Phase Systems

It was shown in discussing the general form of $G(j\omega)$ that only four types of factor are necessary, i.e.

(i) the constant K;

(ii) $j\omega$;

(iii) $1 + j\omega T$;

(iv) $1 + \dfrac{2\zeta}{\omega_n}(j\omega) + \dfrac{1}{\omega_n{}^2}(j\omega)^2.$

If T, ζ and ω_n are positive, the system is called a 'minimum phase system'.

If the numerator contains terms in which T, ζ or ω_n are negative, the system is called a 'non-minimum phase system'.

If the denominator contains terms with negative T, ζ or ω_n, the system is unstable (see Section 6.2).

The system is called minimum phase since all numerator terms will cause phase leads and all denominator terms phase lags.

Example. Consider the minimum phase system:

$$G(j\omega) = \frac{(1 + j\omega)(1 + j\omega2)}{(1 + j\omega0 \cdot 5)(1 + j\omega4)(1 + j\omega3)}$$

$$\therefore \ G(\omega) = \frac{\sqrt{1 + \omega^2} \cdot \sqrt{1 + 4\omega^2}}{\sqrt{1 + 0 \cdot 25\omega^2} \cdot \sqrt{1 + 16\omega^2} \cdot \sqrt{1 + 9\omega^2}}$$

and $\phi(\omega) = \tan^{-1}\omega + \tan^{-1}2\omega - \tan^{-1}0 \cdot 5\omega - \tan^{-1}4\omega - \tan^{-1}3\omega$

Now consider the non-minimum phase system:

$$G(j\omega) = \frac{(1 + j\omega)(1 - j\omega2)}{(1 + j\omega0 \cdot 5)(1 + j\omega4)(1 + j\omega3)}$$

then $\qquad G(\omega) = \dfrac{\sqrt{1 + \omega^2} \cdot \sqrt{1 + 4\omega^2}}{\sqrt{1 + 0 \cdot 25\omega^2} \cdot \sqrt{1 + 16\omega^2} \cdot \sqrt{1 + 9\omega^2}}$

and

$$\phi(\omega) = \tan^{-1}\omega + \tan^{-1}(-2\omega) - \tan^{-1}0 \cdot 5\omega - \tan^{-1}4\omega - \tan^{-1}3\omega$$
$$= \tan^{-1}\omega - \tan^{-1}2\omega - \tan^{-1}0 \cdot 5\omega - \tan^{-1}4\omega - \tan^{-1}3\omega$$

It can clearly be seen that the amplitudes of the two systems are the same, but the phase lag of the non-minimum phase system is greater.

Fortunately, non-minimum phase systems only occur in rare cases.

Interpreting these examples in the s-plane, the minimum phase system:

$$G(s) = \frac{(1 + s)(1 + 2s)}{(1 + 0 \cdot 5s)(1 + 4s)(1 + 3s)}$$

$$= \tfrac{1}{3} \frac{(s + 1)(s + 0 \cdot 5)}{(s + 2)(s + 0 \cdot 25)(s + 0 \cdot 33)}$$

and for the non-minimum phase system:

$$G(s) = \tfrac{1}{3} \frac{(s + 1)(0 \cdot 5 - s)}{(s + 2)(s + 0 \cdot 25)(s + 0 \cdot 33)}$$

Fig. 5.7 shows these two functions plotted on the s-plane.

It is therefore easier to define a minimum phase system as one having no poles or zeros in the R.H. half of the s-plane.

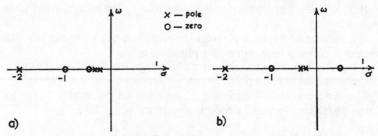

Fig. 5.7 *s*-Plane plot for: (*a*) minimum phase, and (*b*) non-minimum phase systems.

5.5 Logarithmic Plotting

The polar type of plot, which involves the point (0, 0) cannot be expressed in terms of logarithms. Graphs involving scalars only, i.e. amplitude $G(\omega)$, phase $\phi(\omega)$ or frequency, ω, however can, if desired, be expressed logarithmically. In practice, both $G(\omega)$ and ω are commonly plotted logarithmically, although $\phi(\omega)$ never is, for the reasons listed below.

Consider first frequency, ω, which is the horizontal axis of Bode diagrams. For a typical servo the frequency band of particular interest

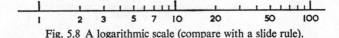

Fig. 5.8 A logarithmic scale (compare with a slide rule).

extends from, say, 1 to 10 Hz; however, in order to ascertain the effect of compensation networks and other additional time constants it is necessary to investigate the frequency range of, say, from 0·1 to 100 Hz. By linear plotting, then, the most important range is limited to less

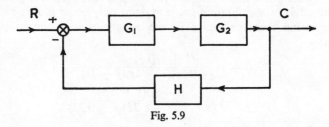

Fig. 5.9

than a tenth of the graph. By using a logarithmic axis as shown in Fig. 5.8 the more important range can be made about one-third of the display. Note, however, that d.c. ($\omega = 0$) is not on this graph as it occurs at log 0 $= -\infty$. The frequency axis can be plotted either in Hz or radians/second.

Next consider $G(\omega)$, as, for example, in plotting the open loop frequency response of the system of Fig. 5.9.

Open loop frequency response $= G_1(j\omega) \,.\, G_2(j\omega) \,.\, H(j\omega)$.

The amplitude $\qquad\qquad G(\omega) = G_1(\omega) \,.\, G_2(\omega) \,.\, H(\omega)$

and the phase angle $\qquad \phi(\omega) = \phi_1(\omega) + \phi_2(\omega) + \phi_H(\omega)$.

Assuming that the plots of $G_1(\omega)$, $G_2(\omega)$, $H(\omega)$, $\phi_1(\omega)$, $\phi_2(\omega)$ and $\phi_H(\omega)$ to a base of ω are known, $\phi(\omega)$ can easily be plotted by adding ordinates of the three separate curves. $G(\omega)$, however, is a much more tedious construction, since ordinates of the three curves $G_1(\omega)$, $G_2(\omega)$ and $H(\omega)$ will have to be multiplied together, at which stage a logarithmic unit (the decibel) suggests itself.

In strict terms the decibel is defined as:

$$\text{Amplitude ratio} = 10 \log_{10} \frac{\text{power out}}{\text{power in}} \text{ db}$$

In its original application to correctly matched transmission lines with equal input and output resistances, R_o, this expression becomes:

$$\text{Amplitude ratio} = 10 \log_{10} \frac{V_o^2}{R_o} \bigg/ \frac{V_i^2}{R_o} \text{ db}$$

$$= 10 \log_{10} \frac{V_o^2}{V_i^2} \text{ db}$$

$$= 20 \log_{10} \frac{V_o}{V_i} \text{ db}$$

This expression has been adapted by the control engineer such that the correctly terminated V_o and V_i have been replaced by *any* f_o and *any* f_i even if f_o and f_i are in different units, so that

$$\text{Amplitude ratio, } N = 20 \log_{10} \frac{f_o}{f_i} \text{ db}$$

If, say, f_o is a velocity (rad/s) and f_i a voltage (volts), it is desirable to define the units thus:

$$N = 20 \log_{10} \frac{f_o}{f_i} \text{ db (rad/s/volt)}$$

Returning to the amplitude equation:

$$G(\omega) = G_1(\omega) \,.\, G_2(\omega) \,.\, H(\omega)$$

Then

$$20 \log_{10} G(\omega) = 20 \log_{10} (G_1(\omega)G_2(\omega)H(\omega))$$

$$= 20 \log_{10} G_1(\omega) + 20 \log_{10} G_2(\omega) + 20 \log_{10} H(\omega)$$

If $G_1(\omega)$, $G_2(\omega)$ and $H(\omega)$ are now plotted in decibels the overall function $G(\omega)$ can be found by *adding* ordinates instead of multiplying.

It is common to calculate $G(\omega)$ in decibels and to plot on linear scales, although the same result would be achieved by plotting $G(\omega)$ direct on logarithmic axes. The former is far easier to manipulate and will be used exclusively from now on in this text.

The diagrams of Figs. 5.3 and 5.4 are ideally suited to this form of plotting as shown in Figs. 5.10 and 5.11, when they are called NICHOLS and BODE diagrams respectively.

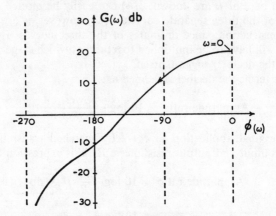

Fig. 5.10 Decibel-phase (Nichols) diagram for:

$$G(s) = \frac{10}{(1 + sT_1)(1 + sT_2)(1 + sT_3)}$$

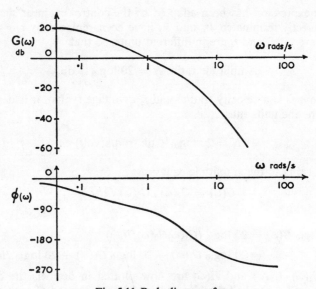

Fig. 5.11 Bode diagrams for

$$G(s) = \frac{10}{(1 + sT_1)(1 + sT_2)(1 + sT_3)}$$

Specially printed paper is available with a linear vertical axis for plotting decibels (and, of course, phase angle), with a logarithmic horizontal axis for plotting ω direct as indicated in Fig. 5.11. This is really a log–log plot. The horizontal axis is usually split into from 1 to 7 decades with 3 or 4 decades the most convenient for servo work.

It must be noted that:

$$20 \log_{10} 1 = 0 \text{ db}$$
$$20 \log_{10} 10 = 20 \text{ db}$$
$$20 \log_{10} \frac{1}{10} = -20 \log_{10} 10 = -20 \text{ db}$$

Hence negative db corresponds to an amplitude ratio less than 1, i.e. an attenuation, not a reversal of phase! In the system of Fig. 5.11 the 'gain' is greater than unity above the 0 db line and less than unity below.

5.6 Asymptotic Approximation on Bode Diagrams

It has been shown that in order to construct frequency response functions only four types of basic functions need be considered; i.e.

   (i) constant K;
   (ii) $j\omega$;
   (iii) $1 + j\omega T$;

   (iv) $1 + \dfrac{2\zeta}{\omega_n}(j\omega) + \dfrac{1}{\omega_n^2}(j\omega)^2$.

Only minimum phase systems will be considered, so that T, ζ and ω_n are positive; for completeness, each factor will be considered raised to the power n, where n is a positive or negative integer.

It must be remembered that complex frequency response functions involving the above functions can be synthesised by adding the decibel amplitudes.

Linear asymptote approximations are used extensively in plotting Bode diagrams, since they can easily be constructed by defining a few points and joining them with lines of specific slope. Corrections can be applied when necessary, although asymptotic constructions are so common that unless accurate values at particular points are required the corrections are ignored.

Let
$$20 \log_{10} G(\omega) = N \text{ db}$$
(i)
$$G(j\omega) = K$$
$$\therefore G(\omega) = K$$
$$N = 20 \log_{10} K \text{ db}$$
and
$$\phi(\omega) = 0$$

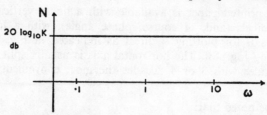

Fig. 5.12 Bode amplitude diagram of $G(j\omega) = K$; phase angle is zero for all ω.

The table of Fig. 5.13 lists some values of decibel equivalents of K.

K	0·1	1/3	1/2	1	$\sqrt{2}$	2	3	4	6	8	10	20	50	100
$N = 20\log_{10}K$ db	−20	−9·5	−6	0	3	6	9·5	12	15·5	18	20	26	34	40

Fig. 5.13 Table of db-amplitude ratio equivalents.

(ii) $G(j\omega) = (j\omega)^k$

therefore $\phi(\omega) = k \cdot 90°$ independent of ω (lead for positive k)

and $G(\omega) = \omega^k$

therefore $N = 20 \log_{10} \omega^k$

$= k \cdot 20 \log_{10} \omega$ db

which is the equation of a straight line if N is plotted against $\log_{10} \omega$. This is the second, possibly most important, reason for logarithmic plotting.

Let $N = N_1$ when $\omega = \omega_1$

and $N = N_2$ when $\omega = 10\omega_1$ i.e. one decade.

The change in $N = N_2 - N_1 = 20 \, k \log_{10} 10\omega_1 - 20 \, k \log_{10} \omega_1$

$= 20 \, k \log_{10} 10$

$= 20 \, k$ db

Therefore, line is of slope 20 k db/decade. Alternatively if ω is doubled, i.e. increased by an octave:

$$N_2 - N_1 = 20 \, k \log_{10} 2$$
$$= 20 \, k \, 0·303$$
$$= 6·06 \, k \text{ db/octave}$$

This is usually quoted as 6 db/octave for $k = 1$. Hence 6 db/octave and 20 db/decade are the same slope.

The line cuts the 0 db axis when $G(\omega) = 1$, i.e. when $\omega = 1$, for all k. No approximations are necessary.

(iii) $G(j\omega) = (1 + j\omega T)^n$

Consider first $n = 1$

Then $\phi(\omega) = \tan^{-1} \omega T$

and $G(\omega) = (1 + \omega^2 T^2)^{\frac{1}{2}}$

$$\therefore N = 20 \log_{10} (1 + \omega^2 T^2)^{\frac{1}{2}}$$
$$= 10 \log_{10} (1 + \omega^2 T^2)$$

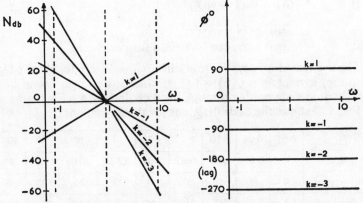

Fig. 5.14 Bode diagrams of $G(j\omega) = (j\omega)^k$.

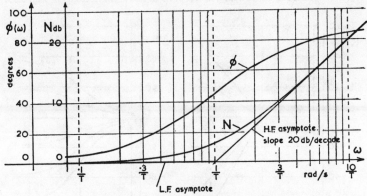

Fig. 5.15 Bode diagrams of $G(j\omega) = 1 + j\omega T$.

$\phi(\omega)$ and $G(\omega)$ can be plotted in full (Fig. 5.15), but straight line approximations can be made to simplify the plotting of N.

At low ω, i.e. $\omega T \ll 1$

$$G(\omega) \simeq 1 \quad \text{i.e. } 0 \text{ db}$$

At high ω, i.e. $\omega T \gg 1$

$$G(\omega) \simeq \omega T$$

It was shown in (ii) that this is a straight line on the db/log ω graph of slope 20 db/decade, passing through the 0 db line, i.e. where the asymptotes cross, when $\omega T = 1\left(\omega = \dfrac{1}{T} \text{ rad/s or } \dfrac{1}{2\pi T} \text{ Hz}\right)$. These asymptotes are shown in relation to the correct curve in Fig. 5.15, while the table of Fig. 5.16 gives the correction factors for curves constructed by asymptotes.

If $G(j\omega) = (1 + j\omega T)^n$

$$G(\omega) = (1 + \omega^2 T^2)^{\frac{n}{2}}$$

and $\phi(\omega) = n . \tan^{-1} \omega T$ (lead for positive n)

The low frequency asymptote is still $G(\omega) \simeq 1$ or 0 db, but the high frequency asymptote is $G(\omega) \simeq (\omega T)^n$, which is a straight line of slope $n \times 20$ db/decade passing through the 0 db line when $\omega T = 1$.

Fig. 5.17 shows the Bode diagrams for these functions.

ωT		0·01	0·04	0·10	0·4	1·0	4	10	40	100
N	Actual	0	0	0·04	0·65	3·01	12·32	20·04	32	40
db	Approx.	0	0	0	0	0	12	20	32	40
ϕ degrees		0·6	2·3	5·7	21·8	45	76	84·3	88·6	89·4

Fig. 5.16 Table for constructing corrections to the asymptotes as in Fig. 5.15.

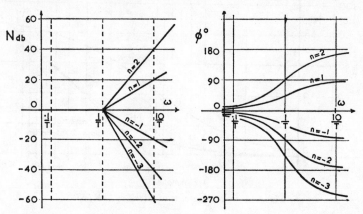

Fig. 5.17 Bode diagrams of $G(j\omega) = (1 + j\omega T)^n$—asymptotic approximation.

(iv)* $G(j\omega) = \dfrac{1}{1 + \dfrac{2\zeta}{\omega_n}(j\omega) + \dfrac{1}{\omega_n^2}(j\omega)^2}$

Putting $\dfrac{1}{\omega_n} = T$

$$G(j\omega) = \frac{1}{1 + j2\zeta\omega T + (j\omega T)^2}$$

* $n = 1$ is the only case for this factor considered here.

The effect of various values of ζ, with T constant, will be considered. All values of ζ will be allowed for, but the only ones of interest are $0 < \zeta < 1$, otherwise the term can be factorised into two terms of type (iii).

$$\phi(\omega) = \tan^{-1} \frac{2\zeta\omega T}{1 - \omega^2 T^2} \text{ lag,}$$

and

$$G(\omega) = [(1 - \omega^2 T^2)^2 + 4\zeta^2\omega^2 T^2]^{-\frac{1}{2}}$$

$$\therefore N = -10 \log_{10} [(1 - \omega^2 T^2)^2 + 4\zeta^2\omega^2 T^2]$$

The asymptotic approximations to this function are of limited use since the accuracy is very dependent upon ζ.

However, when $\omega T \ll 1$ ($\omega \ll \omega_n$)

$$G(\omega) \simeq 1, \text{ i.e. 0 db}$$

and when $\omega T \gg 1$ ($\omega \gg \omega_n$)

$$G(\omega) \simeq \frac{1}{(\omega T)^2}$$

Thus the asymptotes still meet at $\omega T = 1$ ($\omega = \omega_n$), but the high frequency asymptote is of slope 40 db/decade downwards. At the 'break point' it can be calculated

$$G(\omega)\bigg]_{\omega = \omega_n} = \frac{1}{2\zeta}$$

so that the asymptotes can be wildly in error for low ζ.

Another point of interest is the maximum value of $G(\omega)$ and the frequency at which it occurs. From the equation for $G(\omega)$ the maximum value can be found by differentiating and equating to zero, to find the minimum of:

$$x = (1 - \omega^2 T^2)^2 + 4\zeta^2\omega^2 T^2$$

$$\therefore \frac{dx}{d\omega} = 2(1 - \omega^2 T^2) \cdot -2\omega T^2 + 8\zeta^2\omega T^2$$

$$= 0 \text{ for a minimum } x \text{ and } \therefore \text{ maximum } G(\omega).$$

$$\therefore \omega T = \sqrt{1 - 2\zeta^2},$$

i.e. the resonant frequency is $\omega_n\sqrt{1 - 2\zeta^2}$.

From this if $\zeta > 0.707$, there is no maximum other than $\omega = 0$.

Substituting for $G(\omega)$ at this frequency

$$G(\omega)_{\max} = \frac{1}{2\zeta\sqrt{1 - \zeta^2}} \text{ for } \zeta < 0.707$$

$$\therefore N_{\max} = -20 \log_{10} 2\zeta\sqrt{1 - \zeta^2} \text{ db.}$$

If $\quad \zeta = 0, \quad N_{\max} = \infty \text{ db at } \omega = 1/T \quad = \omega_n$

$\quad \zeta = 0.1, \quad N_{\max} = 14 \text{ db at } \omega = 0.99/T = 0.99\omega_n$

$\quad \zeta = 0.5, \quad N_{\max} = 1.2 \text{ db at } \omega = 0.707/T = 0.707\omega_n$

$\quad \zeta = 0.707, N_{\max} = 0 \text{ db} \quad \text{at } \omega = 0$

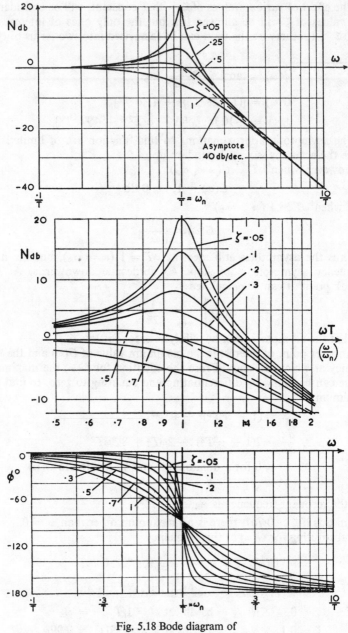

Fig. 5.18 Bode diagram of
$$G(j\omega) = \frac{1}{1 + i2\zeta\omega T + (j\omega T)^2}; \qquad \left(T = \frac{1}{\omega_n}\right)$$

Finally, the phase can be calculated from the original equation so that when

$$\left.\begin{array}{ll} \omega = \dfrac{1}{T} = \omega_n & \phi = 90° \text{ lag} \\[2mm] \omega = 0 & \phi = 0 \\[2mm] \omega = \infty & \phi = 180° \text{ lag} \end{array}\right\} \text{for all } \zeta$$

Fig. 5.18 is a Bode diagram plotted for various values of ζ.

Note that with a second-order term the phase can lag while the amplitude increases.

5.7 Use of Asymptotic Approximation to Construct Bode Diagrams

$$(a) \qquad G(j\omega) = \frac{K(1 + j\omega T_a)}{j\omega(1 + j\omega T_1)(1 + j\omega T_2)}$$

$$= K(1 + j\omega T_a) \cdot \frac{1}{j\omega} \cdot \frac{1}{1 + j\omega T_1} \cdot \frac{1}{1 + j\omega T_2}$$

The curve could have been constructed by *subtracting* curves for $j\omega$, $1 + j\omega T_1$ and $1 + j\omega T_2$ from the K and $1 + j\omega T_a$ curves. The form used above, plotting denominator terms as $\dfrac{1}{1 + j\omega T}$ and *adding* all curves is considered easier. Experience will enable the reader to construct the diagrams by starting with the lowest frequency asymptote and changing the slope (up for numerator and down for denominator) by an extra 20 db/decade at each break frequency, without drawing the component terms.

Hence the asymptotic form of the db–log ω graph can be drawn for each component term and the ordinates added (not multiplied since db scales are used). The use of asymptotes without correction makes the construction particularly easy as can be seen in Fig. 5.19. This graph is constructed for

$$K = 10, \; T_a = 0\cdot5, \; T_1 = 10 \text{ and } T_2 = 2$$

The phase diagram can be constructed in a similar way by summing the phase of each term as in Fig. 5.20. It is important to note that if the amplitude curve is known, then the phase curve can be constructed, particularly easily if the amplitude curve is in asymptotic form, so it is seldom essential to measure both amplitude and phase. This assumes, however, that the system is both linear and minimum phase, so that it is advisable to measure both if possible.

$$(b) \qquad G(j\omega) = \frac{20}{j\omega(1 + 0\cdot1j\omega + 0\cdot25(j\omega)^2)}$$

$$= \frac{20}{i\omega(1 + j0\cdot1\omega + (j0\cdot5\omega)^2)}$$

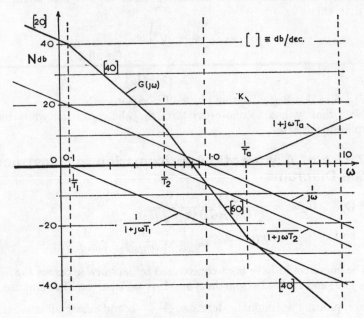

Fig. 5.19 Bode amplitude diagram of

$$G(j\omega) = \frac{10(1 + j\omega 0.5)}{j\omega(1 + j\omega 10)(1 + j\omega 2)}$$

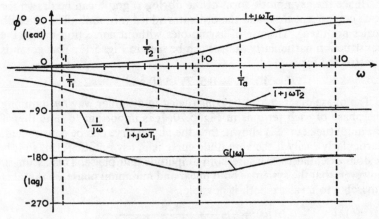

Fig. 5.20 Bode phase diagram of

$$G(j\omega) = \frac{10(1 + j\omega 0.5)}{j\omega(1 + j\omega 10)(1 + j\omega 2)}$$

The quadratic term has complex roots, so that $T = 0.5$ ($\omega_n = 2$) and $\zeta = \dfrac{0.1}{2T} = 0.1$. Here the asymptotes may be drawn, but the second-order term *must* be corrected.

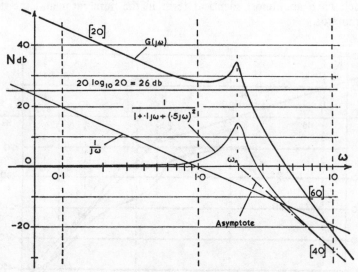

Fig. 5.21 Bode amplitude diagram of
$$G(j\omega) = \frac{20}{j\omega(1 + 0.1j\omega + 0.25(j\omega)^2)}$$

5.8 Use of Measured Frequency Response to Find $G(j\omega)$

Section 5.7 showed the approximate method of constructing the frequency response diagrams in terms of asymptotes; the inverse procedure can be applied to find $G(j\omega)$. Only the modulus curve, $20 \log_{10} G(\omega)$, is considered.

Certain obvious terms can be extracted at first glance, i.e.:

(i) The slope as $\omega \to 0$ will depend upon the type of the system. Thus if the slope is $-k \times 20$ db/decade, then $G(j\omega)$ contains a term $(j\omega)^k$ in the denominator.

(ii) The ultimate slope as $\omega \to \infty$ will be $-n \times 20$ db/decade, indicating that the order of the denominator terms will be n higher than the numerator, including, of course, the k order mentioned in (i).

The general procedure is to draw tangents to the curve of slope $n \times 20$ db/decade and then to interpret the intersections of asymptotes as time constants. Second-order terms will need an estimation for ζ as well as ω_n.

The whole technique is one which is greatly enhanced by experience and it becomes possible to fit curves with good accuracy, but the accuracy is greatly improved by a previous knowledge of the theoretical form of $G(j\omega)$. It is relatively difficult to find $G(j\omega)$ for functions which have an almost identical term in the numerator and the denominator.

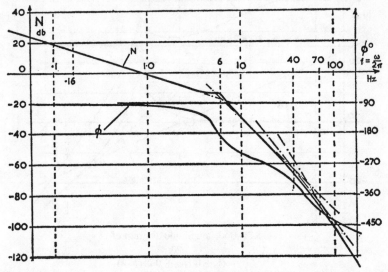

Fig. 5.22 Open loop frequency response of an hydraulic position control servo.

Example. The curve of Fig. 5.22 is the Bode amplitude diagram of the open loop frequency response of a hydraulic, spool valve-controlled position servo. The input is volts applied to the spool valve drive coil and the output is volts from the position transducer.

The theoretical transfer function of this system is derived in Chapter 17 and should be of the form:

$$G(s)H(s) = \frac{K}{s(1 + sT_c)(1 + sT_v)(1 + bs + as^2)}$$

where T_c = coil time constant

T_v = first-order approximation to spool valve characteristic

b and a depend upon the motor, gearing and load.

K = Amplifier gain, valve constant, transducer constant, etc.

It can be seen that:

   (i) The L.F. slope is 20 db/decade, negative, confirming the s term in the denominator, e.g. a type 1 system.

   (ii) The H.F. slope is approx. 100 db/decade indicating a fifth-order system. The phase lag, however, is extending beyond 450°, indicating still further terms at higher frequencies which have had negligible effect on the amplitude over the frequency range covered.

(iii) Break points can be seen at 40 Hz and 70 Hz, i.e. T_c and T_v are $\frac{1}{2\pi 40}$ and $\frac{1}{2\pi 70}$ seconds. Which is which cannot be determined from the graph.

(iv) The remaining term is second-order (40 db/dec H.F. slope) and under-damped due to the 'hump' in the curve. The asymptotes show that $\omega_n = 2\pi$. 6 rad/s, with a peak of 3 db at a frequency just below ω_n. Estimating from Fig. 5.18 this corresponds to a value of ζ of about 0·4.

(v) The gain constant K can be found by noting that at a low frequency, say $\omega = 1$ or $f = 0·16$ Hz, $G(s) H(s) \simeq \frac{K}{s}$ or $G(\omega) H(\omega) = \frac{K}{\omega}$. Thus $20 \log_{10} K = 14$ db and $\therefore K = 5$.

Hence

$$G(s) H(s) = \frac{5}{s\left(1 + s\frac{1}{80\pi}\right)\left(1 + s\frac{1}{140\pi}\right)\left(1 + \frac{2 \cdot 0·4}{12\pi}s + \frac{1}{(12\pi)^2}s^2\right)}$$

$$= \frac{5}{s(1 + 0·004s)(1 + 0·0023s)(1 + 0·0212s + 7·1 \times 10^{-4}s^2)}$$

It is now a good check to plot the phase curve for the calculated $G(j\omega)$ to ensure that it lines up with the measured results.

5.9 Summary

It has now been shown that by measurement or calculation the frequency response function $G(j\omega)$ can be more readily found than other functions, from whence it is a simple step to substitute $j\omega = s$ to find the transfer function $G(s)$. When $G(s)$ is known, the inverse Laplace transform can be applied to find the output as a function of time, $c(t)$, for any given input, $r(t)$, i.e. the transient response.

It should be realised that a given specification in the time domain has a corresponding specification in the frequency domain, as described in Chapter 7. It is thus desired to establish design methods based in the frequency domain without recourse to transformation to the s- or time-domains.

6
Stability Analysis

6.1 Stable and Unstable Systems

The output of a linear system, as a function of time, can be broken down into two parts, steady state terms which are directly related to the input, and transient terms which are either exponential or oscillatory with an envelope of exponential form. If the exponentials decay to zero, the system is said to be STABLE. If any exponential increases, the system is said to be UNSTABLE. Typical examples are shown in Fig. 6.1. In a practical system the exponentials cannot increase to infinity, so that the

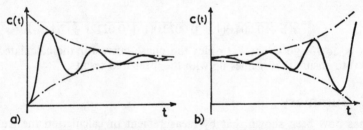

Fig. 6.1 (*a*) a stable system, and (*b*) an unstable system.

system becomes nonlinear, i.e. saturation effects occur and the system settles to a constant amplitude oscillation. Here the system will only be considered in its linear range.

The theoretically 'just stable' system is simple harmonic, in which case the output is a stable amplitude sinusoidal oscillation.

Instability is commonly an unfortunate product of closed loop control. It will be shown in this chapter that the higher the open loop gain, the more likely is instability, so that high precision closed loop control and good stability are incompatible. *The best possible design is a compromise between good stability and as high a loop gain as possible.*

An unstable system does not require an input to start oscillation; the act of closing the loop or small noise signals are sufficient.

There are a number of techniques available for analysing the stability of closed loop systems, many of which are suitable for indicating the degree of stability and possible methods of improving the system, in particular, methods of compensation to make an unstable system stable. This chapter deals with determining whether a given system is stable or unstable and Chapter 8 will cover methods of compensation. In between, Chapter 7 explains what is required of the compensation. The techniques covered here are:

1. Consideration of the closed loop:

   (i) Differential equation approach from transfer functions and the characteristic equation.

   (ii) Routh's criterion for detecting the presence of roots of the characteristic equation with positive real parts.

2. Consideration of the open loop.

   (i) Root loci: plotting the loci of the poles of the closed loop transfer function for various values of gain constant, from a knowledge of the open loop roots.

   (ii) Nyquist criterion of frequency response:

       (*a*) Nyquist diagrams.

       (*b*) Bode diagrams and Bode's theorems.

6.2 Stability Analysis by Pole–Zero Location

6.2.1 The Characteristic Equation

Consider the system represented by the transfer function:

$$\frac{C(s)}{R(s)} = \frac{b_0 + b_1 s + b_2 s^2 + \ldots + b_m s^m}{a_0 + a_1 s + a_2 s^2 + \ldots + a_n s^n}$$

$$\therefore (a_0 + a_1 s + a_2 s^2 + \ldots + a_n s^n)C(s)$$
$$= (b_0 + b_1 s + \ldots + b_m s^m)R(s)$$

To find the transient response of this system the Complementary Function is determined by putting $R(s) = 0$, and since $C(s) \neq 0$,

$$(a_0 + a_1 s + \ldots + a_n s^n) = 0$$

This equation, the denominator of the closed loop transfer function equated to zero, is called the CHARACTERISTIC EQUATION.

The numerator has no bearing on whether a system is stable or not, since it does not affect the complementary function. It will, of course, affect the complete response to a given input.

The transient response to *any* input is given by the complementary function

$$A_1 e^{r_1 t} + A_2 e^{r_2 t} + \ldots + A_n e^{r_n t}$$

where $A_1, A_2, \ldots A_n$ are constants dependent upon the input, the system constants and any initial conditions, and $r_1, r_2, \ldots, r_n$ are the roots of the characteristic equation; these constants and roots may be real or they may be in complex conjugate pairs.

For example, let $n = 5$ with:

$$r_1 = -\sigma_1; \quad r_2 = -\sigma_2 + j\omega_2 \quad \text{and} \quad r_3 = -\sigma_2 - j\omega_2;$$
$$r_4 = \sigma_3 + j\omega_3 \quad \text{and} \quad r_5 = \sigma_3 - j\omega_3$$

where the σ's and ω's are real and positive.

Hence the transient term of $c(t)$ is

$$c_t(t) = A_1 e^{-\sigma_1 t} + A_2 e^{(-\sigma_2 + j\omega_2)t} + A_2^* e^{(-\sigma_2 - j\omega_2)t} + A_3 e^{(\sigma_3 + j\omega_3)t}$$
$$+ A_3^* e^{(\sigma_3 - j\omega_3)t}$$
$$= A_1 e^{-\sigma_1 t} + e^{-\sigma_2 t}(A_2 e^{j\omega_2 t} + A_2^* e^{-j\omega_2 t}) + e^{\sigma_3 t}(A_3 e^{j\omega_3 t} + A_3^* e^{-j\omega_3 t})$$
$$= A_1 e^{-\sigma_1 t} + e^{-\sigma_2 t}(B_1 \cos \omega_2 t + B_2 \sin \omega_2 t)$$
$$+ e^{\sigma_3 t}(B_3 \cos \omega_3 t + B_4 \sin \omega_3 t)$$

The first term is a decaying exponential. The second term is oscillatory, with exponentially decaying amplitude. The third term is

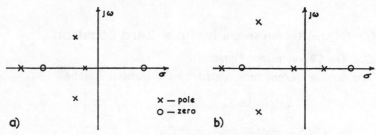

Fig. 6.2 s-plane representation of typical transfer functions of: (a) a stable system, and (b) an unstable system.

oscillatory, but with exponentially increasing amplitude. Thus, this system is unstable. It can now be noticed that the only term which has dictated this instability is σ_3, from which *the presence of one or more roots in the characteristic equation with positive real parts means that the system represented is unstable*. Remember that a root of the characteristic equation is a *pole* of the closed loop transfer function.

This is best and simply represented by plotting the roots of the closed loop transfer function on the complex plane, i.e. the s-plane, as shown in Fig. 6.2, where the presence of poles in the R.H. half plane indicates an unstable system. Remember that zeros in the R.H. half plane indicate a non-minimum phase system, but not an unstable one. A conjugate pair of roots on the $j\omega$ axis indicates a just stable system, i.e. simple harmonic output to a step input.

6.2.2 The Characteristic Equation of a Closed Loop System Derived from the Open Loop Transfer Function

Fig. 6.3 shows a typical closed loop system, for which

$$G(s) = \frac{K_1 N_1(s)}{D_1(s)}$$

and

$$H(s) = \frac{K_2 N_2(s)}{D_2(s)}$$

where N_1, N_2, D_1, D_2, are of the form $s^k(1 + as + bs^2 + \ldots)$, so that K_1 and K_2 are gain constants.

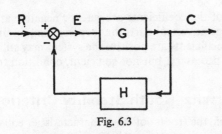

Fig. 6.3

Then

$$\frac{C(s)}{R(s)} = \frac{G(s)}{1 + G(s)H(s)}$$

$$= \frac{\dfrac{K_1 N_1(s)}{D_1(s)}}{1 + \dfrac{K_1 K_2 N_1(s) N_2(s)}{D_1(s) D_2(s)}}$$

$$= \frac{K_1 N_1(s) D_2(s)}{D_1(s) D_2(s) + K N_1(s) N_2(s)}$$

where $K = K_1 K_2$, which is referred to as the OPEN LOOP GAIN CONSTANT.

The characteristic equation is now simply written as

$$D_1(s) D_2(s) + K N_1(s) N_2(s) = 0$$

The closed loop poles (roots of the characteristic equation) depend upon the constant K and all the roots (poles and zeros) of the open loop transfer function.

6.2.3 Some Facts about the Roots of the Characteristic Equation

Consider now the characteristic equation

$$(s - r_1)(s - r_2)(s - r_3) = 0$$

$$\therefore \; s^3 - \underbrace{(r_1 + r_2 + r_3)}_{\text{sum of roots}} s^2 + (r_1 r_2 + r_1 r_3 + r_2 r_3)s - \underbrace{r_1 r_2 r_3}_{\text{product of roots}} = 0$$

Let $r_1 = -\delta$; $r_2 = -\alpha + j\beta$; and $r_3 = -\alpha - j\beta$

Then the equation is

$$s^3 + (\delta + 2\alpha)s^2 + (2\alpha\delta + \alpha^2 + \beta^2)s + \delta(\alpha^2 + \beta^2) = 0$$

It can be seen that if any of the coefficients are either zero or negative, either α or δ must be negative, i.e. a positive real part of a root. Note, however, that β occurs as a plus or minus term and does not affect the polarity of the coefficients.

Thus:

(i) The order of the equation is equal to the number of roots.

(ii) The roots, if complex, occur in conjugate pairs since all coefficients must be real.

(iii) If any of the coefficients are zero or negative at least one root must have a positive real part, thus representing an unstable system.

(iv) If all coefficients are positive the system may still be unstable so that this is a necessary, but not sufficient, condition for stability.

6.3 The Hurwitz–Routh Stability Criterion

Investigation of the roots of the characteristic equation indicates infallibly whether a system is stable or not; it is, however, a difficult problem to find these roots if the equation is of an order higher than two or three. The work of Routh (1884) and Hurwitz (1875) gives a method of indicating the presence and number of unstable roots, but not their value.

This technique has limited application since it will not give any indication of the form of any compensation required to make an unstable system stable.

Let the characteristic equation to be investigated be

$$a_n s^n + a_{n-1} s^{n-1} + \ldots + a_1 s + a_0 = 0.$$

All the 'a' terms are positive, otherwise the system is obviously unstable and need be investigated no further.

The common approach to this technique is to construct 'Routh's array', viz.:

$$
\begin{array}{c|lll}
s^0 & p_1 \\
s & q_1 \\
\vdots & \ldots \\
s^{n-3} & c_1 & c_2 & c_3 \ldots \\
s^{n-2} & b_1 & b_2 & b_3 \ldots \\
s^{n-1} & a_{n-1} & a_{n-3} & a_{n-5} \ldots \\
s^n & a_n & a_{n-2} & a_{n-4} \ldots \\
\end{array}
$$

where

$$b_1 = \frac{1}{a_{n-1}} \begin{vmatrix} a_{n-1} & a_{n-3} \\ a_n & a_{n-2} \end{vmatrix}; \quad b_2 = \frac{1}{a_{n-1}} \begin{vmatrix} a_{n-1} & a_{n-5} \\ a_n & a_{n-4} \end{vmatrix}; \quad b_3 = \text{etc.}$$

and

$$c_1 = \frac{1}{b_1} \begin{vmatrix} b_1 & b_2 \\ a_{n-1} & a_{n-3} \end{vmatrix}; \quad c_2 = \frac{1}{b_1} \begin{vmatrix} b_1 & b_3 \\ a_{n-1} & a_{n-5} \end{vmatrix}; \quad c_3 = \text{etc.}$$

When the array is complete the criterion states that any unstable root will cause a change of sign in the first column. The number of unstable roots is given by the number of times the sign of the first column changes. If all the coefficients of the first column are positive the system is stable.

The following example is applied to a characteristic equation with known roots, i.e. a problem for which the answer is known by inspection. Note, however, that if the problem had been posed in the form of the second line, it would not have been so obvious!

Example 1

$$(s + 4)(s + 1)(s - 2 + j3)(s - 2 - j3)(s + 1 + j2)(s + 1 - j2) = 0$$
$$\therefore s^6 + 3s^5 + 4s^4 + 48s^3 + 135s^2 + 349s + 260 = 0$$

This equation has two unstable roots, $2 \pm j3$. To check this:

1	$f_1[+1]$		
s	$e_1[+1]$		
s^2	$d_1[+1]$ d_2		
s^3	$c_1[+1]$ c_2		
s^4	$b_1[-1]$ b_2	b_3	
s^5	3	48	349
s^6	1	4	135	260

The figures in square brackets are inserted from the following calculations.

$$b_1' = \tfrac{1}{3} \begin{vmatrix} 3 & 48 \\ 1 & 4 \end{vmatrix} = \tfrac{1}{3}(12 - 48) = -12. \quad \text{Put } b_1 = -1$$

$$b_2' = \tfrac{1}{3} \begin{vmatrix} 3 & 349 \\ 1 & 135 \end{vmatrix} = \tfrac{1}{3}(405 - 349) = \tfrac{56}{3} \quad \therefore b_2 = \tfrac{1}{12} \cdot \tfrac{56}{3} = 1 \cdot 56$$

$$b_3' = \tfrac{1}{3} \begin{vmatrix} 3 & 0 \\ 1 & 260 \end{vmatrix} = \tfrac{1}{3}780 = 260 \quad \therefore b_3 = \tfrac{260}{12} = 21 \cdot 6$$

$$c_1' = -1 \begin{vmatrix} -1 & 1 \cdot 56 \\ 3 & 48 \end{vmatrix} = 52 \cdot 67 \quad \text{Put } c_1 = +1$$

$$c_2' = -1 \begin{vmatrix} -1 & 21 \cdot 6 \\ 3 & 349 \end{vmatrix} = 413 \cdot 8 \quad \therefore c_2 = \frac{413 \cdot 8}{52 \cdot 67} = 7 \cdot 88$$

$$d_1' = +1 \begin{vmatrix} 1 & 7 \cdot 88 \\ -1 & 1 \cdot 56 \end{vmatrix} = 9 \cdot 44 \quad \text{Put } d_1 = +1$$

$$d_2' = +1 \begin{vmatrix} 1 & 0 \\ -1 & 21 \cdot 6 \end{vmatrix} = 21 \cdot 6 \quad \therefore d_2 = \frac{21 \cdot 6}{9 \cdot 44} = 2 \cdot 29$$

$$e_1' = +1 \begin{vmatrix} 1 & 2 \cdot 29 \\ 1 & 7 \cdot 88 \end{vmatrix} = 5 \cdot 59 \quad \text{Put } e_1 = +1$$

$$f_1' = +1 \begin{vmatrix} 1 & 0 \\ 1 & 2 \cdot 29 \end{vmatrix} = 2 \cdot 29 \quad \text{Put } f_1 = +1$$

The array shows two changes of sign, i.e. $+3$ to -1 and -1 to $+1$, indicating the two unstable roots.

Note the technique of normalising each row in order to simplify the arithmetic.

Special rules exist to overcome inherent difficulties, e.g. all coefficients in one row being zero, which are covered in detail in Cheng (ref. 7).

Example 2. In the system of Fig. 6.4, find the value of K to just result in instability.

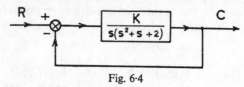

Fig. 6·4

The characteristic equation is $D_1(s)D_2(s) + KN_1(s)N_2(s) = 0$, thus:

$$s(s^2 + s + 2) + K = 0$$
$$\therefore s^3 + s^2 + 2s + K = 0$$

$$
\begin{array}{c|cc}
1 & c & \\
s & b & \\
s^2 & 1 & K \\
s^3 & 1 & 2
\end{array}
$$

where
$$b = 1 \begin{vmatrix} 1 & K \\ 1 & 2 \end{vmatrix} = 2 - K$$

If b is positive, the system is stable (inspection will show that if b is positive, so is c) and if b is negative, the system is unstable; thus the system will just be oscillatory, i.e. roots on the $j\omega$ axis, if $b = 0$,

$$\therefore 2 - K = 0 \quad \text{or} \quad K = 2$$

Since $K = 2$ gives simple harmonic motion $(s^2 + \omega_n{}^2)$ must be a factor of the equation. Bearing this in mind the equation can easier be factorised to

$$s^3 + s^2 + 2s + 2 = (s^2 + 2)(s + 1)$$

Thus $\omega_n{}^2 = 2$ and the frequency of oscillation is $\sqrt{2}$ rad/s.

6.4 Root Loci

6.4.1 The Principle of the Root Locus Method

The techniques previously described are analytical techniques which give an idea of the performance of a system. Problems are involved, first in finding the system equations in a suitable form and then in finding the roots of these equations. It is possible to measure the system equations for a stable system by the methods described in Chapter 13, but it is not possible for an unstable system; it is thus better to base the closed loop design on measurements made with the loop open. The root locus technique, first presented by W. R. Evans in 1948, is such a method.

Root loci are plots on the *s*-plane of the poles of the closed loop transfer function as the open loop gain constant is varied from zero to infinity; the system is then unstable for all values of *K* for which any of the loci pass into the right-hand half of the *s*-plane, i.e. poles with positive real parts.

Direct plotting of the loci, point by point, would be impossibly tedious, but it is possible as outlined below to apply a set of rules to sketch the loci from a knowledge of the roots (poles and zeros) of the open loop transfer function.

The poles of the closed loop transfer function are the zeros of $1 + G(s)H(s)$ which have alternatively been defined as roots of the characteristic equation.

Thus if
$$\frac{C(s)}{R(s)} = \frac{G(s)}{1 + G(s)H(s)} = \frac{K_1 N_1(s) D_2(s)}{D_1(s) D_2(s) + K N_1(s) N_2(s)}$$

where $K = K_1 K_2$ is the open loop gain constant,

then the root locus is the plot of the roots of
$$D_1(s) D_2(s) + K N_1(s) N_2(s) = 0$$

It is important from this to realise that the closed loop *poles* depend upon the open loop poles *and* zeros!

The closed loop zeros are the roots of $N_1(s)D_2(s)$, i.e. the zeros of $G(s)$ and the poles of $H(s)$.

When $K = 0$ the characteristic equation roots are the roots of $D_1(s)D_2(s) = 0$, i.e. the open loop poles:

When $K \rightarrow \infty$ the characteristic equation tends to $K N_1(s) N_2(s) = 0$, and the roots lie at the open loop zeros.

This means that the loci start ($K = 0$) at the open loop poles and finish ($K = \infty$) at the open loop zeros or infinity. It can also be inferred that there will be the same number of loci 'segments' as there are open loop poles.

6.4.2 Some Simple Root Loci

The form $s + a$ for the factor is preferred for the interpretation of root loci, whereas the form $1 + sT$ is preferred for frequency response methods.

Example 1
$$G(s)H(s) = \frac{K'}{s + a}; \quad K = \frac{K'}{a}$$

The characteristic equation is

$$s + a + K' = 0$$

$\therefore$ the closed loop pole is $s = -(a + K')$ and the root locus is the part of the negative real axis from $s = -a$ to $-\infty$, as in Fig. 6.5 (a).

Example 2

$$G(s)H(s) = \frac{K'}{(s + a)(s + b)}; \quad K = \frac{K'}{ab}$$

(N.B. For a type 1 system put $a = 0$.)
The characteristic equation is

$$(s + a)(s + b) + K' = 0$$
$$\therefore \quad s^2 + (a + b)s + ab + K' = 0$$

and $\qquad s = \dfrac{-(a + b) \pm \sqrt{(a + b)^2 - 4(ab + K')}}{2}$

For $K' = 0 \quad s = -a \quad$ and $\quad -b$

For $K' = \dfrac{(a + b)^2}{4} - ab$ the roots are equal at $s = -\dfrac{a + b}{2}$

For $K' > \dfrac{(a + b)^2}{4} - ab$ the roots are imaginary with a real part,

$-\dfrac{a + b}{2}$, increasing to $\pm j\infty$ as $K' \to \infty$, as in Fig. 6.5 (b).

Example 3

$$G(s)H(s) = \frac{K'(s + c)}{(s + a)(s + b)}; \quad K = \frac{K'c}{ab}$$

The characteristic equation is

$$(s + a)(s + b) + K'(s + c) = 0$$
$$\therefore \quad s^2 + (a + b + K')s + K'c + ab = 0$$

and $\qquad s = \dfrac{-(a + b + K') \pm \sqrt{(a + b + K')^2 - 4(K'c + ab)}}{2}$

For $K' = 0 \quad s = -a \quad$ and $\quad -b$

For $K' = \infty, \quad s = \lim_{K' \to \infty} \tfrac{1}{2}(-K' \pm \sqrt{K'^2 - 4cK'})$

$$= \lim_{K' \to \infty} \tfrac{1}{2}\left(-K' \pm K'\left(1 - \frac{4c}{K'}\right)^{\frac{1}{2}}\right)$$

$$= \lim_{K' \to \infty} \tfrac{1}{2}\left(-K' \pm K'\left(1 - \tfrac{1}{2}\frac{4c}{K'} - \tfrac{1}{8}\left(\frac{4c}{K'}\right)^2 - \dots\right)\right)$$

$$= \lim_{K' \to \infty} (-K' + c + \dots) = -\infty$$

and $\qquad \lim_{K' \to \infty} \left(-c - \frac{c^2}{K'} - \dots\right) = -c$

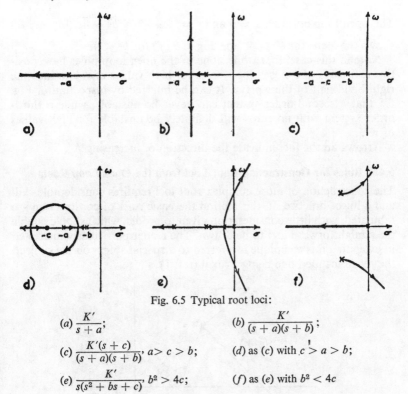

Fig. 6.5 Typical root loci:

(a) $\dfrac{K'}{s + a}$;

(b) $\dfrac{K'}{(s + a)(s + b)}$;

(c) $\dfrac{K'(s + c)}{(s + a)(s + b)}$, $a > c > b$;

(d) as (c) with $c > a > b$;

(e) $\dfrac{K'}{s(s^2 + bs + c)}$, $b^2 > 4c$;

(f) as (e) with $b^2 < 4c$

If $a > c > b$ the two locus segments lie on the negative real axis from $-b$ to $-c$ and from $-a$ to $-\infty$, Fig. 6.5 (c).

If $c > a > b$ the loci are much more complex as shown in Fig. 6.5 (d), the roots being complex conjugate pairs at certain values of K'. This locus could be constructed by point-by-point plotting from the above characteristic equation, but the rules given in the next section simplify the procedure.

Example 4

$$G(s)H(s) = \frac{K'}{s(s^2 + bs + c)}; \quad K = \frac{K'}{c}$$

The characteristic equation is

$$s(s^2 + bs + c) + K' = 0$$
$$\therefore \ s^3 + bs^2 + cs + K' = 0$$

This equation has three roots, so that there are three segments of the root locus. Extracting the roots of this equation is difficult, again stressing the need for graphical aids rather than point by point plotting.

The open loop poles are $s = 0$ and $s = -\dfrac{b}{2} \pm \dfrac{1}{2}\sqrt{b^2 - 4c}$. Fig. 6.5 (e) shows the locus for $b^2 > 4c$ and Fig. 6.5 (f) for $b^2 < 4c$.

Note in this case, that while none of the open loop poles have positive real parts, the closed loop poles for high values of K' go into the right-hand half of the s-plane. It can be inferred by inspection of Fig. 6.5 that a second-order system can never be unstable, while a third-order system with no zeros will definitely be unstable for high values of K'.

Arrows on the loci indicate the direction of increasing K'.

6.4.3 Rules for Constructing Root Loci from the Open Loop Roots

The construction of more complex root loci requires considerable skill and a lot of practice. In this section the basic rules of construction are tabulated, with little consideration given to proofs, which should enable the straightforward loci to be drawn. The reader particularly interested in extending this technique is referred to a specialist text on the subject, the recommended one being Truxal (ref. 1).

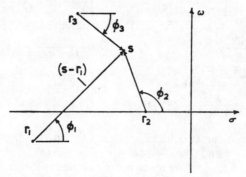

Fig. 6.6 Vector representation of $(s - r)$ factors. ϕ_1 and ϕ_2 are $+$ve, ϕ_3 $-$ve.

All points on the loci satisfy the characteristic equation

$$1 + G(s)H(s) = 0 \quad \text{or} \quad KN_1(s)N_2(s) + D_1(s)D_2(s) = 0$$

This equation can be written in factorised form as

$$G(s)H(s) = \frac{K'(s - z_1)(s - z_2)\ldots}{s^k(s - p_1)(s - p_2)\ldots} = -1$$

where $z_1, z_2, \ldots$, are the open loop zeros and $p_1, p_2, \ldots$, the open loop poles, any complex roots occurring in conjugate pairs. Now consider Fig. 6.6, where s is a point on the s-plane and r_1, r_2 and r_3 are roots; note that the roots are fixed points, whereas s is any point. The vectors representing the factors $(s - r)$ are shown which are of magnitude equal to the distance between the two points and make an angle with the real positive direction equal to ϕ, measured positive anticlockwise.

If all the open loop poles and zeros are plotted on the s-plane and the vectors $(s - z_1)$, $(s - z_2)$, etc., are drawn from any point s, the sum of the angles of the zero factors minus the sum of the angles of the pole factors must be 180°, or an odd multiple, if the value of s is to satisfy the above equation and thus be a point on the root locus. Further, the product of the amplitude of the pole factors divided by the product of the zero factors must equal K'. The technique is to guess points which are likely to lie on the loci, to construct the vectors joining the point to the roots and thus check that the angles add up to 180°, correcting the initial guess as necessary; sufficient points must be located this way to allow the loci to be sketched. This would still be an imposing task except that asymptotes can be constructed which indicate the general shape of the loci, as explained below.

Rules for Constructing Loci Asymptotes

1. Each locus segment starts ($K = 0$) at an open loop pole and finishes ($K = \infty$) at an open loop zero; if there are more poles than zeros the other segments finish at infinity. The number of segments going to infinity is the excess of poles over zeros ($n - m$).

2. The loci are symmetrical about the real axis since complex roots are always in conjugate pairs. The asymptotes to the segments which tend to infinity are equally spaced, i.e. the angle between adjacent asymptotes is $360°/(n - m)$ and to obey the symmetry rule the negative real axis is one asymptote when ($n - m$) is odd.

3. The asymptotes meet at a point on the negative real axis given by

$$\sigma_a = \frac{\Sigma \text{ real part of the poles} - \Sigma \text{ real part of the zeros}}{n - m}$$

4. The parts of the negative real axis which are segments of the loci can be determined by starting from the origin and moving to $-\infty$ when all points which have an odd number of roots (poles + zeros) to the right lie on the loci. Remember to allow for multiple roots and note that complex roots have no effect, since occurring in pairs they always add an even number of roots.

A section which joins two poles must be parts of two segments starting at each pole and moving to a common point from where they follow imaginary, conjugate paths; similarly, a section joining two zeros comprises the ends of two segments.

A section joining a pole and zero is often a complete segment, but may have two intersecting pairs of conjugate paths as in the later example of Fig. 6.8 (*b*). A good check at this point is to add the number of asymptotes and segments on the real axis and see that they add up to n.

5. The point on the negative real axis at which a conjugate pair of segments break away or re-enter, often called the point of emergence from the real axis, σ_e, is given by

$$\frac{1}{p_1 - \sigma_e} + \frac{1}{p_2 - \sigma_e} + \dots + \frac{2(a_1 - \sigma_e)}{(a_1 - \sigma_e)^2 + b_1^2} + \dots$$

$$-\frac{1}{z_1 - \sigma_e} - \frac{1}{z_2 - \sigma_e} - \dots - \frac{2(c_1 - \sigma_e)}{(c_1 - \sigma_e)^2 + d_1^2} = 0$$

where p_1, p_2, ..., z_1, z_2, ..., are the real roots and $a_1 \pm jb_1$, ..., and $c_1 \pm jd_1$, ... are the complex poles and zeros respectively. Remember that for a pole at the origin, $p = 0$. This equation is difficult to solve and is best solved by trial and error. Due to the reciprocal nature of this formula only poles and zeros close to the point of emergence need be considered, points further removed making negligible contributions.

6. The angle at which a locus leaves a complex pole (or approaches a complex zero) is given by 180° − sum of the angles of the vectors from all other poles to the complex pole in question, including its conjugate + sum of the angles of the vectors from all zeros to the complex pole.

Five and six are essential in deciding which segment associates with which asymptote.

The true loci can now be fitted to the above constructions by applying the 'phase angle criterion', point by point, i.e.

The difference between the sum of the angles of the $(s - p)$ vectors and the $(s - z)$ vectors is 180°, or an odd multiple, if s is to be a point on the root locus.

Having constructed the loci the appropriate values of K' for any point on the loci can be calculated from the 'amplitude criterion', i.e.

$$\frac{\text{Product of the modulus of the } (s - p) \text{ vectors}}{\text{Product of the modulus of the } (s - z) \text{ vectors}} = K'$$

If there are no zeros the denominator $= 1$.

The reader should now apply these rules to the simple systems of Fig. 6.5 before considering the following examples.

Example. For the system of Fig. 6.7

$$G(s)H(s) = \frac{K'}{s(s + 1\cdot5)(s + 3)(s^2 + 2s + 2)}$$

There are poles at 0, $-1\cdot5$, -3, $-1 + j1$ and $-1 - j1$; there are no zeros, so that $n = 5$ and $m = 0$ indicating 5 asymptotes going to infinity, one of which is the negative real axis, spaced by 72°.

The asymptotes meet at

$$\sigma_a = \frac{-1\cdot5 - 3 - 1 - 1}{5} = -1\cdot3$$

The parts of the negative real axis which are parts of loci segments are from 0 to $-1\cdot5$ and from -3 to $-\infty$ (on these sections there are always an odd number of roots to the right). The point of emergence from the real axis, σ_e, is given by

$$\frac{1}{0 - \sigma_e} + \frac{1}{-1 \cdot 5 - \sigma_e} + \frac{1}{-3 - \sigma_e} + \frac{2(-1 - \sigma_e)}{(-1 - \sigma_e)^2 + (-1)^2} = 0$$

$$\therefore \frac{1}{\sigma_e} + \frac{1}{1 \cdot 5 + \sigma_e} + \frac{1}{3 + \sigma_e} + \frac{2(1 + \sigma_e)}{(1 + \sigma_e)^2 + 1} = 0$$

This is an extremely difficult equation to solve, but inspection of the asymptotes already drawn will indicate that σ_e lies between 0 and

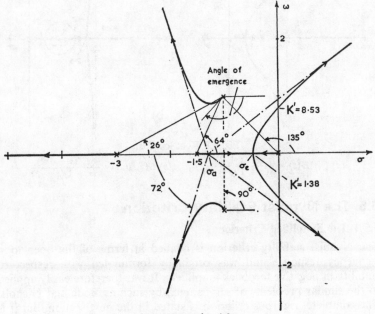

Fig. 6.7 Root loci for:

$$G(s)H(s) = \frac{K'}{s(s + 1 \cdot 5)(s + 3)(s^2 + 2s + 2)}$$

$-1 \cdot 5$ (a pole lies at each end of this section). Let $\sigma_e =$ say, $-1 \cdot 3$, -1 and $-0 \cdot 5$; calculate the value of the L.H.S. of the above equation and plot these values to a base of σ_e so that the value of σ_e for which the L.H.S. is zero can be ascertained. These three values of σ_e gave the L.H.S. as $+4 \cdot 269$, $+1 \cdot 5$ and $+0 \cdot 2$ respectively, with values at the poles of $+\infty$ for $\sigma_e = -1 \cdot 5$ and $-\infty$ for $\sigma_e = 0$. A fourth point at $\sigma_e = -0 \cdot 4$, giving the L.H.S. as $-0 \cdot 321$, was considered necessary to improve the accuracy, resulting in the desired value of σ_e as $-0 \cdot 46$.

Constructing vectors from each root to the complex root $-1 + j1$ gives the angle of emergence of the locus from this pole as

$$180 - (90 + 135 + 63 \cdot 5 + 26 \cdot 5) = -135°$$

Finally, the loci can be accurately fitted to the asymptotes by applying the phase angle criterion as necessary.

Having constructed the loci, typical values of K' can be calculated from the amplitude criterion, e.g. the value of K' for which the loci cuts the $j\omega$ axis is $K' = 8\cdot53$. It is left as an exercise for the reader to repeat the above procedure to verify the loci of Fig. 6.8.

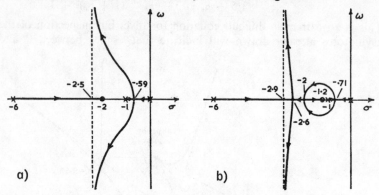

Fig. 6.8 Root loci for:

$$(a)\ G(s)H(s) = \frac{K'(s + 2)}{s(s + 1)(s + 6)}; \qquad (b)\ G(s)H(s) = \frac{K'(s + 1\cdot2)}{s(s + 1)(s + 6)}$$

6.5 The Nyquist Stability Criterion

6.5.1 The Simplified Criterion

The Nyquist stability criterion is defined in terms of the open loop frequency response and was originally formulated with respect to stability of negative feedback amplifiers. It was therefore easily applied to the similar problems of servos, etc., by such as Bode and Nichols. The complete, rigorous criterion is stated in the next section, but it is far easier to understand by a simplified engineering approach, which while not rigorous, is applicable to all except such unusual examples as systems with unstable open loops.

Now
$$\frac{C(s)}{R(s)} = \frac{G(s)}{1 + G(s)H(s)}$$

and
$$\frac{C(j\omega)}{R(j\omega)} = \frac{G(j\omega)}{1 + G(j\omega)H(j\omega)}$$

Considering a basic system with sinusoidal excitation, $E(j\omega)$ will be modified in phase and amplitude by $G(j\omega)H(j\omega)$ and fed back to be subtracted from $R(j\omega)$. If the signal is phase shifted by 180° or more, then the feedback signal will be effectively added to $R(j\omega)$, resulting in positive feedback. If the amplitude of the 180° shifted signal is less than $E(j\omega)$, then the system cannot be self-maintaining and is therefore stable. If, however, the amplitude of the feedback signal is greater than $E(j\omega)$, then $R(j\omega)$ is not needed to produce $E(j\omega)$ and the system will be self-maintaining or unstable. If the feedback signal is 180° phase shifted and equal to $E(j\omega)$, then the system will need an initial input to be just self-maintaining or simple harmonic. If it is realised that any real system will induce phase shifts at certain frequencies greater than 180°, then there must be one frequency at least for which the phase shift due to $G(j\omega)H(j\omega)$ is equal to 180°. If at this frequency the amplitude ratio, i.e. the modulus of $G(j\omega)H(j\omega) = G(\omega)H(\omega)$ is unity or greater, then the system is unstable.

The above explanation is not rigorous, it does not, for example,

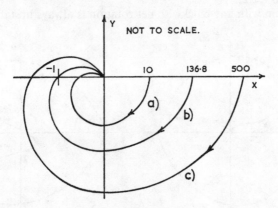

Fig. 6.9 Nyquist diagrams of

$$G(j\omega)H(j\omega) = \frac{K}{(1 + 0\cdot2j\omega)(1 + j\omega)(1 + 10j\omega)}$$

(a) $K = 10$; (b) $K = 136\cdot8$; (c) $K = 500$

account for conditionally stable systems, Section 6.5.4, but by considering the critical value of $G(j\omega)H(j\omega)$, i.e. unity gain, 180° phase shift, which is the point $(-1, j0)$ on an Argand diagram of $G(j\omega)H(j\omega)$, the simplified criterion can be stated:

"The simplified Nyquist stability criterion states that for a closed loop system to be stable the locus of the open loop frequency response function, $G(j\omega)H(j\omega)$, plotted on a 'Nyquist diagram' must not enclose the point $-1,0$ as ω goes from 0 to ∞." Enclosing the point $-1,0$ may be interpreted as passing to the left of the point.

A 'Nyquist diagram' is an Argand diagram of the complex function $G(j\omega)H(j\omega)$.

Fig. 6.9 shows three systems with similar time constants but different gain constants. When the loop is closed the system of (a) is stable, (b) is just stable, or simple harmonic and (c) is unstable.

6.5.2 The Complete Nyquist Stability Criterion

"For a closed loop system to be stable it is necessary and sufficient that the contour of the open loop frequency response $G(j\omega)H(j\omega)$ plotted on an Argand diagram, describes a number of counter clockwise encirclements of the point $(-1, j0)$, as ω varies from $-\infty$ to $+\infty$, not less than the number of poles of $G(s)H(s)$ with positive real parts."

The number of poles of $G(s)H(s)$ with positive real parts can be found by applying Routh's criterion as was described in Section 6.3 for finding the roots of the characteristic equation. Instead of the characteristic equation $D_1D_2 + KN_1N_2 = 0$, the equation $D_1D_2 = 0$ is used (notation of Section 6.2.2).

A system with any clockwise net rotation is always unstable.

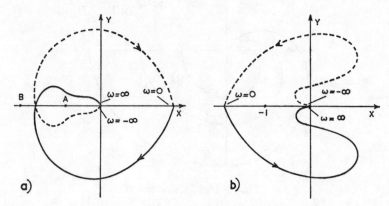

Fig. 6.10 (*a*) Stable open loop:

$$G(j\omega)H(j\omega) = \frac{K}{(1 + j\omega T_1)(1 + j\omega T_2)(1 + j\omega T_3)};$$

(*b*) Unstable open loop:

$$G(j\omega)H(j\omega) = \frac{K(1 + j\omega T_a)}{(j\omega T_1 - 1)(1 + j\omega T_2)(1 + j\omega T_3)}$$

Both systems have stable closed loops.

It must be realised that any system with an open loop pole with a positive real part has an unstable open loop so that the major extension of the complete criterion over the simplified one is to determine the stability of closed loops with unstable open loops.

For Fig. 6.10 (*a*) if the $(-1, j0)$ point is at A there are two clockwise rotations and the system is unstable. If $(-1, j0)$ is at B there are no net rotations. Examination of $G(s)H(s)$ shows that there are no poles with positive real parts so that the system is stable.

For Fig. 6.10 (*b*) the open loop is unstable; there is no need here to apply Routh since the $(j\omega T_1 - 1)$ term has a positive real part. The closed loop makes one counter-clockwise rotation so that the system is stable. This could not have been rigorously determined from the simplified criterion.

6.5.3 Gain and Phase Margins

So far no attempt has been made to define just how stable or unstable a system is. For example, a system which is only just stable will probably have an excessively oscillatory transient response so as to be unsatisfactory. A simple quantitative criterion is to consider the gain and phase margins.

The GAIN MARGIN is the log modulus (db) of the system when the phase shift is 180°, denoted positive for negative db. In other words, if

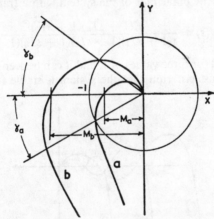

Fig. 6.11 Phase and gain margins on Nyquist diagrams.

For the stable system:

(a) Phase margin, $\gamma_a = +28°$

Gain margin $= -20 \log_{10} M_a = -20 \log_{10} 0{\cdot}75 = +2{\cdot}5$ db

For the unstable system:

(b) Phase margin, $\gamma_b = -37°$

Gain margin $= -20 \log_{10} M_b = -20 \log_{10} 1{\cdot}7 = -4{\cdot}6$ db

the open loop gain is increased by the gain margin the overall gain at 180° phase shift will be unity (0 db) and the system is on the verge of instability. A negative gain margin indicates an unstable system.

The PHASE MARGIN is 180° minus the phase angle at the frequency when the gain is unity (0 db), i.e. the amount the phase lag would have to be increased to make the system unstable. A negative phase margin indicates an unstable system.

The gain margin on the polar plot is given directly by $-20 \log_{10} M$, where M is the value of $G(\omega)H(\omega)$ when the curve cuts the negative $X(\omega)$ axis, since the margin is referred to a gain of 1 (0 db).

The phase margin is found by drawing a circle of unit radius about the origin and drawing the radius from the centre to the point where the curve $G(j\omega)H(j\omega)$ cuts the unit circle. The phase margin is then the angle between this radius and the negative $X(\omega)$ axis, positive below the axis and negative above.

Typical design figures are 3 to 8 db gain margin and 30° to 60° phase margin. It must be stressed, however, that this is a very simple design specification which omits such important considerations as the band width, resonant frequency, etc.

Fig. 6.11 shows the gain and phase margins of two typical systems.

6.5.4 Conditionally Stable Systems

Fig. 6.12 shows the polar plot of the system with a transfer function.

$$G(s)\,H(s) = \frac{K(1 + sT_a)(1 + sT_b)}{s(1 + sT_1)(1 + sT_2)(1 + sT_3)(1 + sT_4)}$$

This is plotted for three values of K and it can be seen by applying the simplified Nyquist criterion that the system is stable for $K = K_b$ and

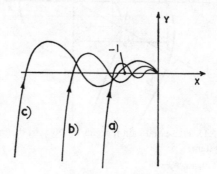

Fig. 6.12 Nyquist diagram of a conditionally stable system:

$$G(j\omega)H(j\omega) = \frac{K(1 + j\omega T_a)(1 + j\omega T_b)}{j\omega(1 + j\omega T_1)(1 + j\omega T_2)(1 + j\omega T_3)(1 + j\omega T_4)}$$

$$T_1 > T_2 > T_a > T_b > T_3 > T_4 \quad \text{and} \quad K_a < K_b < K_c$$

unstable for $K = K_a$ and K_c. The stability of all systems is dependent upon K, but this type of system is particularly so since both increase and decrease of K from K_b can cause instability. Such a system is therefore termed conditionally stable.

While the stability of this type of system can be determined by the application of the simplified criterion it makes the simple explanation of Section 6.5.1 invalid since the stable system of (*b*) cuts the negative real axis at three points, at two of which the modulus is bigger than unity. The rigorous proof follows from the complete Nyquist criterion.

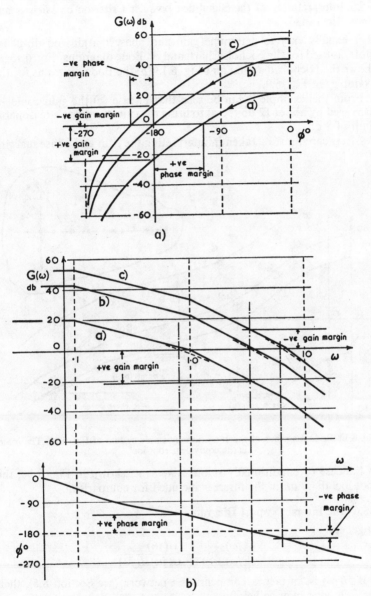

Fig. 6.13 (*a*) Nichols diagram and (*b*) Bode diagrams of:
$$G(j\omega)H(j\omega) = \frac{K}{(1+0\cdot2j\omega)(1+j\omega)(1+10j\omega)};$$
(*a*) $K = 10$; (*b*) $K = 136\cdot8$; (*c*) $K = 500$.

6.5.5 Interpretation of the Simplified Nyquist Criterion on Nichols and Bode Diagrams

It is equally easy to determine gain and phase margins on db-phase plots, as used for the Nichols Chart, and on Bode diagrams. Fig. 6.13 (*a*) shows the Nichols diagram and Fig. 6.13 (*b*) the Bode diagrams for the examples of Fig. 6.9.

From this example it can be seen that in case (*a*) the gain could be increased by about 12 db (× 4) to bring the gain margin to a reasonable limit of 8 db.

Again care must be taken in interpreting the gain and phase margins

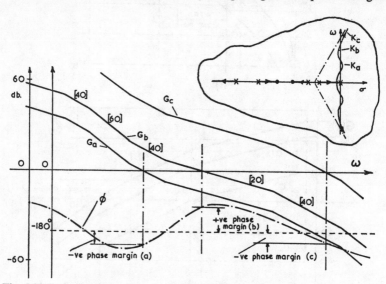

Fig. 6.14 Bode diagrams of the conditionally stable system of Fig. 6.12. The insert is the equivalent root loci.

in the case of conditionally stable systems as shown in Fig. 6.14; the root loci diagram in the insert is included for comparison.

6.5.6 The Inverse Nyquist Diagram

In general

$$\frac{C(j\omega)}{R(j\omega)} = \frac{G(j\omega)}{1 + G(j\omega)H(j\omega)}$$

If $H(j\omega)$ is, in fact, a compensation network (see Section 8.5), then the following may be helpful.

$$\frac{R(j\omega)}{C(j\omega)} = \frac{1 + G(j\omega)H(j\omega)}{G(j\omega)}$$

$$= \frac{1}{G(j\omega)} + H(j\omega)$$

Thus plotting $\dfrac{1}{G(j\omega)}$ on a polar diagram can be a useful guide to find-ing the value of $H(j\omega)$ required. Such a plot is called the inverse Nyquist diagram, as in Fig. 6.15.

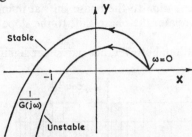

Fig. 6.15 Inverse Nyquist diagram.

6.6 Bode's Theorems

Bode's theorems relate log modulus (db) and phase as functions of real frequency, ω. The application of Bode's work has already been seen in the use of asymptotic approximation in plotting db–log ω and the associated phase–log ω graphs; in fact, these pairs of curves are called Bode diagrams. The db–phase–ω relationship was taken further in Section 5.8, where it was shown that $G(j\omega)$ could be found from a know-ledge of the db–log ω curve.

This stems from two basic theorems, applicable only to linear, minimum phase systems.

(i) Attenuation integral theorem:

$$\int_0^\infty \omega \frac{d}{d\omega} \alpha(\omega) d\omega = -\frac{\pi}{2} \varphi_\infty$$

where
$$\varphi_\infty = -\lim_{\omega \to \infty} \omega\varphi(\omega)$$

and
$$\alpha(\omega) = \log_e G(\omega) \text{ nepers}$$

(1 neper = 8·686 db)

(ii) Phase integral theorem:

$$\varphi(\omega_d) = \frac{\pi}{2} \left[\frac{d}{du} \alpha(u)\right]_{u=0}$$
$$+ \frac{1}{\pi} \int_{-\infty}^{\infty} \left(\frac{d}{du} \alpha(u) - \left[\frac{d}{du} \alpha(u)\right]_{u=0}\right) \log_e \coth \frac{|u|}{2} \, du$$

where $u = \log_e \dfrac{\omega}{\omega_d}$, i.e. $\omega = \omega_d$ when $u = 0$

$\varphi(\omega_d)$ = the phase shift at the desired frequency ω_d.

$\dfrac{d}{du} \alpha(u)$ = the slope of the db–log ω curve (the scale is such that 1 unit $\equiv$ 20 db/decade)

The term $\log_e \coth \frac{|u|}{2}$ is a weighting function of the form shown in Fig. 6.16. The derivation and application of these equations is covered in Chestnut and Mayer, Vol. 1 (ref. 2). Note that the attenuation integral relates the log attenuation to the phase shift at infinite frequency and the phase integral relates the phase shift to the slope of the log amplitude–log ω curve.

The second equation, however, has particular significance to stability

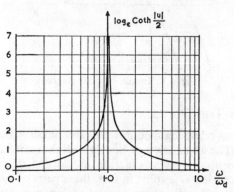

Fig. 6.16 Bode 'weighting function'.

criteria. Consider the second term of the phase integral: if the slope of the curve, $\frac{d}{du}\alpha(u)$ is constant over, say, a decade either side of $\omega_d(u = 0)$, then the integrand is zero over the range $\frac{\omega}{\omega_d} = 0{\cdot}1$ to 10. Outside this range the weighting term is such that a non-zero value of the $\frac{d}{du}\alpha(u) - \left[\frac{d}{du}\alpha(u)\right]_{u=0}$ term has little or no significance. Thus the second term could be considered very small if $\alpha(u)$ is of constant slope a decade above and below the considered frequency, ω_d. In this case, the phase shift is approximately $\frac{\pi}{2}\left[\frac{d}{du}\alpha(u)\right]_{u=0}$, so that phase shift is 180° or more if $\frac{d}{du}\alpha(u) = 2$ nepers/unit of $u = 2 \times 20$ db/decade. If ω_o is the 'cross-over frequency' at which the open loop db/log ω curve cuts the 0 db line, then the phase shift will be 180° or more, i.e. unstable, if the slope is 40 db/decade.

Thus it is a reasonable rule of thumb design figure to say that the open loop db–log ω curve should cut the 0 db axis with a slope of 20 db/decade extending an octave above and below ω_o. Use of 20 db/decade and an octave range allows a practical tolerance of about 45° phase shift. Investigation of the conditionally stable system of Fig. 6.14 provides a good exercise in the application of this technique.

6.7 Methods of Determining the Closed Loop from the Open Loop Frequency Response Function

6.7.1 Introduction

Having utilised open loop frequency response plots to determine system stability by using Nyquist's criterion it would be desirable to use the same plots to find the resulting frequency response when the loop is closed.

The common terminology is to call the closed loop amplitude M, measured either as a ratio or in db, when it is often termed N, and the closed loop phase angle, δ. The technique is to plot contours of constant M and δ on the Nyquist diagram or an equivalent on the db–phase diagram called the Nichols Chart. An approximate method applied to Bode diagrams is also explained.

6.7.2 M and δ Circles

This is applied, on a polar plot (Nyquist diagram), to direct feedback systems only, so that the equivalent direct feedback system should be used as in Section 4.7, i.e.:

$$\frac{C(j\omega)}{R(j\omega)} = \frac{G(j\omega)}{1 + G(j\omega)H(j\omega)}$$

$$= \frac{G(j\omega)H(j\omega)}{1 + G(j\omega)H(j\omega)} \cdot \frac{1}{H(j\omega)}$$

$$= \frac{1}{H(j\omega)} \cdot \frac{C(j\omega)}{R'(j\omega)}$$

The closed loop function found from the M and δ circles, $\dfrac{C(j\omega)}{R'(j\omega)}$, must be multiplied by $\dfrac{1}{H(j\omega)}$ to obtain the complete function.

Put $\qquad G(j\omega)H(j\omega) = G_e(j\omega)$

and $\qquad G_e(j\omega) = X(\omega) + jY(\omega)$

Hence the closed loop amplitude

$$M = \frac{|X + jY|}{|1 + X + jY|}$$

and the closed loop phase angle

$$\delta = \underline{/X + jY} - \underline{/1 + X + jY} \quad \text{(leading)}$$

where M, X, Y and δ are functions of ω.

Hence

$$M^2 = \frac{X^2 + Y^2}{1 + 2X + X^2 + Y^2}$$

$$\therefore \ X^2(M^2 - 1) + 2XM^2 + Y^2(M^2 - 1) = -M^2$$

$$\therefore \ X^2 + \frac{2M^2}{M^2 - 1} \cdot X + Y^2 = -\frac{M^2}{M^2 - 1}$$

Completing the square gives

$$\left(X + \frac{M^2}{M^2 - 1}\right)^2 + Y^2 = -\frac{M^2}{M^2 - 1} + \frac{M^4}{(M^2 - 1)^2}$$

$$= \frac{M^2}{(M^2 - 1)^2}$$

This, for constant values of M, is the equation of a circle of radius $\frac{M}{M^2 - 1}$ with a centre $\left(-\frac{M^2}{M^2 - 1}, 0\right)$. Hence a family of M circles can be drawn which are symmetrical about the X axis. Typical circles are

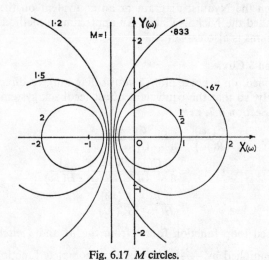

Fig. 6.17 *M* circles.

shown in Fig. 6.17, the plotting of which is simplified by the following examples:

For $Y = 0$

$$X + \frac{M^2}{M^2 - 1} = \pm \frac{M}{M^2 - 1}$$

$$\therefore \; X = -\frac{M}{M^2 - 1}(M \pm 1)$$

$$= -\frac{M}{M + 1} \quad \text{or} \quad -\frac{M}{M - 1}$$

$\therefore$ for $M = 1$ $X]_{Y=0} = -\frac{1}{2}$ and $-\infty$.

The radius $= \infty$, i.e. a vertical line through $X = -\frac{1}{2}$.

For $M = 2$ $X]_{Y=0} = -\frac{2}{3}$ and -2
For $M = \infty$ $X]_{Y=0} = -1$ and -1, i.e. a point
For $M = 0.5$ $X]_{Y=0} = -\frac{1}{3}$ and $+1$

Consider now the equation for the phase angle:

$$\delta = \tan^{-1} \frac{Y}{X} - \tan^{-1} \frac{Y}{1+X}$$

$$= \theta_1 - \theta_2$$

Put $\quad A = \tan \delta = \tan(\theta_1 - \theta_2) = \dfrac{\tan \theta_1 - \tan \theta_2}{1 + \tan \theta_1 \tan \theta_2}$

$$= \frac{\dfrac{Y}{X} - \dfrac{Y}{1+X}}{1 + \dfrac{Y^2}{X(1+X)}} = \frac{Y}{X^2 + X + Y^2}$$

$$\therefore \quad X^2 + X + Y^2 - \frac{Y}{A} = 0$$

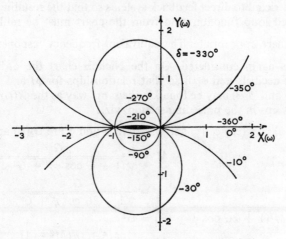

Fig. 6.18 δ circles. The negative sign indicates phase lag.

Completing the square

$$\left(X + \tfrac{1}{2}\right)^2 + \left(Y - \frac{1}{2A}\right)^2 = \frac{1}{4} + \frac{1}{4A^2} = \frac{1}{4}\frac{A^2 + 1}{A^2}$$

This for constant values of δ, and therefore of A, is the equation of a circle of radius $\dfrac{1}{2A}\sqrt{A^2 + 1}$ and centre $\left(-\dfrac{1}{2}, \dfrac{1}{2A}\right)$.

The radius $= \dfrac{\sqrt{\tan^2 \delta + 1}}{2 \tan \delta} = \dfrac{1}{2 \sin \delta}$

Now for $Y = 0$

$$\left(X + \tfrac{1}{2}\right)^2 = \tfrac{1}{4}$$

$$\therefore \quad X = 0 \quad \text{or} \quad X = -1 \text{ for all values of } A.$$

The δ circles lie with their centres on a vertical axis passing through $X = -\frac{1}{2}$. Some confusion occurs since $\tan \delta = \tan (180 + \delta)$, so that there are different values of δ on the same circle above and below the axis, as shown in Fig. 6.18.   The M-and δ circles of Figs. 6.17 and 6.18 can now be used for any system which is stable with known values of $G_e(j\omega)$. The technique is to plot the open loop function $G_e(j\omega)$ and to read off the values of M and δ at the points where the curves intersect. The best method, in practice, is to construct a standard set of M and δ circles on a transparency, which can be superimposed on a normal plot, provided the scales are the same.

6.7 The Nichols Chart

The Nichols chart is the equivalent to the M and δ circles using decibels and is plotted on a db–phase diagram. As with the M and δ circles its use is restricted to direct feedback systems so that the resulting values of the closed loop function read from the chart must be multiplied by $\dfrac{1}{H(j\omega)}$ where appropriate. The equivalent frequency response function $G_e(j\omega)$ is again considered. On the Nichols chart the ordinates are $G_e(\omega)$ in decibels and $\phi(\omega)$ so that relationships for M and δ in terms of $G_e(\omega)$ and $\phi(\omega)$ may be found in the same way as the $X(\omega)$ and $Y(\omega)$ relationships in the preceding section, viz.:

$$G_e(j\omega) = G(\cos \phi + j \sin \phi), \text{ where } G \text{ and } \phi \text{ are } G_e(\omega) \text{ and } \phi(\omega)$$

then
$$M = \frac{G}{\sqrt{(1 + G \cos \phi)^2 + G^2 \sin^2 \phi}}$$
$$= \frac{G}{\sqrt{1 + 2G \cos \phi + G^2}}$$

$$\therefore\ M^2(1 + 2G \cos \phi + G^2) = G^2$$

$$\therefore\ \cos \phi = - \frac{M^2 + G^2(M^2 - 1)}{2M^2G} \tag{6.1}$$

Similarly
$$\delta = \underline{/G_e(j\omega)} - \underline{/1 + G_e(j\omega)} \text{ (leading)}$$

$$= \phi - \tan^{-1} \frac{G \sin \phi}{1 + G \cos \phi} \tag{6.2}$$

Equation 6.1 relates M to $\phi(\omega)$ and $G(\omega)$ so that contours for constant M can be plotted as $20 \log_{10} G(\omega)$ against $\phi(\omega)$. Similarly, equation 6.2 relates δ with $20 \log_{10} G(\omega)$ and $\phi(\omega)$ so that the constant δ contours can also be drawn. Since the logarithm is involved, simple contours do not result so they must be calculated point by point and plotted, preferably on a transparency, to the same scale as the basic db–phase plot. Fig. 6.19 shows a typical Nichols chart. The closed loop function is again derived from the intersects of the open loop function, $G_e(j\omega)$, with the M and δ contours.

Since the results from this can easily be plotted on to a Bode diagram

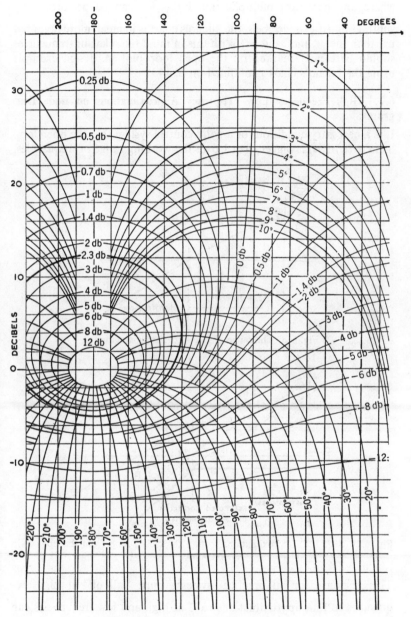

Fig. 6.19 Nichols chart. The angles are labelled as phase lag.

where any necessary multiplication by $\dfrac{1}{H(j\omega)}$ can be performed by subtracting $H(\omega)$ in decibels, the Nichols chart is in much more favoured use than the M and δ circles. With this in mind the relevant frequencies should be clearly marked on the open loop plot. A design example using the Nichols chart is outlined on p. 143.

6.7.4 High Frequency–Low Frequency Approximations on Bode Diagrams

It is a necessary condition for a good control system that the open loop 'gain' $G(\omega)H(\omega)$ is bigger than 1 over the working range of frequencies.

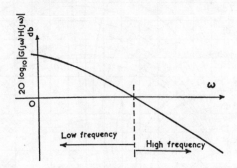

Fig. 6.20 The range of H.F.–L.F. approximations.

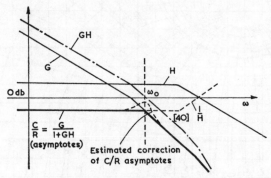

Fig. 6.21 H.F.–L.F. approximations to find the closed loop from the open loop response.

This response for all practical systems must 'fall off' at high frequencies so that approximations to $\dfrac{C}{R}$ at high and low frequencies can be made.

$$\frac{C(j\omega)}{R(j\omega)} = \frac{G(j\omega)}{1 + G(j\omega)H(j\omega)}$$

At low frequency $G(j\omega)H(j\omega) \gg 1$

$$\therefore \frac{C(j\omega)}{R(j\omega)} \simeq \frac{1}{H(j\omega)}$$

At high frequency $G(j\omega)H(j\omega) \ll 1$

$$\therefore \frac{C(j\omega)}{R(j\omega)} \simeq G(j\omega)$$

The terms high and low frequency are relative and should be loosely defined as the regions where the Bode amplitude diagram lies below and above the 0 db line respectively, as in Fig. 6.20.

Note that the results are meaningless for an unstable system. Also, while the low- and high-frequency asymptotes of C/R are reasonably accurate, the curve will be inaccurate around the frequency, ω_0, at which $|GH| = 1$ (0 db). One or two accurate points could be calculated around these frequencies, but an estimate of the shape of the curve near the intersect of the asymptotes can be made from the shape of the GH curve at ω_0. A low phase and gain margin, indicated by a short 40 db/decade slope either side of ω_0, would result in a resonant peak in the C/R curve, while a 20 db/decade slope at ω_0 would result in a smooth roll-off from the low- to the high-frequency C/R asymptotes.

6.8 Summary

The problem of stability of closed loops is the most important single feature of analysis. A variety of methods of determining the stability of a system have been introduced in this chapter in the most logical rather than the most practical order. In fact, the practicality of a method of analysis can often be determined by the ability of the method to indicate the *degree* of stability and possible methods of system improvement.

Because of the impracticality of measuring the closed loop direct, all the important techniques consider the open loop first and interpret the stability of the closed loop by the features of the open loop, e.g. Nyquist's criterion.

Frequency response methods have the virtue of relatively easy measurement of physical systems. Such results can be used directly for frequency domain design techniques. Alternatively the measurements can be used to find the transfer function and hence pole and zeros of the system prior to using s-plane design methods. If data is available in a more basic form e.g. resistances, inertias, aerodynamic coefficients etc., it can be converted via the transfer function to either frequency response or s-plane form. It would be fairly safe to assume that, without the aid of a digital computer in plotting root loci, the frequency response method should be used.

7
Design Procedure and Specifications

7.1 Introduction to Design

It has now been shown that any linear system can be analysed in terms of functions of time; functions of the complex variable, s; or functions of frequency, ω. Design procedure, then, can also be carried out in similar terms.

Any control system is required to do a particular job with all errors kept to a minimum, or at least below a certain level; thus the first step must be to produce a specification.

The next step is to select the necessary hardware to achieve this specification as cheaply and reliably as possible. Part Four of this text is intended to give an overall picture of the scope of components for most servomechanisms, although no consideration is given to components for other types of control systems.

Since the design procedure used here applies only to linear systems, the rating of components is of particular importance. This topic is given special consideration in Chapter 14, the salient feature here being that the use of closed loop control cannot increase the power rating of a component, i.e. a 1 kW motor cannot be expected to develop 10 kW!

While the specification is usually written in terms of transient and steady state performance in real time, an equivalent specification in terms of frequency response or the position of roots on the s-plane can be made. In this book, both frequency domain and s-plane design methods are covered, with the accent on the former. The section on s-plane design, 8.8, can be extended in Truxal (ref. 1). Further, all ideas will be interpreted on Bode diagrams with occasional use of the Nichols chart, the Nyquist diagram being limited to use only as a rapid check of stability if the measured results are in a suitable form.

7.2 Specifications

7.2.1 Requirements of the Specification

The specification must state the maximum allowable errors in the system in the steady state and transient conditions. The only complete answer is to specify the total system error at all instances under actual working conditions, which, however, is impractical for the following reasons:

(i) It would require a completely operative, stable system before any measurement could be made.

(ii) It would be difficult to decide what are the sources of error and how best to reduce them if necessary.

(iii) It is difficult to choose one specific measure of the error as sufficient, e.g. the mean square error is sometimes used, but a system with large infrequent errors could have the same answer as a system with small frequent errors. It then depends upon the application as to which is the better system.

Clearly the nature of the input function has a large bearing on the system errors, so that the best specification may be to define the input by, say, its statistical properties (see Chapter 10) and to design the system from a statistical viewpoint, attempting to produce an optimum system by minimising the error. This, however, is not as practical as could be hoped, due to measuring and interpreting problems. The nature of the input is considered further in Section 14.6 in rating components and, from a transient rather than statistical outlook, in Chestnut and Mayer, Vol. 2 (ref. 3).

The commonest form of specification is to consider the response of the system to a clearly defined (deterministic) input, normally a step function, a ramp function, an impulse or a sinusoid.

7.2.2 Step Input

The following values of the response to a step input can be used for the specification in the time domain. They are shown in Fig. 7.1.

(i) The *peak overshoot* expressed as a percentage of the desired step change. In certain systems, e.g. profile copying, or a tracking system where the output is continually changing, with no step changes in practice, overshoot would be undesirable so that, say, 5% would be the maximum allowable value. In a position control system for a gun or drill, where only the final position is of importance, overshoot up to say 20% could be tolerated. Some overshoot is desirable in any system to minimise the effects of stick–slip, etc.

(ii) The time to settle within a tolerance band either side of the desired value. This is called the *settling time*. The acceptable tolerance band must also be specified and will depend upon the required accuracy of the system; ±5%, or ±2%, of the step is reasonable. The final steady state accuracy may of course be required better than 2%.

(iii) The *rise time* which is defined usually as the time for the system to change from 5% to 95% of its final value in the initial part of the curve. 10% and 90% are also commonly used, although it would seem reasonable to this author to use 0 to 90%. It should be noted that all these terms are quoted as percentages since the time of response for a linear system is independent of the amplitude.

These specifications do not directly include any mention of the frequency of transient oscillations (it is in order to assume that there will be an oscillation since it has already been pointed out that slight under-

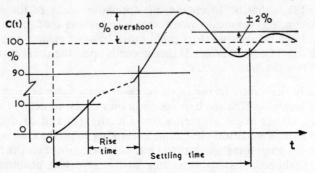

Fig. 7.1 Transient response terms.

damping is desirable). The frequency should be as high as possible in order to reduce the rise time, but a practical limit should be considered since high frequencies involve high accelerations and therefore high mechanical stressing. Equally important, frequencies near to any resonant frequencies of associated structures must be avoided; it is fortunate that transient oscillation frequencies of the servo are nearly always lower than mechanical resonances.

7.2.3 Ramp, or Constant Velocity Input

The specification with respect to a step input is the most common in practice, but the important error known as velocity lag, which is often present when the input is varied at a constant rate, is not brought into

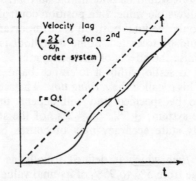

Fig. 7.2 Ramp or constant velocity input.

consideration. This error is brought out when the input is a ramp function, as shown in Fig. 7.2. The magnitude of the velocity lag is a function of the rate of change of input so that the magnitude of the ramp input must be specified. This error is not important in a simple position control, but is critical in a tracking or contouring system, The maximum velocity lag error allowable should then be specified, corresponding to the maximum anticipated velocity of the input.

7.2.4 Parabolic, or Constant Acceleration Input

In a few cases the specification is made even more stringent by defining the allowable errors during acceleration. For the type 1 system of Fig. 7.3 it can be seen that the error increases with time, so that a more

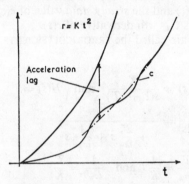

Fig. 7.3 Parabolic or constant acceleration input.

realistic specification is, say, for a machine tool, that it can contour at a constant tangential velocity, e.g. 10 m/min, around a circular path of radius 25 mm, with the total error not exceeding 0·02 mm. Thus the input can be calculated as a function of time and the total error calculated from the equations of the system. This form of input has the added advantage that some idea can be obtained of the magnitudes of the accelerations involved so that a power rating for the drive can be estimated.

7.2.5 Error Coefficients

It is instructive to analyse the actuating signal, E, in terms of the input, R, thus:

$$\frac{C(s)}{R(s)} = \frac{G(s)}{1 + G(s)H(s)} \quad \text{and} \quad C(s) = G(s) E(s)$$

$$\therefore \quad \frac{E(s)}{R(s)} = \frac{1}{1 + G(s)H(s)}$$

$E(s)/R(s)$ can be expressed as a rational polynomial, which will have the same denominator as $C(s)/R(s)$:

$$\frac{E(s)}{R(s)} = \frac{B_0 + B_1 s + B_2 s^2 + \dots}{A_0 + A_1 s + A_2 s^2 + \dots}$$

$$= \frac{B_0}{A_0} + \left(B_1 - \frac{A_1 B_0}{A_0}\right)\frac{s}{A_0}$$

$$+ \left[\left(B_2 - \frac{A_2 B_0}{A_0}\right) - \frac{A_1}{A_0}\left(B_1 - \frac{A_1 B_0}{A_0}\right)\right]\frac{s^2}{A_0} + \dots$$

(by long division)

$$\therefore \ E(s) = C_0 \cdot R(s) + C_1 \cdot sR(s) + C_2 \cdot s^2 R(s) + \dots$$

where $\qquad C_0 = \dfrac{B_0}{A_0}, \quad C_1 = \dfrac{1}{A_0}\left(B_1 - \dfrac{A_1 B_0}{A_0}\right)$, etc.

For a particular value of $r(t)$ the $s^x R(s)$ terms can be interpreted as the xth derivative of $r(t)$ and the *steady state* value of $e(t)$ can be found by summing the terms $C_x \cdot x$th derivative of $r(t)$.

C_0, C_1, C_2, etc., are called the ERROR COEFFICIENTS.

Example

$$G(s) = \frac{K}{s(1 + sT)} \quad \text{and} \quad H(s) = 1$$

then

$$\frac{C(s)}{R(s)} = \frac{1}{1 + \dfrac{2\zeta}{\omega_n}s + \dfrac{1}{\omega_n^2}s^2}$$

where

$$\frac{2\zeta}{\omega_n} = \frac{1}{K} \quad \text{and} \quad \frac{1}{\omega_n^2} = \frac{T}{K}$$

$$\frac{E(s)}{R(s)} = \frac{1}{1 + G(s)H(s)} = \frac{s + Ts^2}{K + s + Ts^2}$$

$$\therefore \ A_0 = K, \quad A_1 = 1 \quad \text{and} \quad A_2 = T$$
$$B_0 = 0, \quad B_1 = 1 \quad \text{and} \quad B_2 = T$$

so that the first three error coefficients are

$$C_0 = 0; \quad C_1 = \frac{1}{K}; \quad \text{and} \quad C_2 = \frac{1}{K^2}(TK - 1)$$

For a unity magnitude, constant acceleration input $s^2 R(s) \equiv 1$, $sR(s) \equiv t$ and $R(s) \equiv \frac{1}{2}t^2$, all higher derivatives being zero.

$$\therefore \ e(t) = 0 + \frac{1}{K}t + \frac{1}{K^2}(TK - 1) \cdot 1 \quad \text{in the steady state}$$

$$= \frac{2\zeta}{\omega_n}t + \frac{1}{\omega_n^2}(1 - 4\zeta^2) \quad \text{(cf. Fig. 7.3)}$$

An alternative notation akin to the error coefficients are the *gain constants*.

For a type 0 system

$$G(s)H(s) = \frac{K_p(1 + b_1 s + \dots)}{1 + a_1 s + a_2 s^2 + \dots}$$

where K_p is the dimensionless 'position gain constant'—the open loop gain constant of a type 0 system. The name can be misleading since a type 0 system would not be a position control system.

For a type 1 system

$$G(s)H(s) = \frac{K_v(1 + b_1s + \ldots)}{s(1 + a_1s + a_2s^2 + \ldots)}$$

where K_v is the 'velocity gain constant'—the open loop gain constant of a type 1 system which has the dimensions of second^{-1}.

For a type 2 system

$$G(s)H(s) = \frac{K_A(1 + b_1s + \ldots)}{s^2(1 + a_1s + a_2s^2 + \ldots)}$$

where K_A is the 'acceleration gain constant'—the open loop gain constant of a type 2 system, which has the dimensions of second^{-2}.

It can be shown that for a

type 0 system $\qquad C_0 = \dfrac{1}{1 + K_p}$

type 1 system $\qquad C_0 = 0 \quad$ and $\quad C_1 = \dfrac{1}{K_v}$

type 2 system $\qquad C_0 = C_1 = 0 \quad$ and $\quad C_2 = \dfrac{1}{K_A}$

This notation is not so extensive as the error coefficients, e.g. it does not define the acceleration and higher order errors for a type 1 system.

7.2.6 Frequency Response Specifications

For a system analysed in the frequency domain it would be desirable if the specification where written in the same terms.

While few systems are of second order, many can be approximated by second-order systems, at least over the bandwidth. N.B. This is only appropriate for the closed loop specification, since the higher frequency components of an open loop thus neglected could contribute markedly to the resulting closed loop. The following are the prominent relationships between transient and frequency responses of a second-order system:

$$\frac{C(j\omega)}{R(j\omega)} = \frac{1}{1 + \dfrac{2\zeta}{\omega_n}j\omega + \dfrac{1}{\omega_n^2}(j\omega)^2}$$

From Section 5.6 (iv), the resonant frequency, ω_R, and the peak value M_p are given by

$$\left.\begin{array}{l} \omega_R = \omega_n\sqrt{1 - 2\zeta^2} \\ M_p = -20\log_{10}2\zeta\sqrt{1 - \zeta^2} \end{array}\right\} \text{for } \zeta < 0.707$$

and

From Section 3.5 (*a*) and Fig. 3.8, the peak overshoot for a unit step input ($\zeta < 1$) occurs when

$$t_p = \frac{\pi}{\omega_t} = \frac{\pi}{\omega_n \sqrt{1 - \zeta^2}}$$

giving the maximum percentage overshoot $= 100 \cdot e^{-\frac{\zeta\pi}{\sqrt{1-\zeta^2}}}$ and the transient frequency of oscillation as

$$\omega_t = \omega_n \sqrt{1 - \zeta^2}$$

$$\therefore \frac{\omega_R}{\omega_t} = \frac{\sqrt{1 - 2\zeta^2}}{\sqrt{1 - \zeta^2}}$$

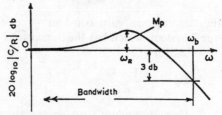

Fig. 7.4 Frequency response terms.

The number of oscillations for the envelope to fall within a $\pm 5\%$ settling band, with a settling time, t_s, is given by

$$e^{-\zeta\omega_n t_s} = 0 \cdot 05$$

$$\therefore t_s = \frac{3}{\zeta\omega_n}$$

and the number of oscillations,

$$N = \frac{t_s}{2\pi/\omega_t} = \frac{t_s \omega_n}{2\pi} \sqrt{1 - \zeta^2}$$

$$= \frac{1 \cdot 5\sqrt{1 - \zeta^2}}{\pi\zeta}$$

Finally the bandwidth, ω_b, is that frequency at which the log amplitude of C/R is -3 db.

$$\therefore -3 = -20 \log_{10} \sqrt{\left(1 - \left(\frac{\omega_b}{\omega_n}\right)^2\right)^2 + 4\zeta^2\left(\frac{\omega_b}{\omega_n}\right)^2}$$

from which $\quad \omega_b = \omega_n \sqrt{1 - 2\zeta^2 + \sqrt{2 - 4\zeta^2 + 4\zeta^4}}$

Some of these relationships are shown graphically in Fig. 7.5.

The bandwidth, having been defined, must be high enough to encompass the frequency content of the input, but should be no higher than necessary, so as to minimise noise problems (this is an instinctive statistical method of design), since an increase in bandwidth above that required by the input will only include extra noise in the output for no improvement in the response to the signal. Velocity and acceleration lags can be determined by first ascertaining the type of the system from the slope of the open loop Bode diagram as $\omega \to 0$, which will be $-20 \times$ the type db/decade, and then finding the gain constant K.

The reciprocal of K is C_0 for a type 0 system or C_1 for a type 1 system, etc. The higher order coefficients are not so obvious and must be calculated as in Section 7.2.5.

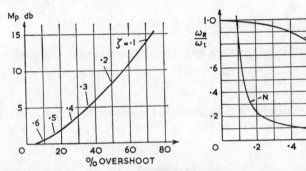

Fig. 7.5 Relationships between transient and frequency response terms for the second order system:

$$G(s) = \frac{1}{1 + \frac{2\zeta}{\omega_n}s + \frac{1}{\omega_n{}^2}s^2}$$

7.2.7 *s*-Plane Specifications

It is commonly claimed that the advantage of *s*-plane analysis and design, e.g. root loci, is that the transient performance can easily be interpreted in terms of the pole–zero locations of the system transfer function. Thus interpreting the second-order example of the previous section in terms of pole locations will enable transient, frequency response and *s*-plane specifications to be related:

(*a*) A single pole at $s = -a$ corresponds to a transient term of the form

$$Ae^{-at} \ldots \quad \text{(see Section 6.2.1)}$$

(*b*) A second-order term with complex poles at $s = -\sigma \pm j\omega$ corresponds to a transient term of the form

$$B_1 e^{(-\sigma + j\omega)t} + B_2 e^{(-\sigma - j\omega)t} = e^{-\sigma t}(C_1 \cos \omega t + C_2 \sin \omega t)$$

Consider the term $\dfrac{1}{1 + \dfrac{2\zeta}{\omega_n}s + \dfrac{1}{\omega_n{}^2}s^2}$, where $\zeta < 1$

The poles are
$$s = -\zeta\omega_n \pm j\omega_n\sqrt{1 - \zeta^2}$$
$$= -\sigma \pm j\omega_t$$

Thus for any complex conjugate pair of poles in the L.H. half of the *s*-plane, the corresponding transient response term will be damped sinusoidal, of frequency equal to the imaginary part of the pole, with an exponential envelope of decay rate equal to the magnitude of the real part. It is noted in Fig. 7.6 that poles which lie on a horizontal line have the same frequency of oscillation, and poles which lie on a vertical line have the same decay rate.

The length of the vector from the origin to one of the complex poles is given by

$$\sqrt{\sigma^2 + \omega_t{}^2} = \sqrt{\zeta^2\omega_n{}^2 + \omega_n{}^2(1 - \zeta^2)} = \omega_n$$

and the angle this vector makes with the negative real axis, θ, is given by

$$\cos\theta = \frac{\sigma}{\omega_n} = \frac{\zeta\omega_n}{\omega_n} = \zeta \qquad (\zeta < 1)$$

Thus poles which lie on the same radial line from the origin have the same damping factor. Texts such as Truxal (ref. 1, Chapter 5) and

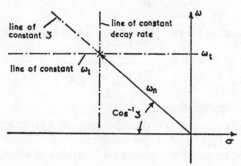

Fig. 7.6 Relationship between the location of a complex pole and the transient response.

Towill (ref. 5) extends the relationships between the transient and the *s*-plane specifications to higher order systems and systems with zeros.

Error Coefficients

$$\frac{E(s)}{R(s)} = C_0 + C_1 s + C_2 s^2 + \ldots$$

and since $\qquad C(s) = R(s) - E(s), \qquad$ assuming that $\qquad H(s) = 1$:

$$\frac{C(s)}{R(s)} = 1 - C_0 - C_1 s - C_2 s^2 - \ldots$$

In terms of the closed loop poles and zeros

$$\frac{C(s)}{R(s)} = \frac{K'(s + z_1)(s + z_2)\ldots(s + z_m)}{(s + p_1)(s + p_2)\ldots(s + p_n)}$$

(note that this is not the usual $(1 + sT)$ form of the equation so that K' is not the gain constant), so that long division and manipulation of this function enable C_0, C_1, etc., to be expressed in terms of the poles and zeros, viz.:

$$1 - C_0 = \frac{K' z_1 z_2 z_3 \ldots z_m}{p_1 p_2 p_3 \ldots p_n}$$

This factor is unity for all direct feedback systems of above type 0, giving $C_0 = 0$ as expected.

The general relations for C_1 and C_2 are more complex, but for type 1 or higher systems they simplify, with $C_0 = 0$, to

$$C_1 = \sum_{i=1}^{n} \frac{1}{p_i} - \sum_{i=1}^{m} \frac{1}{z_i}$$

and

$$C_2 = -C_1 \sum_{i=1}^{n} \frac{1}{p_i} + \tfrac{1}{2}\left[\sum_{1}^{n} \frac{1}{p_i p_j} - \sum_{1}^{m} \frac{1}{z_i z_j} \right]$$

where

$$\sum_{1}^{n} \frac{1}{p_i p_j} = \frac{1}{p_1 p_2} + \frac{1}{p_1 p_3} + \ldots$$
$$+ \frac{1}{p_1 p_n} + \frac{1}{p_2 p_3} + \ldots + \frac{1}{p_2 p_n} + \ldots + \frac{1}{p_{n-1} p_n}$$

7.3 Sensitivity Functions

The objective of closed loop control is to reduce the effects of unwanted variations in system parameters and of disturbances on the system

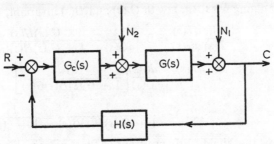

Fig. 7.7 A compensated closed loop system with disturbances.

(noise). Sensitivity functions are quantitative measures of the effectiveness of control.

Consider the system of Fig. 7.7, in which $G_c(s)$ and $H(s)$ are compensation networks and can be considered free from variations. The plant transfer function $G(s)$ is prone to variation. N_1 and N_2 are disturbances (noise).

With $N_1 = N_2 = 0$, let

$$T(s) = \frac{C(s)}{R(s)} = \frac{G_c(s)G(s)}{1 + G_c(s)H(s)G(s)} = \frac{G_c(s)G(s)}{1 + L(s)}$$

$L(s)$ is the open loop transfer function.

With $R = N_2 = 0$, let

$$T_{N_1}(s) = \frac{C(s)}{N_1(s)} = \frac{1}{1 + L(s)}$$

With $R = N_1 = 0$, let

$$T_{N_2}(s) = \frac{C(s)}{N_2(s)} = \frac{G(s)}{1 + L(s)}$$

The two transfer functions $T_{N_1}(s)$ and $T_{N_2}(s)$ define the effects of noise on the system. To determine the effect of variation in $G(s)$, the sensitivity function is defined as the ratio of the per-unit change in the dependent function to the per-unit change in the independent function. The particular example of interest here is the effect of change in $G(s)$ on the closed loop transfer function $T(s)$. Thus

$$S_G^T(s) = \frac{\dfrac{dT(s)}{T(s)}}{\dfrac{dG(s)}{G(s)}} = \frac{d(\log_e T(s))}{d(\log_e G(s))}$$

where $T(s)$ and $G(s)$ are taken as the nominal values.

For the system under consideration

$$\log_e T(s) = \log_e G_c(s) + \log_e G(s) - \log_e (1 + G_c(s)H(s)G(s))$$

$\therefore$ differentiating w.r.t. $G(s)$, with $G_c(s)$ and $H(s)$ invariant,

$$\frac{1}{T(s)}\frac{dT(s)}{dG(s)} = 0 + \frac{1}{G(s)} - \frac{G_c(s)H(s)}{1 + G_c(s)H(s)G(s)}$$

$$\therefore \frac{dT(s)}{T(s)} = \frac{dG(s)}{G(s)}\left[\frac{1}{1 + G_c(s)H(s)G(s)}\right]$$

$$\therefore S_G^T(s) = \frac{1}{1 + G_c(s)H(s)G(s)} = \frac{1}{1 + L(s)}$$

Note now that the disturbance transfer functions are directly related to the sensitivity function defined. Thus S should be chosen to satisfy the noise specification as well as the plant variation specification; S can be considered a fundamental concept of feedback control.

If a specific parameter, p, of $G(s)$ is considered to vary independently, then individual sensitivity functions can be found. In this case $dG(s)$

becomes $\dfrac{\partial}{\partial p}\, G(s)\,.\, dp$.

$$\therefore \qquad \frac{dT(s)}{T(s)} = \frac{\partial G(s)}{\partial p} \cdot \frac{dp}{G(s)} \left[\frac{1}{1 + L(s)} \right]$$

Hence

$$S_p^T(s) = \frac{\dfrac{dT(s)}{T(s)}}{\dfrac{dp}{p}} = \frac{p \dfrac{\partial G(s)}{\partial p}}{G(s)\,(1 + L(s))}$$

If the nominal plant transfer function is factorised as

$$G(s) = \frac{K(1 + sT_a) \ldots}{(1 + sT_1) \ldots}$$

then

$$S_K^T = \frac{K \cdot \dfrac{G(s)}{K}}{G(s)\,(1 + L(s))} = \frac{1}{1 + L(s)}$$

$$S_{T_a}^T = \frac{T_a \cdot \dfrac{sG(s)}{1 + sT_a}}{G(s)\,(1 + L(s))} = \frac{sT_a}{1 + sT_a} \cdot \frac{1}{1 + L(s)}$$

and

$$S_{T_1}^T = \frac{T_1 \dfrac{-s\,G(s)}{1 + sT_1}}{G(s)\,(1 + L(s))} = \frac{-sT_1}{1 + sT_1} \frac{1}{1 + L(s)}$$

The extended use of sensitivity functions as a design method is proposed by Horowitz (ref. 18).

7.4 Design Procedure

The following is a recommended line of approach to a design problem:

1. Write a clear specification in transient and frequency response, or acceptable regions of the *s*-plane.

2. Determine the rating and type of component to be used for the actuator (see Chapter 14).

3. Calculate the initial value of the loop gain from the specification for static errors, i.e. a radar aerial is to be within 5 minutes of arc of the desired value under wind loading conditions equivalent to say 400 N m at the output shaft. The d.c. loop gain is calculated so that with 5 minutes of error the torque at the output is 400 N m. It would be well to allow here another say 100 N m for the effects of static friction. The actual magnitudes of these terms must, of course, be included in the specification. The gain calculated here may not be sufficient to keep the velocity lag within limits, but it is a good starting-point.

4. Calculate, or preferably measure, the open loop frequency response with this gain and check for stability.

5. Introduce any necessary compensation to make the system stable and to 'pull' the closed loop frequency response up to the specification.

6. Close the loop on the system and check for any undesirable effects, which often occur in practice due to nonlinearity.

7. Check the transient response to step and velocity inputs and make any corrections to the compensation which may prove necessary.

An analogue computer is of great help in (5) and in anticipating (7).

There are systems, particularly those with marked nonlinear and noise problems, for which this approach is not sufficient, so that more complex design procedures must be adopted.

8
Compensating Techniques

8.1 Methods of Compensation

It has previously been explained that most systems will require some form of compensation in order to produce a stable system which will meet the specification.

In this chapter many possible methods of compensation are described.

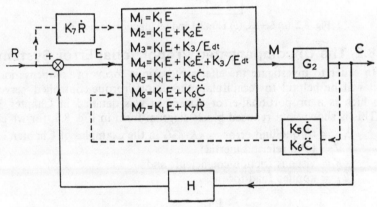

Fig. 8.1 The basic methods of compensation.

The approach used is to initially analyse their effect on a simple second-order system, to expand this application to any system in terms of Bode diagrams and finally, to review design techniques using the s-plane.

In discussing the Nyquist stability criterion it was made clear that stability could be achieved by reduction of gain constant, except for a conditionally stable system. Here, however, other methods must be devised since reduction of gain will also adversely affect the perform-ance of the system. These methods can be termed 'frequency conscious' gain and phase modification.

A summary of all these possibilities is given in Fig. 8.1, all of which will be investigated.

The various possibilities expressed in Fig. 8.1 can be put into three categories, viz.:

131

(i) Series or error compensation (Fig. 8.2 (*a*)).
(ii) Parallel or feedback compensation (Fig. 8.2 (*b*)).
(iii) Input or feedforward compensation (Fig. 8.2 (*c*)).

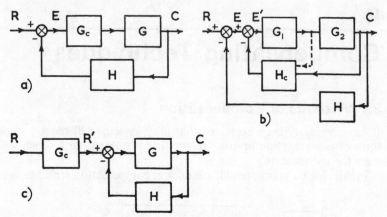

Fig. 8.2 (*a*) Series, (*b*) parallel and (*c*) input compensation.

8.2 The Uncompensated Proportional Error System

In order to investigate the effects of various forms of compensation, it will be helpful to recapitulate the simple 'torque-controlled' servo, which is a proportional-error system as was detailed in Chapter 3. This is shown in a reduced block diagram form in Fig. 8.3, for which

K_T = torque/unit error = nK_1K_2G in the examples of Chapter 3;
J = total referred inertia;
F = viscous friction coefficient; and
T_L = applied load torque.

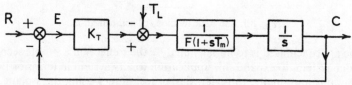

Fig. 8.3 A position servo.

If $J/F = T_m$ and the open loop gain constant $K = K_T/F$, then

$$K_T E(s) = s(sJ + F)C(s) + T_L(s) = sF(1 + sT_m)C(s) + T_L(s)$$

and $E(s) = R(s) - C(s)$

$$\therefore C(s) = \frac{K_T}{K_T + Fs(1 + sT_m)} \cdot R(s) - \frac{1}{K_T + Fs(1 + sT_m)} \cdot T_L(s)$$

$$= \frac{1}{1 + s\frac{1}{K} + s^2\frac{T_m}{K}} \cdot R(s) - \frac{1}{KF}\frac{1}{1 + s\frac{1}{K} + s^2\frac{T_m}{K}} \cdot T_L(s)$$

In the steady state, if R and T_L and therefore C are constant,

$$C = R - \frac{1}{KF}T_L$$

or $$E = R - C = \frac{1}{KF}T_L \qquad (8.1)$$

Further if $$T_L = 0$$

$$\frac{C(s)}{R(s)} = \frac{1}{1 + s\frac{1}{K} + s^2\frac{T_m}{K}}$$

which is usually written as

$$\frac{C(s)}{R(s)} = \frac{1}{1 + \frac{2\zeta}{\omega_n}s + \frac{1}{\omega_n^2}s^2}$$

where $$\omega_n = \sqrt{\frac{K}{T_m}} \qquad (8.2)$$

and $$\zeta = \frac{1}{2\sqrt{T_m K}} \qquad (8.3)$$

Consider now a constant velocity input, with $T_L = 0$, i.e. $r(t) = Qt$, the complete solution of which was developed in Section 3.5; the steady state error can be found from the particular integral of this solution, or alternatively solved in terms of error coefficients (Section 7.2.5) or the final value theorem (Appendix). Using error coefficients

$$\text{velocity error coefficient, } C_1 = \frac{1}{K_v} = \frac{1}{K}$$

and the velocity lag error in the steady state when $r(t) = Qt$ is

$$E_{\text{s.s.}} = \frac{1}{K} \cdot Q \qquad (8.4)$$

Summarising, with T_m constant:

(i) The 'stiffness' of the system, the load required to produce unit error, is $KF = K_T$ from (8.1).

(ii) The steady state position error when R is constant and $T_L = 0$ is zero, from (8.1), corresponding to $C_0 = 0$ for the type 1 system.

(iii) ω_n, increase in which corresponds to an increase in the speed of response, is proportional to $\sqrt{K}$ from (8.2).

(iv) The damping is proportional to $1/\sqrt{K}$ from (8.3).

(v) The velocity lag is proportional to $1/K$ from (8.4).

These results show that in all respects except damping, the system is improved by increasing the open loop gain constant, K. The practical limit, as indicated by (iv) lies in instability, i.e. the transient errors worsen while the steady state improves, so that additional methods of improving the system are required, particularly methods of increasing K without reducing the damping.

8.3 Proportional Error System with Derivative of Output (Velocity) Feedback

This is a form of parallel compensation which is considered separately here because of its major significance as a practical form of damping. It is enlarged upon in Section 8.5.

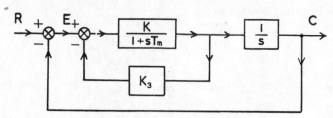

Fig. 8.4 A position servo with velocity feedback.

In Fig. 8.4 K_3 = velocity feedback constant and the error detector constant defined in chapter 3, $K_1 = 1$.

Here

$$C(s) = \frac{1}{1 + s\left(K_3 + \frac{1}{K}\right) + s^2 \frac{T_m}{K}} \cdot \left(R(s) - \frac{1}{KF} T_L(s)\right)$$

Comparing these results with the previous, uncompensated case:

(i) the stiffness is unaltered;

(ii) the steady state position error when R and $T_L = 0$ is still zero;

(iii) ω_n is unaltered;

(iv) the damping is increased to $\zeta = \dfrac{1 + KK_3}{2\sqrt{KT_m}}$; if the bulk of the damping is provided by velocity feedback and the friction F minimised, then $KK_3 \gg 1$, so that ζ is proportional to $K_3\sqrt{K}$.

(v) the velocity lag error is $\dfrac{1 + KK_3}{K}$. Q when $r(t) = Qt$, which is approximately K_3Q if F is minimised.

Thus increase of K will give improvement in all respects. It must be realised, however, that a simple second-order system such as this does not exist in practice, so that K is once again limited by stability considerations. Further, to provide adequate damping K_3 must be large which may lead to undesirably large velocity lag errors.

In summary, velocity feedback is a good, simple means of producing a satisfactory system, provided the specification is not too demanding. If the input is continuously varying, i.e. a tracking radar or a continuous contouring machine tool, then the velocity lag errors will almost certainly be too big, so that additional compensation is necessary.

8.4 Series Compensation

8.4.1 The General Case

Fig. 8.5 shows the general class of system to be considered under this heading. Since the compensation is made on the actuating signal, loosely termed the error signal, such phrases as derivative of error, integral of error, etc., are common.

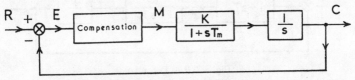

Fig. 8.5 Series compensation.

8.4.2 Proportional + Derivative Control

This is commonly referred to simply as derivative control; it in fact means proportional plus derivative action, i.e. to the normal actuating signal will be added some of its own derivative. Thus

$$M(s) = (1 + sT_d) \cdot E(s)$$

where T_d is a constant.

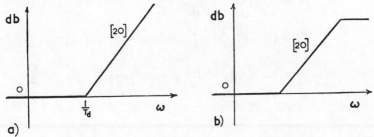

Fig. 8.6 (*a*) Proportional plus derivative, (*b*) phase lead.

The frequency response of this function is shown in Fig. 8.6 (*a*), from which it is clear, since the infinite gain required at high frequency can never be achieved, that such a network is not feasible. An approximation, however, is feasible, having the form shown in Fig. 8.6 (*b*). This is better known as a phase lead network and is discussed later from a frequency response viewpoint. First, however, consider the ideal case of Fig. 8.6 (*a*), as shown in Fig. 8.7.

$$\frac{C(s)}{R(s)} = \frac{(1 + sT_d)K}{K(1 + sT_d) + s(1 + sT_m)}$$

$$= \frac{1 + sT_d}{1 + \left(T_d + \dfrac{1}{K}\right)s + \dfrac{T_m}{K}s^2}$$

The T_d term introduces the required damping in the coefficient of the s term in the denominator as for velocity feedback. The velocity lag error

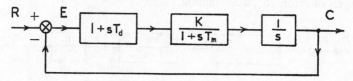

Fig. 8.7 P + D control.

can be found from the particular integral to an input $r(t) = Qt$ (Section 3.5), by application of the final value theorem to $E(s)$ for $R(s) = Q/s^2$ (Appendix), or from the velocity error coefficient, C_1 (Section 7.2.5) viz.:

$$\frac{E(s)}{R(s)} = \frac{1}{1 + G(s)H(s)} = \frac{1}{1 + \dfrac{K(1 + sT_d)}{s(1 + sT_m)}}$$

$$= \frac{s + T_m s^2}{K + (1 + KT_d)s + T_m s^2}$$

$$\therefore C_1 = (1 - 0) \cdot \frac{1}{K} = \frac{1}{K} \quad \text{and the velocity lag is } \frac{1}{K} \cdot Q$$

OR $\quad \lim_{t \to \infty} e(t) = \lim_{s \to 0} sE(s) = \lim_{s \to 0} s \dfrac{s + T_m s^2}{K + (1 + KT_d)s + T_m s^2} \cdot \dfrac{Q}{s^2}$

$$= \frac{1}{K} \cdot Q$$

Thus the system has been damped as with velocity feedback, but the velocity lag is still only that associated with the viscous friction. The magnitude of T_d is chosen to give the required degree of damping.

The disadvantage with any type of differentiating network is that the high frequency content of any noise present in the actuating signal is emphasised.

As with velocity feedback, the static stiffness is unaffected.

8.4.3 Proportional + Integral Control

As with derivative compensation, integral compensation or control means proportional plus integral; the proportional signal must in all cases be retained. Thus from Fig. 8.5:

$$M(s) = \left(1 + \frac{1}{T_i s}\right) \cdot E(s) = \frac{1 + sT_i}{sT_i} \cdot E(s)$$

where T_i is a constant.

The frequency response of this network is shown in Fig. 8.8 (*a*), along with Fig. 8.8 (*b*), the related phase lag network. Note that while

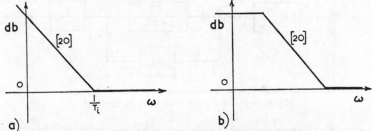

Fig. 8.8 (*a*) Proportional plus integral, (*b*) phase lag.

the addition of true derivative action is impossible, a good approximation to integral action can be achieved by using active networks, Section 8.4.7. The phase lag network will be considered later, along with the phase lead network, but first consider simple integral control applied

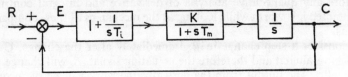

Fig. 8.9 P + I control.

to a second-order system as in Fig. 8.9. Note that the integral compensation has converted the basic type 1 system into a type 2 system, for which

$$C(s) = \frac{1 + sT_i}{1 + sT_i + s^2 \frac{T_i}{K} + s^3 \frac{T_i T_m}{K}} R(s)$$

$$- \frac{1}{KF} \cdot \frac{sT_i}{1 + sT_i + s^2 \frac{T_i}{K} + s^3 \frac{T_i T_m}{K}} T_L(s)$$

Since the denominator is a cubic, the simple term 'damping factor' is no longer applicable and the transient performance must be interpreted more rigorously. The velocity lag, however, is completely eliminated as this is a type 2 system for which both C_0 and C_1 are zero. The choice of a value for T_i cannot be so easily interpreted and depends upon the time allowable for the error to be reduced without too adversely affecting the damping. The phase lag network will not reduce the velocity lag error to zero, but it will enable K to be increased so as to appreciably reduce the velocity lag. If velocity feedback is used for damping, then additional integral control, Fig. 8.10, can eliminate the velocity lag. Note that the integral network must not be included inside the velocity feedback loop!

Finally, due to the s term in the numerator of the torque term, the usual reasoning will show that the actuating signal produced to provide

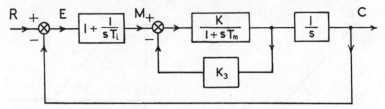

Fig. 8.10 P + I control with velocity feedback.

the reaction torque to a load torque will be zero ($E = 0$, M is finite), resulting in infinite static stiffness. Note, however, that it is the steady state stiffness that is improved, and steady state velocity lag error, the dynamic properties being worsened.

8.4.4 Physical Interpretation of Derivative and Integral Control

Before any quantitative analysis of derivative and integral control is attempted it is worth while looking at the mechanics of these compensations.

Consider a step change in R. Immediately after the change, C will not have altered and therefore the actuating signal, E, will change as a step also. The output from the derivative network will be the proportional step signal plus a pulse due to the derivative action. This will possibly saturate the system, but in any case it will cause large negative torques at periods of rapid changes in E. In the steady state condition it will have no effect, leaving only the velocity lag error due to F.

On the other hand, the integral network will give no improvement to sudden changes in error and in most cases will prove detrimental with a reduction in stability, i.e. the extra root in the now third-order characteristic equation of the previous example. However, with a constant velocity input (and output in the steady state) or a steady disturbing load, the necessary actuating signal from the network will be produced by integrating E for a short period. When sufficient actuating signal, M, has accumulated, the actuating signal, E, will have decayed to zero.

It is common, in practice, to use combinations of derivative plus integral action, P + D + I control, the so-called three-term controller.

8.4.5 Phase Lead (Advance) Compensation

The Characteristics of the Network

It was pointed out in Section 8.4.2 that phase advance is the practical form of derivative of error control, the effect of which is to provide damping with no increase in velocity lag error. Using Bode diagrams, the effect of such a network on any system, not just a second-order

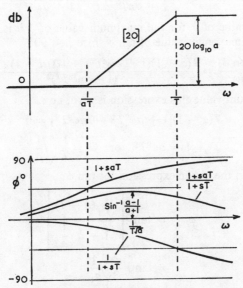

Fig. 8.11 Phase lead network:

$$G(s) = \frac{1 + saT}{1 + sT}; \quad a > 1$$

system, will be investigated. The transfer function of the phase lead network is:

$$G(s) = \frac{1 + saT}{1 + sT}, \quad \text{where} \quad a > 1$$

Fig. 8.11 shows the Bode diagrams of this transfer function which makes clear the phrase 'phase lead' network.

If
$$G(s) = \frac{1 + saT}{1 + sT}$$

then
$$G(j\omega) = \frac{1 + j\omega aT}{1 + j\omega T}$$

∴ the amplitude

$$G(\omega) = \sqrt{\frac{1 + \omega^2 a^2 T^2}{1 + \omega^2 T^2}}$$

and the phase angle

$$\phi(\omega) = \tan^{-1} \omega aT - \tan^{-1} \omega T \quad \text{leading}$$

$$= \phi_1 - \phi_2$$

$$\therefore \tan \phi = \frac{\tan \phi_1 - \tan \phi_2}{1 + \tan \phi_1 \tan \phi_2}$$

$$= \frac{\omega aT - \omega T}{1 + \omega aT \cdot \omega T} = \frac{\omega T(a - 1)}{1 + a\omega^2 T^2}$$

To find the value of ω for the maximum value of ϕ it is necessary to find only the maximum value of $\tan \phi$ since $\phi < 90°$.

$$\therefore \frac{d(\tan \phi)}{d\omega} = \frac{T(a - 1)(1 + a\omega^2 T^2) - \omega T(a - 1) \cdot 2a\omega T^2}{(1 + a\omega^2 T^2)^2}$$

For a maximum value this expression is zero, i.e.:

$$T(a - 1)[1 + a\omega^2 T^2 - 2a\omega^2 T^2] = 0$$

$$\therefore 1 = a\omega^2 T^2 \quad \text{or} \quad \omega = \frac{1}{T}\sqrt{\frac{1}{a}}$$

Substituting in the above expression for $\tan \phi$

$$\phi_{\max} = \tan^{-1} \left[\frac{\frac{1}{\sqrt{a}}(a - 1)}{1 + a \cdot \frac{1}{a}} \right]$$

$$= \tan^{-1} \frac{a - 1}{2\sqrt{a}}$$

$$= \sin^{-1} \frac{a - 1}{a + 1}$$

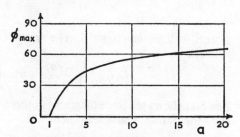

Fig. 8.13 Relationship between maximum phase shift and the ratio of the break frequencies for a phase lead or phase lag network.

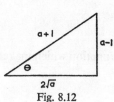

Fig. 8.12

by using the construction of Fig. 8.12. Thus $\phi_{\max} = \sin^{-1} \dfrac{a - 1}{a + 1}$ at a frequency, $\omega = \dfrac{1}{\sqrt{a}T}$. In practice, a should be limited to a maximum of 20 and more usually 5 to 15 because of the problem of increasing the high-frequency noise content to unacceptable proportions. Fig. 8.13 shows the relationship of ϕ_{max} to a. Phase shift of more than 60° is best achieved by cascading networks, e.g. 90° phase shift is obtained by two networks having $a = 5\cdot8$, provided they are arranged not to interact.

Fig. 8.14 shows a passive phase lead network. If the source impedance is assumed to be zero and the load impedance infinite (or part of R_1), then:

$$\frac{V_o(j\omega)}{V_i(j\omega)} = \frac{R_1}{R_1 + \dfrac{R_2 \cdot \dfrac{1}{j\omega C}}{R_2 + \dfrac{1}{j\omega C}}}$$

$$= \frac{R_1(1 + j\omega C R_2)}{R_1 + R_2 + j\omega C R_1 R_2}$$

$$= \frac{R_1}{R_1 + R_2} \cdot \frac{1 + j\omega C R_2}{1 + j\omega C \dfrac{R_1 R_2}{R_1 + R_2}}$$

$$= \frac{1}{a} \cdot \frac{1 + j\omega aT}{1 + j\omega T}$$

where $\qquad a = \dfrac{R_1 + R_2}{R_1} \quad$ and $\quad T = C\dfrac{R_1 R_2}{R_1 + R_2}$

Note, however, that this network has an attenuation, $\dfrac{1}{a}$, so that the open loop gain of a system using this network must be increased by a to compensate for this. This is known as the insertion loss of a filter.

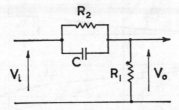

Fig. 8.14 A passive phase lead network.

Phase Lead Compensation

The simplest interpretation of phase lead stabilisation is to consider how to modify the open loop frequency response of the given system. Consider the given system with the gain constant K adjusted so that the resulting closed loop system would just be unstable, i.e. zero gain and phase margins. Such a system is represented in Fig. 8.15. Note that K may be lower than required by the steady state specification, e.g. stiffness; this will be corrected by adding phase lag compensation, in

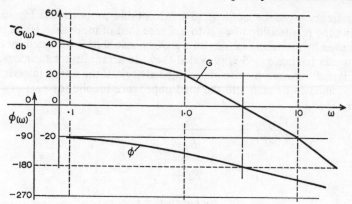

Fig. 8.15 Bode diagrams for the open loop transfer function of a system which will be unstable when the loop is closed.

$$G(s)H(s) = \frac{11}{s(1 + 0.1s)(1 + s)}$$

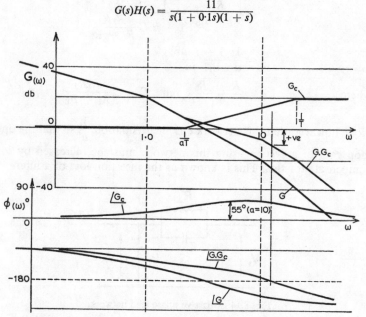

Fig. 8.16 Bode diagrams for the system of Fig. 8.15 with active series phase lead compensation:

$$G_c(s) = \frac{1 + saT}{1 + sT}; \quad a = 10 \quad \text{and} \quad T = \frac{1}{20}$$

addition to the phase lead considered here, as discussed in the next section.

Fig. 8.16 shows the resulting Bode diagrams for the system with compensation. The choice of the compensation constants, a and T, is rather arbitrary: $a = 10$ is a sensible practical value, and a good initial guess is to choose T so that the network 'break-up frequency' is approximately equal to the 'cross-over' frequency (it is shown slightly lower on Fig. 8.16, for clarity). Often, better results can be achieved by trial and error adjustment of a and T; further phase lead can be introduced by using more than one network if necessary.

Use of Nichols Chart for Selection of Phase Lead Network

In using the Nichols chart (see Section 6.7.3) it must be remembered that $H = 1$, and if H is not unity a modification must be made by multiplying the closed loop response obtained from the chart by $1/H(j\omega)$. Since for most systems stabilised by series compensation H is a constant, this is but a small disadvantage.

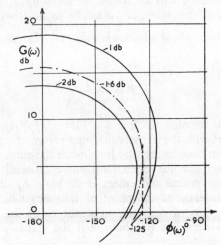

Fig. 8.17 Extract from Nichols chart (Fig. 6.19).

This method is particularly suited to problems for which the specification is given as a peak in the frequency response of, say, M db occurring at a resonant frequency, ω_R.

Let the uncompensated open loop frequency response function be

$$G(j\omega)H(j\omega) = \frac{K}{j\omega(1 + j\omega T_1)(1 + j\omega T_2)}$$

Since the compensation may also involve some change in system gain let the compensation frequency response function be:

$$G_c(j\omega) = \frac{K_c(1 + j\omega aT)}{1 + j\omega T} \qquad a > 1$$

The problem is to find K_c, a and T to give the specified values of

M_{max} and the corresponding resonant frequency, ω_R. The procedure is as follows:

(i) From the Nichols chart, Fig. 6.19, find the value of phase shift, ϕ, corresponding to the peak of the specified M_{max} contour, e.g. for an overshoot, in response to a step, of 18%, $M_{max} = 1\cdot6$ db, from Fig. 7.5, then $\phi = 125°$ lag, as indicated in Fig. 8.17.

(ii) Find the value of phase shift for the uncompensated system when $\omega = \omega_R$, say ϕ_R lag.

(iii) The required value of phase lead then is:

$$\phi_c = \phi_R - \phi \quad \text{lead}$$

If compensation is required, $\phi_R > \phi$. The number of stages of compensation required can be found by allowing a maximum phase shift of, say, 60° per stage, and a can be found from Fig. 8.13.

(iv) Maximum phase advance is achieved when $\omega = \dfrac{1}{\sqrt{aT}}$. Thus if

$$\omega = \omega_R$$

$$T = \frac{1}{\sqrt{a}\,.\,\omega_R}$$

(v) It now remains to find K_c. Plot $G_c(j\omega)G(j\omega)H(j\omega)$ on the Nichols chart with the above calculated values of a and T, but with $K_c = 1$. It can now be seen by how much the curve must be moved vertically to make the open loop curve tangential to the M_{max} contour. The vertical movement is 20 $\log_{10} K_c$ db, hence K_c. In fact, the frequency at the point of tangency will not be exactly ω_R, as the point of tangency may not be at the extreme of the contour as assumed in (i) above, but the values obtained will be close.

8.4.6 Phase Lag Compensation

The Characteristics of the Network

The very name, 'phase lag' indicates that such networks should be introduced into systems with caution, otherwise the extra phase shift will cause instability. The phase lag compensation is used to reduce velocity lag errors and to increase stiffness by effectively increasing the loop gain constant, K. In order to maintain good stability the gain is only increased at low frequencies and is removed at higher frequencies in the region of 180° phase shift. Since the gain is increased only at the low frequency end where GH is much greater than 0 db in any case, the resultant closed loop frequency response is unaffected, being approximately $1/H$ in each case. The Bode diagrams of the network are shown in Fig. 8.18.

The required transfer function of the network is

$$G_c(s) = b \cdot \frac{1 + sT}{1 + sbT}, \quad \text{where} \quad b > 1$$

$$\therefore G_c(j\omega) = b \cdot \frac{1 + j\omega T}{1 + j\omega bT}$$

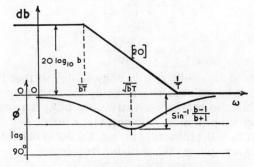

Fig. 8.18 Phase lag network:

$$G(s) = b \cdot \frac{1 + sT}{1 + sbT}; \quad b > 1$$

The same manipulations can be applied as to the phase lead network in Section 8.4.5, so that the maximum phase shift is

$$\phi_{max} = \sin^{-1} \frac{b - 1}{b + 1} \text{ lag}$$

occurring at

$$\omega = \frac{1}{\sqrt{b}T}$$

It is also possible to use the curve of Fig. 8.13 to find the value of ϕ_{max} for a given b, but there is no practical limit to the maximum value of b,

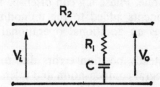

Fig. 8.19 A passive phase lag network

although for $b > 20$, approaching pure integral control, an active network is preferred. Consider now the passive phase lag network of Fig. 8.19, in which the source impedance is zero and the load impedance infinite.

$$\frac{V_o(j\omega)}{V_i(j\omega)} = \frac{R_1 + \dfrac{1}{j\omega C}}{R_2 + R_1 + \dfrac{1}{j\omega C}}$$

$$= \frac{1 + j\omega C R_1}{1 + j\omega C(R_1 + R_2)}$$

$$= \frac{1 + j\omega T}{1 + j\omega bT}$$

where $\qquad\qquad\qquad T = CR_1$

and $\qquad\qquad\qquad b = \dfrac{R_1 + R_2}{R_1}$

Note that the d.c ($\omega = 0$) gain is unity (0 db) so that the constant factor, b, must be included by increasing the open loop gain factor, K, to bK. This also proved necessary when using a passive network for phase lead compensation. Also note that a simple R–C network, i.e. Fig. 8.19 with $R_1 = 0$, produces phase lag, but does not fulfil the requirements here of zero phase shift at higher frequencies.

Application of Phase Lag Compensation

The design of a phase lag compensator, i.e. the choice of the constants b and T, is simple. The open loop transfer function must indicate that the resulting closed loop system will be stable, i.e. the gain and phase margins *must* be positive. This can be achieved by adjustment of K, but more desirably by using a phase lead compensator as previously explained. There is no limit on the choice of b, and T is simply chosen so that the higher break frequency of the phase lag compensator is below the 'cross-over frequency' by a factor of around 10. This is shown in Fig. 8.20, from which it can be seen from the low frequency end of the diagram that K has been effectively increased to bK.

Phase Lag Versus Integral Compensation

The choice between integral compensation or phase lag compensation is mainly a practical one. Phase lag can often be incorporated for the cost of a few components added to the unmodified system. Integral compensation requires an additional operational amplifier (see next section).

Velocity lag and steady position errors due to external disturbing loads are zero for integral compensation and tolerably small, but finite, for phase lag compensation. With a constant acceleration input, $r(t) = Pt^2$, the actuating signal, $e(t)$, after transients have died away, i.e. the acceleration lag, is finite but constant with integral compensation but with phase lag has an additional term proportional to time. Thus the error could become excessive if the constant acceleration input is maintained for any length of time, an unusual but not impossible situation, which would make integral compensation essential.

8.4.7 Active Networks

Networks comprising resistors, capacitors and inductors only are called 'passive networks'. Networks utilising amplifiers with feedback are called 'active networks'. Using active networks more complex transfer functions can be synthesised, including networks with gain at certain frequencies. Inductances, which are always inconvenient, can be avoided in active networks.

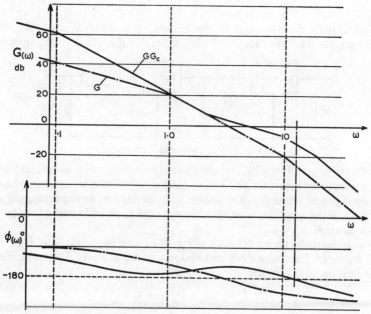

Fig. 8.20 System of Fig. 8.15 with (active) phase lead and (active) phase lag compensation. Note that the effective gain constant has been increased to bK.

$$G(s)H(s) = \frac{11}{s(1 + 0.1s)(1 + s)}; \quad G_{\text{lead}}(s) = \frac{1 + s\frac{10}{20}}{1 + s\frac{1}{20}} \quad \text{and} \quad G_{\text{lag}}(s) = \frac{10\left(1 + s\frac{1}{1}\right)}{1 + s\frac{10}{1}}$$

High gain d.c. amplifiers with high input impedance and low output impedance are called 'operational amplifiers'. They originated as amplifiers for analogue computers for which they are generally connected as summing amplifiers or integrators, as detailed in Section 13.2.2. This was one of the last uses of valve amplifiers, which employed 'chopper stabilisation' to achieve the necessary low drift values. Transistorised units are now available but, more importantly, integrated circuit amplifiers have brought the cost down. While not up to analogue computer standards these miniature amplifiers are ideal for compensation networks.

Operational amplifiers have an odd number of stages so that there is an inherent phase reversal. This enables negative feedback to be achieved by *adding* the output voltage, via an impedance Z_0, to the input. The phase reversal introduced when using active networks must be correct at another part of the loop, e.g. by reversing field connections.

As an example consider a combined lead–lag transfer function,

$$G(s) = b \cdot \frac{1 + sT_b}{1 + sbT_b} \cdot \frac{1 + saT_a}{1 + sT_a}$$

This transfer function can be synthesised by using R–C networks, an amplifier and a buffer stage as in Fig. 8.21. A buffer stage is an amplifier

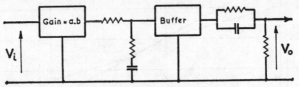

Fig. 8.21 A cascade lead–lag network.

with unity gain but with an infinite input impedance so as not to load the preceding network, and a zero source impedance so as to correctly feed the next stage. Such a system may be easy to incorporate, particularly if the amplifier is required in any case to provide part of the gain constant, K.

However, the active network of Fig. 8.22 may be simpler and is more likely to be free from drift and distortion, particularly if the amplifier open loop gain is high.

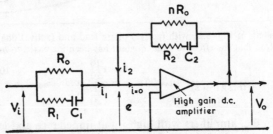

Fig. 8.22 An active lead–lag network.

If the amplifier gain is high enough, e is negligible, so that the current $i_1 = - i_2$ (see Section 13.2.2).

$$\therefore \frac{V_i(s)}{Z_i(s)} = -\frac{V_o(s)}{Z_o(s)}$$

where $Z_i(s) = \dfrac{R_0\left(R_1 + \dfrac{1}{sC_1}\right)}{R_0 + R_1 + \dfrac{1}{sC_1}} = \dfrac{R_0(1 + sC_1R_1)}{1 + sC_1(R_1 + R_0)}$

and $\quad Z_o(s) = \dfrac{nR_0(1 + sC_2R_2)}{1 + sC_2(R_2 + nR_0)}$

$\therefore \dfrac{V_o(s)}{V_i(s)} = -\dfrac{Z_o(s)}{Z_i(s)}$

$\qquad = -n \cdot \dfrac{1 + sC_2R_2}{1 + sC_2(R_2 + nR_0)} \cdot \dfrac{1 + sC_1(R_1 + R_0)}{1 + sC_1R_1}$

It is worth showing that individual phase lead or phase lag networks can be obtained by putting $R_1 = R_2 = 0$ and appropriately choosing n, R_0, C_1 and C_2.

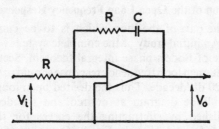

Fig. 8.23 An active integral plus proportional network.

As a second example, Fig. 8.23, the integral plus proportional control suggested in Section 8.4.3 can be quite accurately approximated:

$$\frac{V_o(s)}{V_i(s)} = -\frac{Z_o(s)}{Z_i(s)} = -\frac{R + \dfrac{1}{sC}}{R}$$

$$= -\left(1 + \frac{1}{CR} \cdot \frac{1}{s}\right)$$

8.5 Parallel Compensation

8.5.1 The General Case

Fig. 8.24 shows a general case of the application of parallel compensation, or as it is often called, feedback compensation. H_c is the com-

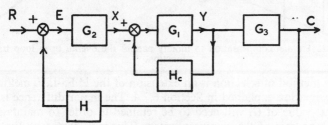

Fig. 8.24 General case of parallel compensation.

pensation network, G_1, G_2, G_3 and H are the usual transfer functions, some of which may be unity.

In this case H_c and G_1 are considered as a minor loop, altering the transfer function, Y/X, and hence reshaping the open loop functions, i.e. referring to Fig. 8.24 the uncompensated open loop transfer function

$$= G_1(s) \cdot G_2(s) \cdot G_3(s) \cdot H(s)$$

and the compensated open loop transfer function

$$= G_2(s) \cdot G_3(s) \cdot \frac{G_1(s)}{1 + G_1(s)H_c(s)} \cdot H(s)$$

8.5.2 Modification of the Open Loop Frequency Response

Consider only the part of the loop which is to be compensated, i.e. G_1 in Fig. 8.24. An initial study of the complete system will be required. Use can be made of Bode's phase integral theorem, Section 6.6, which shows that the net open loop frequency response should pass through 0 db with a slope of 20 db/decade. Thus the desired open loop response can be drawn on a Bode diagram as desired and the desired value of $Y(s)/X(s)$ established by subtracting the curves for the unmodified portions, $G_1(s)$, $G_2(s)$ and $H(s)$. In Fig. 8.25 a value of $G_1(s)$ is shown together with the desired value of $Y(s)/X(s)$ which can be achieved by selection of $H_c(s)$.

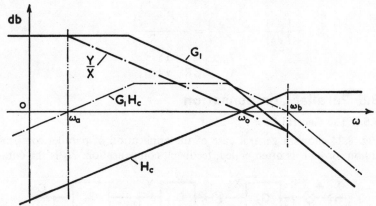

Fig. 8.25 Parallel compensation to modify part of the overall open loop transfer function.

The method of selection is an extension of the H.F.–L.F. method of approximation explained in Section 6.7.4. The major difference is that the d.c. value of G_1 will need to be retained in order to maintain the open loop gain of the complete system. This is indicated by Fig. 8.25 where typical values of G_1 and the desired Y/X are plotted.

The technique used is to consider three frequency bands in the approximation, low, mid and high frequency, with the compensation only to be effective in the mid frequency band. Thus

$$\frac{Y}{X} = \frac{G_1}{1 + G_1 H_c}$$

At L.F. $\qquad G_1 H_c \ll 1$

so that $\qquad \frac{Y}{X} \simeq G_1$ $\qquad\qquad$ (a)

at Mid-F. $\qquad G_1 H_c \gg 1$

so that $\qquad \frac{Y}{X} \simeq \frac{1}{H_c}$ $\qquad\qquad$ (b)

and at H.F. $\qquad G_1 H_c \ll 1$

and $\qquad \frac{Y}{X} \simeq G_1$ $\qquad\qquad$ (c)

The frequency ranges termed low, mid and high are relative and depend upon the particular system; they can be easily defined by the range of the desired compensation, i.e. from Fig. 8.25 L.F. is $\omega < \omega_a$, M.F. is $\omega_a < \omega < \omega_b$ and for H.F. $\omega > \omega_b$. To fulfil condition (a) it is necessary that the numerator of H_c shall have at least one more common s term than the denominator of G_1. Thus in the example quoted of

$$G_1(j\omega) = \frac{K_1}{(1 + j\omega T_1)(1 + j\omega T_2)}$$

$$H_c(j\omega) = \frac{j\omega . K_H(1 + b_1(j\omega) + b_2(j\omega)^2 + \ldots)}{(1 + a_1(j\omega) + a_2(j\omega)^2 + \ldots)}$$

From condition (b) the form of $1/H_c$ can be defined as the desired value of Y/X. In the example of Fig. 8.25, H_c has been chosen as $H_c = K_H j\omega$ for $\omega_a < \omega < \omega_b$. Finally to fulfil condition (c) H_c should not increase with increase in frequency so that $H_c = K$ or $H_c = K/j\omega$. In Fig. 8.25 the H.F. value has been indicated as a constant.

In this example, then, the L.F. value of H_c can be $j\omega K_H$, as in the Mid-F. range. Remembering that by L.F. or H.F., etc., value of H_c is meant the approximate value in that range, the complete function for H_c is

$$H_c(j\omega) = \frac{K_H . j\omega}{1 + j\omega T_b}, \qquad \text{where} \quad T_b = \frac{1}{\omega_b}$$

A value for K_H can be found easiest from the graph of Fig. 8.25, i.e.

$$\frac{1}{K_H \omega_a} = G_1(\omega)]_{\omega = \omega_a}$$

or $K_H \omega_0 = 1$, since the effect of T_b can be considered negligible at ω_0 and ω_a.

8.5.3 First Derivative of Output (Velocity) Feedback

The principle of velocity feedback stabilisation of servomechanisms was outlined originally in Section 3.2.3 and extended in Section 8.3, but with limited application to second-order systems.

Fig. 8.26 shows the construction, in two stages, of the open loop frequency response of the velocity feedback system of Fig. 8.4 showing the effect of the feedback in reducing the L.F. gain and pushing the 'break' to a higher frequency, thereby improving the gain margin but increasing the velocity lag. What is not shown on these diagrams is the fact that the stiffness is not reduced, even though the apparent value

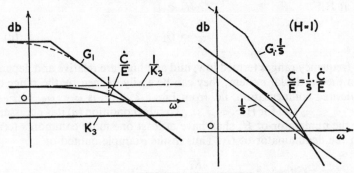

Fig. 8.26 Open loop response of velocity feedback stabilised servo of Fig. 8.4.

of K is reduced; this could be shown by constructing the Bode diagrams for C/T_L with $R = 0$, but was shown adequately in Section 8.3. The H.F.–L.F. approximation technique has been used to construct Fig. 8.26, although it could have been done more accurately with a Nichols chart, and the closed loop response, C/R, can be constructed from the resulting open loop response in similar fashion. Note from the construction of Fig. 8.26 that systems of higher order than second can just as easily be analysed, but some care must be taken over the stability of the minor loop.

8.5.4 Second Derivative of Output (Acceleration) Feedback

Second derivative feedback can be used in order to provide damping with no velocity lag. Feedback, and therefore damping is only applied while the first derivative is changing, so that with a steady velocity input and output there is no damping applied, nor needed. Second derivative feedback can be approximated by feeding the tachometer signal through a differentiating network as in Fig. 8.27, for which:

$$\text{Tacho voltage} = KsC(s)$$

and the output from the network:

$$V_o(s) = \frac{R}{R + \frac{1}{sC}} \cdot KsC(s)$$

$$= \frac{sCR}{1 + sCR} \cdot KsC(s)$$

$$= K_a \frac{s^2}{1 + sT_a} C(s)$$

where $$CR = T_a \quad \text{and} \quad KCR = K_a$$

This is not true second derivative feedback, but T_a can be made small enough to make $sT_a \ll 1$, as shown in Fig. 8.28. The tachometer constant K may need to be increased, possibly by amplification, in order to achieve the required value of the constant K_a.

Fig. 8.29 shows the circuit of this type of feedback as applied to a simple position control system, which is then analysed by interpreting

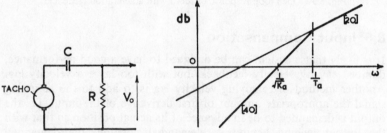

Fig. 8.27 Acceleration feedback using tachogenerator and R–C network.

Fig. 8.28 Characteristics for acceleration feedback.

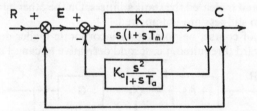

Fig. 8.29 Position servo with acceleration feedback.

the effect on the open loop frequency response, as in Fig. 8.30, constructed by applying the H.F.–M.F.–L.F. approximations to the minor loop.

Comparing this result with the first derivative feedback example of Fig. 8.26, it can be seen that the high frequency responses are similar but the low frequency gain is maintained by using second derivative feedback, thus eliminating velocity lag errors. The complete closed loop response and stability can be calculated in the usual ways.

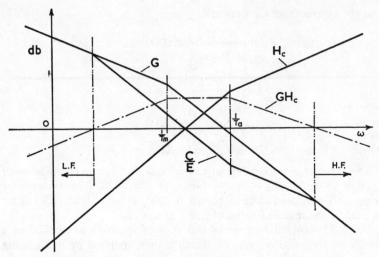

Fig. 8.30 Open loop response of servo with acceleration feedback.

8.6 Input Compensation

It is likely that a system can be designed to have a good performance, damped say by velocity feedback, but with too large a velocity lag. Another method of removing velocity lag is to add to the actuating signal the appropriate amount of first derivative of the input, i.e. the output is demanded to be in advance of its actual position so that with the output lagging behind its demanded position, the demanded advance and the inherent lag will cancel. Note, however, that the stiffness is unaffected whereas, for example, integral control reduced the velocity lag and increased the static stiffness. On the other hand, stability and dynamic stiffness are unimpaired.

It would, of course, be possible to incorporate higher derivatives in the feedforward if required, i.e. second derivative to cancel acceleration

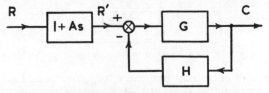

Fig. 8.31 Input compensation.

lags. Fig. 8.31 shows the circuit for first derivative feedforward. Assuming as an example that a second-order system is to be compensated

$$R'(s) = \left(\frac{1}{\omega_n^2} s^2 + \frac{2\zeta}{\omega_n} s + 1 \right) C(s)$$

and $$R'(s) = (1 + As)R(s)$$

$$\therefore \left(\frac{1}{\omega_n^2} s^2 + \frac{2\zeta}{\omega_n} s + 1 \right) C(s) = (1 + As)R(s)$$

If $r(t) = Qt$ where Q is a constant, then in the steady state $c(t) = Qt -$ velocity lag.

If, however, $A = \frac{2\zeta}{\omega_n}$, then in the steady state the constants due to the s terms will be equal on both sides of the equation, leaving no velocity lag.

The following points must be noted.

(i) Using $R–C$ networks the compensation element will not have a simple derivative term but will be of the form $\frac{1 + As}{1 + Bs}$, but B can be made much smaller than A (see section 8.4.5).

(ii) If $A > \frac{2\zeta}{\omega_n}$, then the output will lead the input for steady state velocity inputs. This is the derivative effect on the input and not a defiance of nature.

(iii) Since this form of compensation has no effect on the characteristic equation it has no effect on stability. This will be equally clear from Fig. 8.31 since the compensation is outside the loop.

(iv) This method is limited, in practice, to systems with very clear input signals, e.g. processed information for the input to a machine tool, otherwise the differentiation effect will increase the input noise level to unacceptable proportions.

8.7 Compensation for Two Inputs

In the system of Fig. 8.32 a specification has been set for the relations C/R_1 and C/R_2 so that some form of compensation is required for each function.

If $R_2 = 0$ $$C = \frac{G_1 G_2}{1 + G_1 G_2 H} \cdot R_1$$

If $R_1 = 0$ $$C = \frac{G_2}{1 + G_1 G_2 H} \cdot R_2$$

In order to compensate both these equations two variables are required. Unfortunately, there are not two independent variables so that any two of G_1, G_2 or H can be used, practical considerations very often eliminating one possibility, and sometimes two, in which case design for both C/R_1 and C/R_2 would have to be a compromise. It would probably be best to compensate G_1 and H, either by series or parallel compensation as appropriate. Input compensation may help in some cases like this.

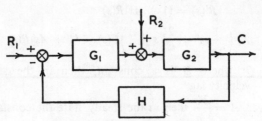

Fig. 8.32 A system with two inputs.

8.8 Design Techniques Using the *s*-Plane

Earlier chapters of this book showed that a system can be analysed in terms of transient response, transfer functions or frequency response functions; Chapter 7 indicated methods of specifying the requirements of the system in similar terms. Design technique, however, has been limited to frequency response methods. An equally extensive coverage of design using Laplace transforms and transfer functions, i.e. *s*-plane design, is not attempted here, but the following résumé will give the reader an introduction.

There are two principal approaches to *s*-plane design:

(i) *Network Synthesis.* Following on Darlington's procedure for synthesising filters by first determining the transfer function required to meet the specification and then realising a network which has the pole–zero configuration suggested by the transfer function, Guillemin in 1947 proposed that a closed loop system could be similarly treated. Guillemin's procedure can be summed up as:

(*a*) Determine the required closed loop transfer function from the specification.
(*b*) Determine the corresponding open loop transfer function.
(*c*) Synthesise the appropriate compensation networks.

An extension of this work to optimum design of systems was suggested by Weiner, Bode and Shannon (see Section 10.1) and the topic as a whole is extensively covered by Truxal (ref. 1).

(ii) *Interpretation of Conventional Series and Parallel Compensation Techniques, Using Root Loci.* While they have no significance in terms of pole–zero locations, terms such as phase lead and phase lag are retained. Fig. 8.33 shows the location of the poles and zeros of typical phase lead and phase lag networks.

The dynamic performance of a system is determined by the location of the poles and zeros. The closed loop transfer function is given by

$$\frac{C(s)}{R(s)} = \frac{K_1 N_1(s) D_2(s)}{K N_1(s) N_2(s) + D_1(s) D_2(s)} \quad \text{(see 6.4.1)}$$

The root loci, Section 6.4, are constructed to determine the closed loop poles given the open loop poles and zeros; the closed loop zeros are the zeros of $G(s)$ and the poles of $H(s)$.

The static performance of the system is a function of the type number of the system and the open loop gain constant, K.

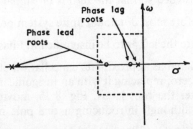

Fig. 8.33 Pole–zero location of phase lag and phase lead networks. The dotted zone indicates the region of the dominant roots of the system, relative to the network roots.

Fig. 8.34 shows the root loci of the uncompensated system to be considered in the following examples. The open loop transfer function is

$$G(s)H(s) = \frac{K_1 K_2}{s(1 + sT_f)(1 + sT_m)} = \frac{K'}{s\left(s + \dfrac{1}{T_f}\right)\left(s + \dfrac{1}{T_m}\right)}$$

The open loop gain constant, $K = K_1 K_2$ and the gain constant in the second form, as used in constructing root loci, is $K' = \dfrac{K_1 K_2}{T_m T_f}$.

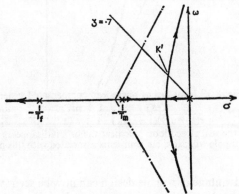

Fig. 8.34 Root loci of uncompensated third order system:

$$G(s)H(s) = \frac{K_1 K_2}{s(1 + sT_f)(1 + sT_m)}; \qquad K' = \frac{K_1 K_2}{T_f T_m}$$

Example 1. Phase lead compensation.

The phase lead network has the transfer function

$$G_c(s) = \frac{1 + saT_d}{1 + sT_d}$$

The network is designed so that the zero is of similar magnitude to the smallest and therefore most dominant finite system pole, i.e. $-\dfrac{1}{T_m}$; the pole $-\dfrac{1}{T_d}$ is so far to the left as to be insignificant. Phase lead compensation can be considered as approximate cancellation of a dominant pole with the network zero, replacing it with an insignificant pole. The effect is to distinctly alter the root locus, Fig. 8.35, moving the oscillatory poles to the left although introducing a real pole nearer the origin. $\left(\text{It is suggested that the reader should inspect the loci resulting if the network zero is to the left of } -\dfrac{1}{T_m}\right)$.

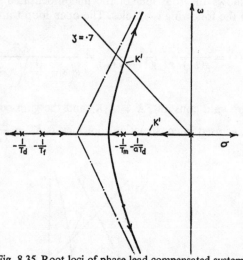

Fig. 8.35 Root loci of phase lead compensated system:

$$G(s)H(s) = \frac{K_1K_2}{s(1 + sT_f)(1 + sT_m)} \cdot \frac{1 + saT_d}{1 + sT_d}; \qquad K' = \frac{aK_1K_2}{T_fT_m}$$

If $aT_d = T_m$ the real closed loop pole near the origin disappears.
If $aT_d \simeq T_m$ a pole will exist, but transients associated with this pole will be small in amplitude.

A major advantage of *s*-plane design can now be seen; while there are no terms equivalent to damping ratio and undamped frequency of oscillation for third and higher order systems, the boundaries on the *s*-plane set by minimum allowable values of ζ (radial lines) and ω_n (a circle of radius ω_n) together with maximum oscillation frequency and minimum decay rate (see Chapter 7, Fig. 7.6) apply to any order system, since the dominant closed loop poles will be immediately obvious.

Thus comparing Figs. 8.34 and 8.35 it can be seen that for the same damping ratio for the dominant terms, the decay rate is appreciably increased by the compensation. By applying the amplitude rule it can also be ascertained that the value of K' required to give the same ζ is roughly 'a' times higher in the compensated case.

Now for the compensated system

$$G(s)H(s) = \frac{K_1 K_2}{s(1 + sT_f)(1 + sT_m)} \cdot \frac{1 + saT_d}{1 + sT_d}$$

so that $K' = \dfrac{aK_1 K_2}{T_m T_f}$, while K is still $K_1 K_2$.

Thus as K' is approximately 'a' times bigger from the root loci, K is largely unchanged so that the steady state performance of the system is unaffected.

It can be envisaged that this form of compensation which 'pushes' the loci to the left is essential to stabilise a type 2 system or a system which goes into the right-hand half s-plane for the desired value of K.

Example 2. Phase lag compensation.

The object of phase lag compensation is to improve the steady state performance of a system by increasing K without appreciably altering the dynamic performance. Intuitively the pole and zero of the network must nearly cancel to maintain the dynamics; however, the zero must be bigger than the pole to achieve the increase in K; careful consideration will show that both these conditions can be met only by placing the pole–zero pair near to the origin on the s-plane.

The phase lag network transfer function is

$$G_c(s) = \frac{b(1 + sT_i)}{1 + sbT_i}$$

$$\therefore \; G(s)H(s) = \frac{bK_1 K_2 (1 + sT_i)}{s(1 + sT_f)(1 + sT_m)(1 + sbT_i)}$$

$$= \frac{K'\left(s + \dfrac{1}{T_i}\right)}{s\left(s + \dfrac{1}{T_f}\right)\left(s + \dfrac{1}{T_m}\right)\left(s + \dfrac{1}{bT_i}\right)}$$

where $K' = \dfrac{K_1 K_2}{T_f T_m}$ as in the uncompensated case. Thus while K' is similar for both systems, the open loop gain constant is $K_1 K_2$ for the uncompensated case and $bK_1 K_2$ for the compensated case.

Inspection of the loci of Fig. 8.36 shows such a system. For K' to be similar for compensated and uncompensated cases the distance from the pole to a point on the locus and from the zero to the same point must be similar, which, as previously indicated, can only be achieved with the network pole and zero close together.

The disadvantage of phase lag compensation is equally clear in

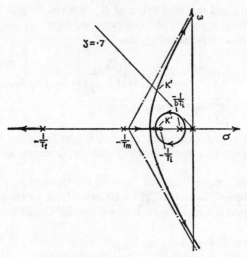

Fig. 8.36 Root loci of phase lag compensated system:

$$G(s)H(s) = \frac{K_1 K_2}{s(1 + sT_f)(1 + sT_m)} \cdot \frac{b(1 + sT_i)}{(1 + sbT_i)}; \qquad K' = \frac{K_1 K_2}{T_f T_m}$$

In practice the network pole and zero can be closer to the origin, making less effect on the complex poles compared to the uncompensated system, but placing the real closed loop pole closer to the origin.

Fig. 8.36 with the introduction of the real closed loop pole close to the origin. The dynamics have thus not strictly been maintained; the oscillatory terms are similar, but there is a slowly decaying term now included in the transient response.

Example 3. Integral compensation.

Integral compensation can be considered from the previous example by letting the network pole move to the origin, i.e.

$$G_c(s) = \frac{b(1 + sT_i)}{1 + sbT_i}$$

$$= \frac{s + \dfrac{1}{T_i}}{s}$$

as required, by fixing the zero and letting $b \longrightarrow \infty$, i.e. the pole $= 0$. The shape of the root loci will be practically unaltered and the value of K' will also remain unchanged. Thus the effective open loop gain constant, $bK_1K_2 \longrightarrow \infty$, resulting in zero velocity lag and infinite stiffness. This is an unusual interpretation of the type 2 (integral compensated) system as being a type 1 system with infinite open loop gain constant at d.c.!

Example 4. Open loop transfer functions with complex poles.

Systems with complex open loop poles can be compensated in the normal manner as explained above. An alternative approach is to introduce a network having complex zeros of such a value as to coincide with and (approximately) cancel the complex open loop poles, introducing new poles in less significant positions. This can be achieved with bridged T networks, discussed briefly in Section 16.5, Fig. 16.7.

Example 5. Velocity feedback.

The velocity feedback example considered in Section 8.5.3 is a

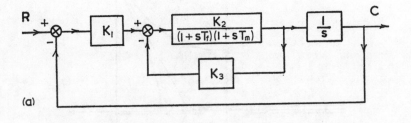

(a)

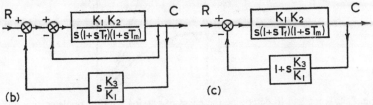

(b)   (c)

Fig. 8.37 A third order position control system with velocity feedback, expressed in alternative forms.

common example of parallel compensation. Fig. 8.37 shows the basic system, reduced in two stages.

(a) For Fig. 8.37 (a), opening only the position feedback loop, the open loop transfer function is

$$G(s)\,H(s) = K_1 \cdot \frac{K_2}{K_2K_3 + (1 + sT_f)(1 + sT_m)} \cdot \frac{1}{s} \qquad (8.5)$$

$$\therefore \; K' = \frac{K_1K_2}{T_fT_m}$$

(b) For Fig. 8.37 (b), opening only the velocity feedback loop, the open loop transfer function is

$$G(s)\,H(s) = \frac{K_1 K_2}{K_1 K_2 + s(1 + sT_f)(1 + sT_m)} \cdot s\frac{K_3}{K_1} \qquad (8.6)$$

$$\therefore\ K' = \frac{K_1 K_2}{T_f T_m} \cdot \frac{K_3}{K_1} = \frac{K_2 K_3}{T_f T_m}$$

(c) For Fig. 8.37 (c), opening effectively both position and velocity loops, the open loop transfer function is

$$G(s)\,H(s) = \frac{K_1 K_2\left(1 + s\dfrac{K_3}{K_1}\right)}{s(1 + sT_f)(1 + sT_m)} \qquad (8.7)$$

$$\therefore\ K' = \frac{K_2 K_3}{T_f T_m}$$

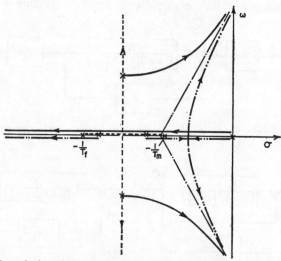

Fig. 8.38 Root loci to determine the effect of varying K_1 after predetermining the minor loop poles for various values of K_3 (Fig. 8.37 (a)).
- - - -   loci of minor loop poles as K_3 increases.
— · · —   loci of major loop poles as K_1 increases (K_3 small).
———   loci of major loop poles as K_1 increases (K_3 large).
— · —   asymptote.

While each of the above systems corresponds to the same closed loop the open loop functions are wildly different.

(a) must be used to determine the effect of varying K_1 for fixed values of K_3. The root locus is plotted in Fig. 8.38. Two sets of loci are drawn, first for the minor loop with varying K_3 to determine the poles of equation (8.5) and then typical loci for varying K_1 for two specific values of K_3.

(*b*) must be used to determine the effects of varying K_3 for fixed values of K_1. The root locus is plotted in Fig. 8.39. Again two sets of loci are drawn, first for the uncompensated system varying K_1 to determine the poles of equation (8.6), and then typical loci for varying K_3 for a specific value of K_1.

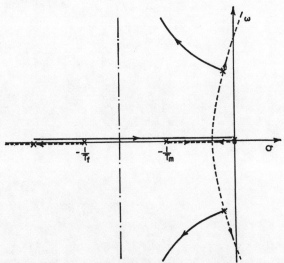

Fig. 8.39 Root loci to determine the effect of varying K_3. The minor loop in this case is the uncompensated system (Fig. 8.37 (*b*)).

- - - -   loci of uncompensated system poles as K_1 increases.
———   loci of major loop poles as K_3 increases for a fixed value of K_1.
— · —   asymptote.

(*c*) is the easiest case and can be constructed straightforwardly. It is, however, unrealistic, corresponding in fact to adjusting K_2 with K_1 and K_3 fixed; in other words, affecting both minor and major loops simultaneously. Sketching of these loci is left to the reader.

Part Three

AN INTRODUCTION
TO SOME
FURTHER TECHNIQUES

9
Nonlinear Systems

9.1 Introduction

What is a nonlinear system?* Previously a linear system has been considered as one which can be represented by linear, constant coefficient, differential equations; however, any system to which the principle of superposition† applies is defined as linear, so that systems described by linear, time-varying coefficient, differential equations are also linear. All other systems for which the coefficients of the differential equations depend upon the variable, e.g. $a(x)\dfrac{dx}{dt} + x = y$ are classed as *nonlinear*.

Most time-invariant (autonomous) systems have been discussed already; an exception, covered briefly below, is a finite time delay.

Time-variant parameter (nonautonomous) systems can be adequately analysed by conventional linear techniques if the rate of change of coefficients is slow. If the range of variation is large, then the automatic controller (compensation) parameters may have to be varied to maintain adequate control, i.e. an adaptive control. If the rate of change of parameters is appreciable, as say in a rocket re-entry control, then optimal, time-varying control laws must be used.

Nonlinear systems analysis‡ occupies a large percentage of present-day research work. No one approach is acceptable, and a number of techniques have been developed which are applicable only in a limited range of cases.

The following are some of the methods at present in use:

(a) the frequency response method known as 'describing function' analysis;

(b) the 'phase plane' analysis, a graphical method which is commonly only applied to first and second order systems;

---

* Gibson (ref. 19) is a most thorough text devoted to the modern methods of analysis of nonlinear control systems.

† If $f(x_1) = a_1$ and $f(x_2) = a_2$, then $f(x_1 + x_2) = a_1 + a_2$.

‡ It is only necessary to consider nonlinearities which are appreciable in magnitude by these methods. If the system is subjected only to small variations in signal amplitude, then good approximate linear equations can usually be derived. Typical examples of this technique of 'linearisation for small departures' are included in Section 16.3 for the a.c. motor torque-control current characteristic and in Section 17.5.1 for the hydraulic spool valve flow–load pressure relationship.

(c) numerical methods, e.g. digital computer solution of time response;

(d) analogue computer simulations.

Here only (a) and (b) are covered. (d) is the most practical and can be studied in Jackson (ref. 23).

A *finite time delay* (transportation lag) is included here simply because it does not fit conveniently earlier. A simple delay in which the output reproduces, exactly, the input at a later time is easily analysed by conventional means, e.g. in determining the temperature at a point A of a moving fluid, a reading is taken at a point B farther downstream from A. Thus

$$f_o(t) = f_i(t - T_d)H(t - T_d)$$

where $T_d =$ time delay and $H(t - T_d)$ is the delayed unit step function,

then $\quad F_o(s) = F_i(s)e^{-sT_d}$

and $\therefore \quad G(s) = e^{-sT_d}$

This can be expanded as a power series, the first few terms giving an approximation to a normal transfer function. Frequency-response analysis is more exact, since $G(j\omega) = e^{-j\omega T_d}$ and therefore $G(\omega) = 1$ and $\phi(\omega) = \omega T_d$ lagging.

In practice, such delays are often more troublesome, since: (a) the response $f_o(t)$ may be related to the function $f_i(t)$ by a differential equation, as well as a time delay, e.g. in the above example due to heat loss in the fluid moving from A to B; and (b) the time delay is often variable, e.g. due to change in velocity of flow in the above example.

It is a common fallacy to consider all nonlinearities undesirable since they always present problems of analysis. Backlash and stick–slip are on their own objectionable, but if in a system backlash is unavoidable, then stick–slip will have the *desirable* effect of minimising jitter in the backlash zone. Saturation is unavoidable in practice and can usually be tolerated for large error conditions. The economic penalty for increasing the saturation level is often not justified in improved performance.

9.2 The Common Types of Nonlinearities

The commonly encountered nonlinearities in control systems are:

(a) Continuous functions such as a nonlinear amplifier (Fig. 9.1) for which, say,

$$f_o = af_i + bf_i^3$$

(b) Dead zones (Fig. 9.2), e.g. the output torque of a relay servo is zero for small signals while the relay stays in the centre position, or the output position of a position control remaining zero, due to static friction, until the torque exceeds a certain value.

(c) Saturation (Fig. 9.3) occurring due to a limitation on the amount of energy available in any system, e.g. an amplifier output

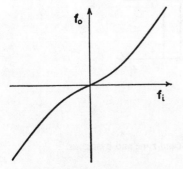

Fig. 9.1 A continuous nonlinearity.　　　Fig. 9.2 Dead zone.

reaching the limit set by the H.T. supplies; torque saturation of an hydraulic motor due to the limit set by the available supply pressure. Saturation is often deliberately introduced for protection, i.e. acceleration limiting to avoid damage to mechanical structures. Such

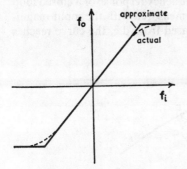

Fig. 9.3 Saturation.　　　Fig. 9.4 Approximate hysteresis characteristic.

limits are usually introduced electronically in the control stages of a system.

(d) Hysteresis (Fig. 9.4). Stick–slip and backlash mechanically, and magnetic hysteresis electrically are the common sources. This form of nonlinearity depends upon the past history of the function, i.e. which direction of approach, and therefore effects phase as well as amplitude, as will be shown later.

(e) Combinations of the above, as, for example, Fig. 9.5, which is the characteristic of an on–off servo with dead zone and hysteresis.

Nonlinearities such as (*b*) to (*e*), which can be considered as a series of linear portions are called 'PIECEWISE LINEAR' elements.

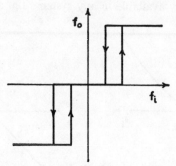

Fig. 9.5 Relay characteristic with dead zone and hysteresis.

9.3 Some Effects of Nonlinearities in Closed Loop Control Systems

In some practical closed loop systems certain effects are noticed which cannot be explained by linear theory but which can to some extent be related to the nonlinearities. Some of these features are listed below.

(i) *Jump Resonance.* Fig. 9.6 is the frequency response of a closed loop system involving a nonlinear element, measured with the input magnitude, $R(\omega)$, a constant.* As ω is increased from d.c. the curve reaches

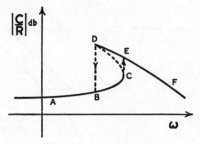

Fig. 9.6 Jump resonance.

point C and then jumps to point E and continues along EF. If ω is then steadily reduced to zero the curve reaches point D and then jumps to point B and continues along BA. This phenomena is usually associated with saturation.

(ii) *Limit Cycle Oscillation.* According to linear theory the output of a system with a constant input will either settle at the desired value or

* Change in $R(\omega)$ will produce a different characteristic with a nonlinear system.

exponentially (or sinusoidally with an exponential envelope) increase to infinity. However, a phenomenon occurs where a closed loop system can oscillate at a fixed frequency with constant amplitude which, since it cannot be explained by linear theory, must be due to the nonlinearity in the loop. The amplitude of oscillation can be very small, but is often objectionably large. The linear unstable system is not practical since the amplitude cannot extend to infinity; saturation therefore occurs and the oscillation settles at a constant amplitude which is in fact an extreme case of a limit cycle oscillation. Van Der Pol showed that an electronic oscillator (positive feedback) must possess nonlinear properties in order to achieve a stable amplitude oscillation and formulated the nonlinear differential equation

$$\ddot{q} - \varepsilon(1 - q^2)\dot{q} + q = 0$$

where q is the output and ε a positive constant.

This equation indicates that the damping is negative (unstable) when $q^2 < 1$ and positive when $q^2 > 1$, so that a stable amplitude of oscillation is reached.

Note that the basic waveform of a limit cycle oscillation is not sinusoidal.

(iii) *Sub-harmonic Content.* If the input signal to a linear system contains, say, two component frequencies ω_1 and ω_2, the superposition theorem shows that the output contains just these two frequencies. For a nonlinear system, however, cross-modulation of these signals produce components in the output at frequencies $m\omega_1 \pm n\omega_2$, where m and n are integers. Undesirable components can thus exist at frequencies below those of the applied signals.

9.4 Describing Functions

9.4.1 The Describing Function, $G_N(j\omega)$

Describing functions,* as the name suggests, are approximate frequency response functions. The analysis is treated only in terms of sinusoidal input signals so that there is not an equivalent transfer function.

A sinusoidal input signal to a nonlinear device produces a distorted output signal which is, however, of the same fundamental frequency. The output can then be expressed by a Fourier series as the sum of an infinite number of sinusoidal functions of frequency ω, 2ω, 3ω, etc., and phase ϕ_1, ϕ_2, ϕ_3, etc. It is then assumed that the fundamental component of the output adequately describes the system response. The describing function is thus the ratio of the fundamental component of the output to the input.

---

* Describing function analysis has stemmed mainly from work on relay servo-mechanisms (see Section 15.7) by such as Kochenburger and Tustin.

Let $\qquad\qquad |G_N(j\omega)| = G_N(\omega)$

and $\qquad\qquad \underline{/G_N(j\omega)} = \phi_N(\omega)$

$$\therefore \; G_N(\omega) = \frac{\text{Fundamental amplitude of the output}}{\text{Amplitude of the input}}$$

and $\quad \phi_N(\omega) =$ Phase shift between the fundamental component of the
output and the input

9.4.2 Validity of the Application of Describing Functions

For the simple system of Fig. 9.7, $G_N(j\omega)$ represents the nonlinear part
and $G_p(j\omega)$ the linear part of the system.

Let the system excitation $R(j\omega)$ be sinusoidal. For the describing
function to apply, $E(j\omega)$ must also be sinusoidal, but $M(j\omega)$ is not. The

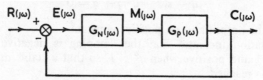

Fig. 9.7 A closed loop system with a single nonlinearity.

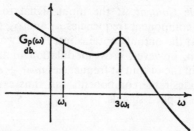

Fig. 9.8 A source of error in the describing function at the frequency ω_1.

basic assumption, then, is that $G_p(j\omega)$ acts as a low-pass filter so that the
harmonics of $M(j\omega)$ are attenuated to negligible proportions, resulting
in a near sinusoidal value for $E(j\omega)$ as required.

This method then is applicable to many systems because:

(a) the amplitude of the harmonics of $M(j\omega)$ are usually much
smaller than the fundamental; and

(b) most open loop functions, $G_p(j\omega)$, inherently attenuate the
higher frequencies.

Difficulty and gross inaccuracy of the method occurs in systems for
which $G_p(j\omega)$ has the form of Fig. 9.8, when certain harmonics will
clearly be amplified and not attenuated, i.e. the third harmonic of a
frequency ω_1. Also

(i) The system must be time invariant, i.e. $G_N(j\omega)$ cannot be a
function of time.

(ii) $G_N(j\omega)$ is the only nonlinear element. Techniques for dealing with more than one nonlinear element are difficult since the input to the second nonlinear element is not sinusoidal.

9.4.3 Characteristics of Describing Functions

It has been pointed out that $G_N(j\omega)$ cannot be a function of time.

$G_N(j\omega)$ is always a function of the input amplitude and therefore for, say, the system of Fig. 9.7 should have been written $G_N(j\omega, |E|)$.

In general, $G_N(j\omega)$ is also a function of frequency and is called a 'frequency variant' describing function. If, however, no energy storage is involved, i.e. backlash or saturation of the input (as distinct from a derivative, i.e. velocity limiting), the describing function is dependent only upon the amplitude of the input. Such describing functions are termed 'frequency invariant'.

In order to calculate the describing function the waveform of the output must be found and then Fourier analysed.

Thus if

$$e(t) = E \sin \omega t$$

then

$$m(t) = \frac{M_0}{2} + M_1 \sin(\omega t + \phi_1) + M_2 \sin(2\omega t + \phi_2)$$
$$+ M_3 \sin(3\omega t + \phi_3) + \ldots$$

The describing function is given by

$$G_N(\omega) = \frac{M_1}{E} \quad \text{and} \quad \phi_N(\omega) = \phi_1$$

In general terms the Fourier expansion of a periodic function $m(t)$ of period $T = 2\pi/\omega$ is

$$m(t) = \frac{A_o}{2} + \sum_{n=1}^{n=\infty} A_n \cos n\omega t + \sum_{n=1}^{n=\infty} B_n \sin n\omega t$$

where

$$A_n = \frac{2}{T} \int_0^T m(t) \cos n\omega t \, dt \quad \text{for} \quad n = 0 \text{ to } \infty$$

and

$$B_n = \frac{2}{T} \int_0^T m(t) \sin n\omega t \, dt \quad \text{for} \quad n = 1 \text{ to } \infty$$

From this $M_n = \sqrt{A_n^2 + B_n^2}$

and

$$\phi_n = \tan^{-1} \frac{A_n}{B_n}$$

The d.c. component $\frac{M_o}{2}$ would occur for an asymmetrical non-linearity and would cause steady state errors in the system in a similar fashion to drifting.

9.4.4 Examples of Describing Functions

As a worked example the relatively simple case of saturation will be considered. The describing functions of a few common nonlinearities are also listed, while extensive tables can be found in Gibson, ref. 19, p. 358.

Let $e(t) = E \sin \omega t$

then $m(t) = KE \sin \omega t$ for $0 < t < t_1$

                        $= KE \sin \omega t_1$ for $t_1 < t < \dfrac{T}{2} - t_1$

                        $= KE \sin \omega t$ for $\dfrac{T}{2} - t_1 < t < \dfrac{T}{2} + t_1$

                        $= -KE \sin \omega t_1$ for $\dfrac{T}{2} + t_1 < t < T - t_1$

                        $= KE \sin \omega t$ for $T - t_1 < t < T$

where $T = \dfrac{2\pi}{\omega}$

If S is the value of $e(t)$ at which the output saturates, then

$$S = E \sin \omega t_1$$

Since it is an odd function the output $m(t)$ will contain no cosine, d.c. or even harmonic terms. Conversely inspection of the waveform shows

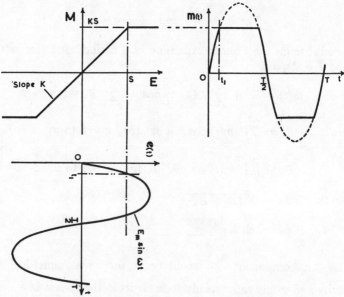

Fig. 9.9 Input and output relationships for a saturation characteristic.

that it is essentially in phase with the input and therefore contains only odd sine terms.

$$\therefore\ m(t) = \sum_{n=1,\,3,\,5}^{\infty} B_n \sin n\omega t$$

where

$$B_n = \frac{2}{T}\int_0^T m(t) \sin n\omega t\, dt = \frac{8}{T}\int_0^{\frac{T}{4}} m(t) \sin n\omega t\, dt \text{ for an even function}$$

$$= \frac{8}{T}\left[\int_0^{t_1} KE \sin \omega t \,.\, \sin n\omega t\, dt + \int_{t_1}^{\frac{T}{4}} KE \sin \omega t_1 \,.\, \sin n\omega t\, dt\right]$$

$$= \frac{8KE}{T}\left[\int_0^{t_1} \frac{\cos (n-1)\omega t - \cos (n+1)\omega t}{2}\, dt \right.$$
$$\left. - \sin \omega t_1\left[\frac{\cos n\omega t}{n\omega}\right]_{t_1}^{\frac{T}{4}}\right]$$

$$= \frac{8KE\omega}{2\pi}\left[\tfrac{1}{2}\left[\frac{\sin (n-1)\omega t}{(n-1)\omega} - \frac{\sin (n+1)\omega t}{(n+1)\omega}\right]_0^{t_1} \right.$$
$$\left. + \frac{\sin \omega t_1}{n\omega} \,.\, \cos n\omega t_1\right]$$

$$= \frac{8KE}{2\pi}\left[\frac{\sin (n-1)\omega t_1}{2(n-1)} - \frac{\sin (n+1)\omega t_1}{2(n+1)} + \frac{\sin \omega t_1 \,.\, \cos n\omega t_1}{n}\right]$$

$$= \frac{2KE}{\pi}\left[\frac{\sin (n-1)\omega t_1}{n-1} - \frac{\sin (n+1)\omega t_1}{n+1} \right.$$
$$\left. + \frac{\sin (n+1)\omega t_1}{n} - \frac{\sin (n-1)\omega t_1}{n}\right]$$

$$= \frac{2KE}{n\pi}\left[\frac{\sin (n-1)\omega t_1}{n-1} + \frac{\sin (n+1)\omega t_1}{n+1}\right]$$

To find G_N it is necessary only to calculate B_1, but some idea as to the accuracy of the approximation given by G_N can be obtained by comparing the relative amplitudes of the first two terms, in this case the third harmonic to the fundamental—the smaller the ratio the more accurate the result.

Thus
$$B_1 = \frac{2KE}{\pi}\left[\omega t_1 + \frac{\sin 2\omega t_1}{2}\right]$$

and
$$B_3 = \frac{2KE}{3\pi}\left[\frac{\sin 2\omega t_1}{2} + \frac{\sin 4\omega t_1}{4}\right]$$

Now $\sin \omega t_1 = \dfrac{S}{E}$ which is independent of ω. Thus put $\omega t_1 = \theta_1$ so as to stress the independence of B_n on ω (changes in ω change t_1, θ_1 being constant).

$$\therefore\ G_N = \frac{B_1}{E} = \frac{2K}{\pi}\left[\theta_1 + \tfrac{1}{2}\sin 2\theta_1\right]$$

and the 'goodness' factor

$$\frac{B_3}{B_1} = \frac{1}{6} \,.\, \frac{2\sin 2\theta_1 + \sin 4\theta_1}{2\theta_1 + \sin 2\theta_1}$$

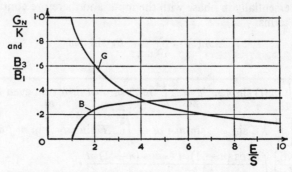

Fig. 9.10 Describing function for a saturation characteristic.

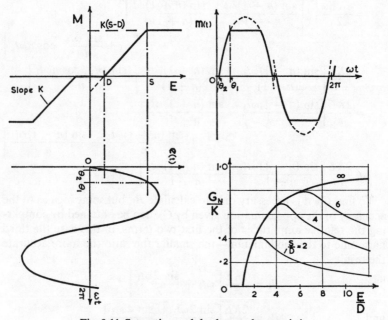

Fig. 9.11 Saturation and dead zone characteristic.

$D = E \sin \theta_2$ and $S = E \sin \theta_1$.

$$
\begin{aligned}
m(t) &= 0 & \text{for } 0 < \omega t < \theta_2 \\
&= KE (\sin \omega t - \sin \theta_2) & \text{for } \theta_2 < \omega t < \theta_1 \\
&= KE (\sin \theta_1 - \sin \theta_2) & \text{for } \theta_1 < \omega t < \frac{\pi}{2}
\end{aligned}
$$

This is an even function for which $A_n = 0$ and

$$
B_n \atop n \text{ odd} = \frac{2KE}{n\pi} \left[\frac{\sin (n-1)\theta_1 - \sin (n-1)\theta_2}{n-1} + \frac{\sin (n+1)\theta_1 - \sin (n+1)\theta_2}{n+1} \right]
$$

$$
G_N = \frac{B_1}{E} = \frac{2K}{\pi} [\theta_1 - \theta_2 + \tfrac{1}{2}(\sin 2\theta_1 - \sin 2\theta_2)]
$$

These functions are plotted in Fig. 9.10. Note that this is a frequency invariant describing function. For $E < S$ no distortion occurs and $G_N = K$; for $E > S$, G_N gets smaller and the accuracy decreases.

Fig. 9.11 and 9.12 show the describing functions of two more frequency invariant functions, the first being a combination of a dead zone

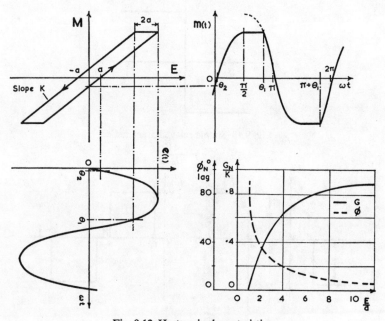

Fig. 9.12 Hysteresis characteristic.

$$\sin \theta_1 = \frac{E - 2a}{E} \quad \left(\text{N.B. } \theta_1 > \frac{\pi}{2}\right)$$

$$
\begin{aligned}
m(t) &= K(E \sin \omega t - a) &&\text{for } \theta_1 - \pi < \omega t < \frac{\pi}{2} \\
&= K(E - a) &&\text{for } \frac{\pi}{2} < \omega t < \theta_1 \\
&= K(E \sin \omega t + a) &&\text{for } \theta_1 < \omega t < \frac{3\pi}{2} \\
&= -K(E - a) &&\text{for } \frac{3\pi}{2} < \omega t < \pi + \theta_1
\end{aligned}
$$

$$A_1 = -\frac{2KE}{\pi} \cos^2 \theta_1$$

$$B_1 = \frac{2KE}{\pi} \left[\frac{3\pi}{4} - \frac{\theta_1}{2} - \frac{\sin 2\theta_1}{4}\right]$$

$$G_N = \frac{\sqrt{A_1{}^2 + B_1{}^2}}{E} \quad \text{and} \quad \phi_N = \tan^{-1} \frac{A_1}{B_1}$$

with saturation and the second, which exhibits phase shift, i.e. has both sine and cosine terms, being hysteresis. Hysteresis is a typical example of a nonlinearity said to have 'memory'. Note that the phase shift is a function of amplitude and not of frequency as is the case for a simple linear element.

9.4.5 Analysis of Relay Servomechanisms

Relay systems are discussed from a hardware viewpoint in Section 15.7. In order to analyse such systems a means of evaluating the relay as a circuit element must be devised; here a describing function will be calculated.

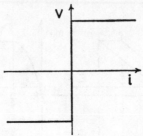

Fig. 9.13 Simple relay characteristic:

$$G_N = \frac{4V}{\pi I_{max}}$$

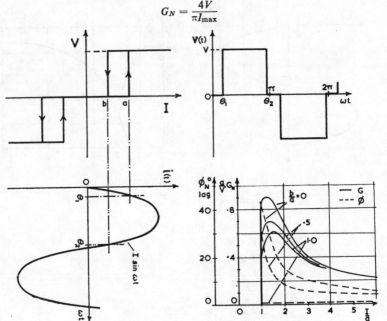

Fig. 9.14 Relay characteristic, with dead zone and hysteresis.
Pull-in current = a, drop out current = b.
Input current, $i = I \sin \omega t$.

$$a = I \sin \theta_1 \text{ and } b = I \sin \theta_2. \left(\text{N.B. } \theta_2 > \frac{\pi}{2} \right)$$

$$A_1 = -\frac{2V}{\pi} (\sin \theta_1 - \sin \theta_2)$$

$$B_1 = \frac{2V}{\pi} (\cos \theta_1 - \cos \theta_2)$$

$$G_N = \frac{\sqrt{A_1{}^2 + B_1{}^2}}{I} \text{ and } \phi_N = \tan^{-1} \frac{A_1}{B_1}$$

The simplest form of relay has one polarity for a positive input and reverse polarity for negative inputs as in Fig. 9.13, which shows the switched output voltage against relay coil current. The output voltage magnitude is a function of the power source and not the relay. System stability and the relay life can be greatly improved if the relay is of the polarised type with a small dead zone around zero input. A further sophistication lies in allowing a linear conventional mode of operation for small errors in place of the dead band, a technique which is not analysed here. A further complication arises since electromagnetic relays requiring a certain value of pull-in current will not drop-out until a lower value of current, thus exhibiting hysteresis, i.e. a memory type nonlinearity which will involve amplitude dependent phase shift. Again the describing function is frequency invariant although the relay coil time constant must be included in the linear part of the system. The general case is shown in Fig. 9.14.

9.5 Stability Analysis Using Describing Functions

Since the rules of superposition do not apply to nonlinear systems and the system cannot be represented by pole and zero locations (a linear system can be represented by its poles and zeros, independent of the input), the normal Routh and Nyquist criteria cannot be directly

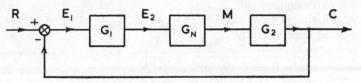

Fig. 9.15

applied. The Liapounov method (see Gibson (ref. 19)) is rather akin to Routh's method in that it attempts to say yes or no to stability without attempting a complete solution. The less accurate describing function is akin to the Nyquist method, when applicable, in that it gives more quantitative results. As with linear systems the results can be interpreted on a Nyquist diagram and a db-phase diagram (as for the Nichols chart), but not very readily on a Bode diagram.

Consider the system of Fig. 9.15, for which

$$\frac{C(j\omega)}{R(j\omega)} = \frac{G_N(j\omega)G_1(j\omega)G_2(j\omega)}{1 + G_N(j\omega)G_1(j\omega)G_2(j\omega)}$$

Put the linear part $G_1(j\omega)G_2(j\omega) = G(j\omega)$.

Note the following:

(i) It has been assumed that the describing function is valid.

(ii) These terms cannot be expressed as functions of s.

(iii) Any gain constant in $G_N(j\omega)$ can often be conveniently incorporated in $G(j\omega)$ although it makes little difference.

Thus
$$\frac{C(j\omega)}{R(j\omega)} = \frac{G_N(j\omega)G(j\omega)}{1 + G_N(j\omega)G(j\omega)}$$

A stability rule, equivalent to Nyquist's criterion is to consider the equation,

$$1 + G_N(j\omega)G(j\omega) = 0$$

i.e.
$$-G_N(j\omega) = \frac{1}{G(j\omega)}$$

The technique is to solve the equation graphically by plotting $\dfrac{1}{G(j\omega)}$ and $-G_N(j\omega)$ on a polar diagram $\left(\dfrac{1}{G(j\omega)}\right.$ is effectively an inverse Nyquist plot). For a linear system to be stable the inverse Nyquist plot, $\dfrac{1}{G(j\omega)}$, must pass *outside* the -1 point (see Section 6.5.6). This approach is similar, except that the locus $\dfrac{1}{G(j\omega)}$ must pass outside the point given by $-G_N(j\omega)$ and not -1. The point, $-G_N(j\omega)$, which is not necessarily on the negative real axis (see Fig. 9.16), will vary with signal amplitude, E, so that a system may be stable for one value of E and unstable for another.

Alternatively $G(j\omega)$ and $-\dfrac{1}{G_N(j\omega)}$ can be plotted on a db-phase chart and interpreted similarly.

Whether $G(j\omega)$ and $-\dfrac{1}{G_N(j\omega)}$ or $\dfrac{1}{G(j\omega)}$ and $-G_N(j\omega)$ are plotted is insignificant, the two examples quoted being common practice.

Plots of $G(j\omega)$, or $\dfrac{1}{G(j\omega)}$, must be clearly labelled with the parameter ω; $G_N(j\omega)$ must be marked with the parameter E, (the amplitude) and in the case of frequency dependent describing functions with a second parameter, ω. If $G_N(j\omega)$ is frequency invariant, then one curve only exists with E as a parameter; for frequency variant cases a family of curves exists.

Fig. 9.16 is a Nyquist-type plot for the function $-G_N(j\omega)$ for:

(*a*) Saturation (Fig. 9.10 with $K = S = 1$).

(*b*) Dead zone (Fig. 9.11 with $K = D = 1$ and $S = \infty$).

(*c*) Saturation and dead zone (Fig. 9.11 with $K = D = 1$ and, say, $S = 4$).

(*d*) Hysteresis (Fig. 9.12 with $K = a = 1$).

(*e*) On-off relay (Fig. 9.13 with $V = 1$).

(*f*) Relay with hysteresis and dead zone (Fig. 9.14 with $a = V = 1$ and, say, $b = \frac{1}{2}$).

Consider first a simple case where $G(j\omega) = \dfrac{K}{s(1 + sT_1)(1 + sT_2)}$ and $G_N(j\omega)$ is a dead zone. Fig. 9.17 is a plot of $\dfrac{1}{G(j\omega)}$ and $-G_N(j\omega)$ for various values of K.

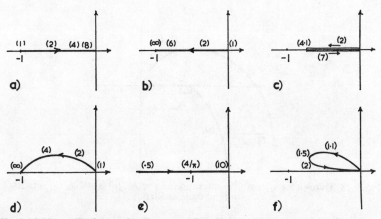

Fig. 9.16 Normalised Nyquist diagrams of $-G_N(j\omega)$. The parameters shown are values of e (or i).

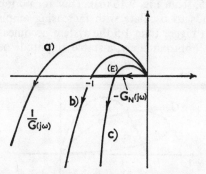

Fig. 9.17 Third-order system with a dead zone: $K_a < K_b < K_c$. (a) is stable, (b) just stable and (c) amplitude conditionally stable.

Case (a) is stable; case (c) may or may not be stable, as follows. The value of E marked on the diagram is the value of the parameter of $-G_N(j\omega)$ (from Fig. 9.16 (b), E is approximately 6 in this case) at which the two curves intersect. If E is less than 6, $\dfrac{1}{G(j\omega)}$ passes outside the point $-G_N(j\omega)$ and the system is stable; if E is bigger than 6 the system is unstable and the increasing amplitude of E resulting drives the point on the $-G_N(j\omega)$ curve further into the unstable region, so that the amplitude of the oscillations build up to infinity. Note that such a system is **amplitude conditionally stable** which is not the same as conditional stability previously defined in Section 6.5.4.

Oscillations must be bounded so that a practical system will exhibit saturation as well as the dead zone. First, however, examine the case of saturation only as shown in Fig. 9.18.

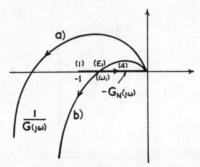

Fig. 9.18 Third-order system with saturation: $K_a < K_b$. (*a*) is stable, (*b*) is unstable with limit cycle oscillation.

In this example, case (*a*) is again stable. Case (*b*), however, indicates instability unless E is *bigger* than E_1 (for the curve sketched, E_1 is approximately 1·5, from Fig. 9.16 (*a*)). Thus for no input the system is unstable and tends to oscillate with increasing amplitude; when the amplitude of E is bigger than 1·5 the system becomes stable so that E falls again. Thus, approximately, a stable amplitude oscillation sets up

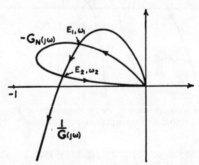

Fig. 9.19 Third-order system with relay having dead zone and hysteresis. ω is the parameter of $1/G(j\omega)$ and E the parameter of $-G_N(j\omega)$.

of amplitude E_1 at a frequency ω_1. This is the special case of instability known as a 'limit cycle oscillation'.

Any nonlinearity for which G_N gets smaller as E increases can result, under certain conditions, in limit cycle oscillations.

Now consider the example of Fig. 9.19 involving a plot of $-G_N(j\omega)$

which is a loop, i.e. dead zone with saturation, or in the case taken, a relay with dead zone and hysteresis.

On the diagram $E_2 > E_1$ (see Fig. 9.16 (f)). If $E < E_1$ the system is stable. If $E > E_1$ the system is unstable and so the amplitude of E builds up. When $E > E_2$, however, the system becomes stable again so

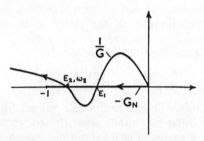

Fig. 9.20 Dead zone with $G(s) = \dfrac{K(1 + sT_a)}{s(1 + sT_1)(1 + sT_2)}$. A limit cycle oscillation occurs of amplitude E_2 and frequency ω_2, excited by $E > E_1$.

that the amplitude begins to fall; thus a stable limit cycle occurs at a frequency ω_2 of amplitude E_2. E_1, ω_1 is not a stable oscillation.

The major difference between this example and simple saturation is that, provided the gain is high enough for oscillation in both cases, the relay system requires an excitation of amplitude $> E_1$ (this could be any

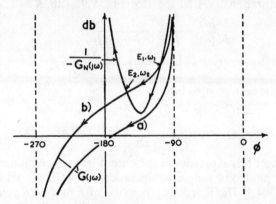

Fig. 9.21 Db-phase diagram for a third-order system with a relay having dead zone and hysteresis (see Fig. 9.19). (a) is stable, (b) has a limit cycle oscillation of amplitude E_2 and frequency ω_2, excited by $E > E_1$.

frequency, including d.c.) whereas the saturation case oscillates in any condition.

Finally, limit cycle oscillations can occur with dead zone only, if $G(j\omega)$ is a conditionally stable system of the form shown in Fig. 9.20.

As an alternative method of presentation Fig. 9.21 is the equivalent of Fig. 9.19 plotted on a db-phase chart.

9.6 Closed Loop Response of a Nonlinear System

In the preceding section a method was outlined of determining whether or not a system is stable and of suggesting one method of stabilising an unstable system, i.e. by shaping the linear part, $G(j\omega)$, in a conventional manner. Having determined that a system is stable it is desirable to find the output–input relationship of the closed loop, realising that the linear system M and δ type contours do not apply.

Assuming the describing function to be valid*

$$\frac{C(j\omega)}{R(j\omega)} = \frac{G(j\omega)G_N(j\omega)}{1 + G(j\omega)G_N(j\omega)}$$

Being nonlinear, this will not be a single valued function as it will depend upon the input magnitude $R(\omega)$. It is therefore necessary to produce a family of curves of C/R for various values of $R(\omega)$. Further, the function cannot be plotted directly since $G_N(j\omega)$ is a function of E and yet E is a function of $G_N(j\omega)$, with $R(\omega)$ a constant. The technique is to expand the relationship between E and G_N and to solve graphically.

As an example consider a saturating system for which

$$G(j\omega) = \frac{250}{j\omega(1 + j\omega)}$$

and take $R = 5$†

Let the saturation level be 0·4 (see Fig. 9.9 with $K = 1$, $S = 0·4$).

Then $\dfrac{E(j\omega)}{R(j\omega)} = \dfrac{1}{1 + G(j\omega)G_N(j\omega)}$

$$\therefore E = \left| \frac{j\omega(1 + j\omega)}{j\omega(1 + j\omega) + 250G_N} \right| R$$

which can be manipulated into

$$E^2((250G_N - \omega^2)^2 + \omega^2) = \omega^2(1 + \omega^2)R^2$$

$$\therefore G_N E = \frac{\omega^2 E}{250} \pm \frac{\omega}{250} \sqrt{(1 + \omega^2)R^2 - E^2}$$

The L.H.S. of this expression is the output from the nonlinear element and can be plotted by simple manipulation of the G_N against E curve as in Section 9.4.4. The R.H.S. is a function of E and ω (R constant), so that a family of functions for constant ω must be calculated. The solution is obtained by the intersects of plots to a base E of

(a) $x = G_N E$‡

* It is well to remember here that the foregoing describing function analysis is an approximate, and not infallible method.

† This is the example used in Gibson, p. 393, but solved in the manner suggested by Truxal, p. 582, after Levinson.

‡ Note that while in this example the magnitude of the output of the nonlinearity is saturated at S, the describing function output, $G_N E$, increases to about 1·25 S, since it is given by the amplitude of the fundamental.

(b) $$y(\omega) = \frac{\omega^2 E}{250} \pm \frac{\omega}{250} \sqrt{(1 + \omega^2)R^2 - E^2}$$

These functions are plotted on Fig. 9.22.

To find and plot $C(\omega)$ with $R(\omega)$ constant it is easiest to read off $G_N E$ from the above graph over a range of frequencies, given by the intersects of the two curves and to multiply this by the linear part

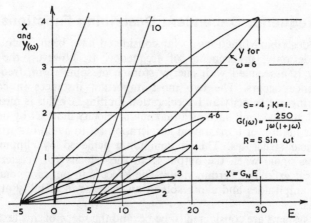

Fig. 9.22 Graphical solution to find C/R for system with saturation.

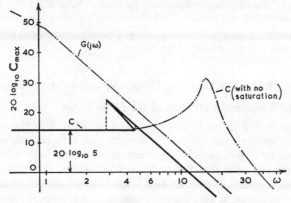

Fig. 9.23 Output of system of Fig. 9.22 ($R = 5 \sin \omega t$).

$G(\omega)$, which is best achieved on a logarithmic axis. However, inspection of Fig. 9.22 will show that for certain frequencies the two curves intersect at three points. For ω from d.c. up to about 3 rad/s there is only one intersect, the maximum of y being less than $G_N E$. For ω from 3 to 4·6 rad/s there are three intersections. For $\omega > 4·6$ there is only one intersect, this time at the higher values of E. This gives rise to the phenomena known as 'jump resonance' expounded in Section 9.3.

Fig. 9.23 is the resulting curve for $C(\omega)$ for this example; this is not a frequency response function, it is an amplitude; another curve could be calculated for another value of $R(\omega)$.

Finally, note that jump resonance does not occur for all nonlinearites but is often associated with saturation.

9.7 Frequency Dependent Describing Functions

All the describing functions so far considered have been functions of amplitude only, independent of frequency. If, however, the non-linearity is associated with energy storage of some form, frequency dependence occurs. The two most important instances in control engineering are velocity and acceleration limiting. By this is meant an artificial applied limit such as a maximum velocity limit set by bearing and gear failure, or a maximum acceleration set to avoid distorting of mechanical structures. This is commonly achieved by clipping the actuating signal when the output from the velocity (or acceleration) transducer exceeds a certain value. The effect is to put more damping at high amplitudes and can result in an increase in stability at these extremes.

Such systems are considered to be beyond the scope of this text since the analysis proves appreciably more difficult. The above-mentioned problems, however, where covered by R. J. Kochenburger, one of the pioneers of the describing function, in *Trans. A.I.E.E.*, Vol. 72, Part 2, pp. 180–94, 1953.

In practice, mathematical analysis is often best avoided, the solution being achieved by simulation techniques.

9.8 An Introduction to Phase-Plane Analysis

9.8.1 The Scope and Definition of the Phase-Plane

The short introduction to nonlinear system analysis contained in this chapter is based on the frequency response method since this is the most applicable to engineering, and particularly servomechanism, problems. It should have been made clear by now, however, that this technique cannot be successfully applied to a large number of problems.

The most common alternative technique is a graphical one called the 'phase-plane' analysis; no approximation is made, but the method has other severe limitations, viz.:

(i) It is only successfully applicable to systems, the linear part of which can be described by a second-order differential equation.*

* Attempts to plot a three-dimension phase-space to allow for third-order systems are possible but extremely difficult to interpret.

(ii) The method is as yet unapplicable to systems with time-varying parameters.

(iii) The form of the input signal $r(t)$ is limited to initial conditions only. This in fact allows step or ramp inputs which can be incorporated as constants, but sinusoidal or random inputs are not allowable.

The phase-plane is a plot of $\dfrac{dx}{dt}$ vertically against x, where x is a function of time. It is common to make the output, $c(t)$, the variable x, or alternatively the actuating signal, $e(t)$. The plotting of $e(t)$ is almost essential for a ramp input, so as to limit the size of the plot, since $c(t)$ will continually increase.

The general equation which can be studied under the above rules is the equation of an unexcited second-order system:

$$a\ddot{x} + b\dot{x} + cx = k, \text{ a constant*}$$

where a, b and c need *not* be constant coefficients, but can be dependent upon the magnitude (x) or the first derivative of magnitude ($\dot{x}$), thus covering a wide range of nonlinear systems.

Any initial conditions of x and $\dot{x}$ are possible.

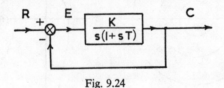

Fig. 9.24

Time is involved in the phase-plane only implicitly, as both x and $\dot{x}$ are functions of time, but not directly. For any given initial conditions, time is a parameter of the locus of $\dot{x}$ against x. The references should be consulted for methods of calculating time from a given plot.

Any points on the locus for which $\dfrac{d\dot{x}}{dt}$ and $\dfrac{dx}{dt}$ are zero are called 'singular points', the significance of which is that a stable system will eventually settle at one of them, while an unstable system will diverge.

Consider first the linear type 1 system of Fig. 9.24. For this system:

$$\frac{C}{R} = \frac{K}{s^2 T + s + K} \quad \text{and} \quad \frac{E}{R} = \frac{s(1 + sT)}{s^2 T + s + K}$$

Thus $$T\ddot{c} + \dot{c} + Kc = Kr$$

and $$T\ddot{e} + \dot{e} + Ke = \dot{r} + T\ddot{r}$$

* Truxal suggests a change of variable $x_1 = x - \dfrac{k}{c}$ in order to give a general equation with zero R.H.S.

The necessity for the R.H.S. to be a constant helps to decide the best equation to use.

(a) $r(t) = RH(t)$, a step function.

∴ $r = R$; $\dot{r} = \ddot{r} = 0$. Thus either equation can be used.

(b) $r(t) = Qt$, a ramp function.

∴ $r = Qt$; $\dot{r} = Q$; and $\ddot{r} = 0$. Thus the error equation only can be used.

As an example, consider the output equation with a step input:

$$T\ddot{c} + \dot{c} + Kc = KR$$

Rewriting the equation in the common form

$$\frac{1}{\omega_n^2}\ddot{c} + \frac{2\zeta}{\omega_n}\dot{c} + c = R$$

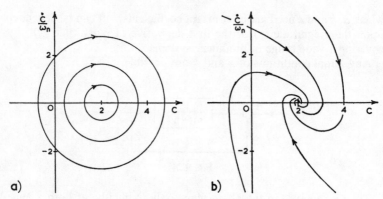

Fig. 9.25 Typical phase portraits for a second-order system:
(a) $\zeta = 0$, (b) $\zeta = 0.5$.

A change of variable here, putting $\tau = \omega_n t$, gives a simplification in plotting on the phase-plane, since it makes the lengths of the axes required roughly equal.

With $\tau = \omega_n t$; $\dfrac{dc}{dt} = \omega_n \dfrac{dc}{d\tau}$ and $\dfrac{d^2 c}{dt^2} = \omega_n^2 \dfrac{d^2 c}{d\tau^2}$

$$\therefore \frac{d^2 c}{d\tau^2} + 2\zeta \frac{dc}{d\tau} + c = R$$

The phase-plane plot is now $\dfrac{\dot{c}}{\omega_n}$ against c. For any 'phase portrait' the system constants ω_n and ζ must be fixed, any change requiring a complete new portrait. The portrait, or complete set of curves, is necessary to allow for *any* initial conditions, $\dot{c}(0)$ and $c(0)$.

A typical portrait is plotted in Fig. 9.25, for (a) $\zeta = 0$, and for (*b*) $\zeta = 0.5$. The initial conditions determine which locus to start on: (*a*) since it neither approaches nor diverges from the singular point is just stable, S.H.M. (the singular point here is taken as 2, 0 for a step input of magnitude 2); (*b*) is stable and settles at the singular point. The direction of the arrows indicate increasing time.

When $\dot{x}$ is positive, x is increasing positively and vice versa; there is thus a definite limit to the direction of the contours as indicated in Fig. 9.26.

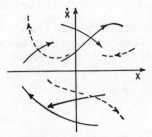

Fig. 9.26 ——▶—— possible trajectories, ---▶--- impossible.

9.8.2 Construction of the Phase Portrait

The phase portrait can be constructed in the following ways:

    (i) direct from an analogue computer;
    (ii) by numerical methods;
    (iii) solution of the differential equations;
    (iv) by graphical construction, in particular the 'isocline method'.

Consider (iv) in particular, but with passing reference to (iii). Method (iii) could be achieved the hard way by solving for $\dot{x}$, and then x as functions of t, from which the portrait could be plotted, with t as a parameter.

A more direct solution, however, can be achieved thus:

put
$$\frac{dx}{d\tau} = y$$

$$\therefore \frac{dy}{d\tau} + 2\zeta\frac{dx}{d\tau} + x = k$$

$$\therefore \frac{1}{y}\frac{dy}{d\tau} + 2\zeta + \frac{x - k}{y} = 0$$

$$\therefore \frac{dy}{dx} + 2\zeta + \frac{x - k}{y} = 0$$

This is a first-order differential equation relating y to x, and can be solved to plot $y = \dfrac{dx}{d\tau}$ to a base of x.

A graphical construction, however, can now be developed from this equation. Rearranging

$$y = \frac{-1}{2\zeta + \dfrac{dy}{dx}} \cdot x + \frac{k}{2\zeta + \dfrac{dy}{dx}}$$

For constant values of $\dfrac{dy}{dx}$ this is the equation of a straight line, which when plotted is called an 'isocline'. The isocline for, say, $\dfrac{dy}{dx} = 2$ is drawn crossed with construction lines of actual slope 2, thus the trajectory must cut the isocline at the same slope as the marks. Note that the slope of the isocline is not equal to the slope of the contour.

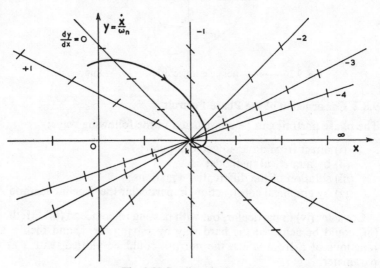

Fig. 9.27 Isoclines for $\zeta = 0.5$.

A typical set of isoclines is shown in Fig. 9.27, plotted for $\zeta = 0.5$ and $k = 2$. New sets of isoclines must be plotted for different values of ζ, but different values of k simply shift the whole phase portrait to a new singular point.

Starting from *any* point on the plot the trajectory is drawn by freehand estimating of the direction indicated by the predetermined slope given by the isoclines.

9.8.3 Application to a Simple Nonlinear Problem

Two examples are considered here:

   (i) a continuous mathematically defined nonlinearity;
   (ii) a piecewise-linear nonlinearity.

Example 1

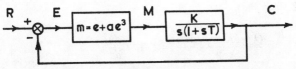

Fig. 9.28

$$c = r - e = \frac{K}{D(1 + DT)} \cdot (e + ae^3)^*$$

$$\therefore rD(1 + DT) = K(e + ae^3) + D(1 + DT)e$$

$$\therefore \ddot{r} + T\ddot{r} = K(e + ae^3) + y + T\frac{dy}{dt}$$

where $y = \dfrac{de}{dt}$. This could be normalised, putting $\sqrt{\dfrac{K}{T}} = \omega_n$, substituting $\tau = \omega_n t$ and putting $y = \dot{e}/\omega_n$.

As an example, plot the phase-plane trajectory for a ramp input, $r = Qt$, with the system initially dead, i.e. $y = e = 0$ when $t = 0$.

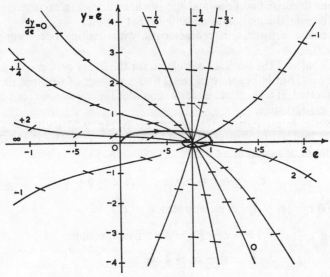

Fig. 9.29 Isoclines for system of example 1 with $Q = 2$; $T = 4$; $a = 0.5$; and $K = 2$.

Thus $\dot{r} = Q$ and $\ddot{r} = 0$, which substituting in the above gives

$$Q = K(e + ae^3) + y + T\frac{dy}{dt}$$

* The D, used here instead of s, can be interpreted as short-hand for $\dfrac{d}{dt}$, the nonlinearity making the Laplace transform invalid.

dividing by $y = \dfrac{de}{dt}$

$$\frac{Q - K(e + ae^3)}{y} = 1 + T\frac{dy}{de}$$

$$\therefore \; y = \frac{-K(e + ae^3)}{1 + T\dfrac{dy}{de}} + \frac{Q}{1 + T\dfrac{dy}{de}}$$

For constant values of $\dfrac{dy}{de}$ a set of isoclines can be drawn as in Fig. 9.29 and the trajectory for the given initial conditions sketched in as shown.

A useful exercise for the student is to construct the phase portrait for Van Der Pol's equation, given in Section 9.3, when discussing limit cycles, and to show the existence of the stable amplitude oscillation.

Example 2. For piecewise linear systems the phase portrait is split into various linearised sections and the isoclines drawn in each section for the differential equation applicable in that section.

Consider a saturation characteristic with limiting occurring when $|m| = 2$.

For $|m| < 2$ the isoclines calculated for the linear system of Fig. 9.27 will apply, but $|m|$ cannot exceed 2 and a new set of isoclines must be constructed. It is therefore easier to consider the trajectories for c rather than e, if possible.

Let the linear part of the system be as Fig. 9.24 with $K = 2$ and $T = 4$, and let the saturation level be $m = \pm 2$. Consider the case of a step input of magnitude 6:

$$e = r - c = 6 - c \qquad \therefore \; c = 6 - e$$

(i) For $|e| < 2$, i.e. unsaturated, $8 > c > 4$:

$$T\ddot{c} + \dot{c} + Kc = Kr, \quad \text{from Section 9.8.1}$$

$$\therefore \; 4\ddot{c} + \dot{c} + 2c = 2 \cdot 6$$

Put $$y = \frac{dc}{dt}$$

$$\therefore \; 4\frac{dy}{dt} + y + 2c = \left(4\frac{dy}{dc} + 1\right)y + 2c = 12$$

Hence $$y = \frac{-2}{4\dfrac{dy}{dc} + 1}\,c + \frac{12}{4\dfrac{dy}{dc} + 1}$$

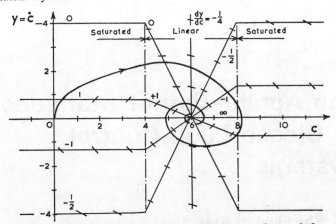

Fig. 9.30 Isoclines for the piecewise linear system of example 2.

(ii) Saturated, $e > +2$, $\therefore$ $m = +2$ and $c < 4$.

then
$$c = \frac{K}{D(1 + DT)} \cdot 2$$

$$\therefore T\ddot{c} + \dot{c} - 2K = 0$$

$$\therefore 4\ddot{c} + \dot{c} - 4 = 0$$

Hence
$$4\frac{dy}{dt} + y - 4 = 0$$

$$\therefore 4\frac{dy}{dc} + 1 - \frac{4}{y} = 0$$

$$\therefore y = \frac{4}{4\frac{dy}{dc} + 1}$$

(iii) Saturated, $e < -2$, $\therefore$ $m = -2$ and $c > 8$.

$$\therefore y = \frac{-4}{4\frac{dy}{dc} + 1}$$

These isoclines are plotted on Fig. 9.30, the trajectory shown being for initial conditions of $\dot{c} = 0$ and $c = 0$. Again these trajectories could have been normalised, putting $\omega_n = \sqrt{\frac{K}{T}}$ and putting $y = \frac{\dot{c}}{\omega_n}$.

10
The Application of Statistics to Closed Loop Control Systems

10.1 The Field of Statistical Application

This chapter is included with the object of presenting to readers who have a working knowledge of classical control system design, the basic principles and applications of statistics. No attempt is made to make this an authoritative article on design using statistical means, but more to present the background necessary for reading the many papers now being published. The backbone of statistical design was laid down by Wiener and Phillips,* and extended by such as Bode, Shannon, Westcott, Kalman, etc., although the text on basic statistical principles recommended here is Papoulis (ref. 20), which is one of many examples of the spread of communications theory into automatic control.

The application of statistics can be split into three categories, viz.:

(i) As a measuring technique for finding system transfer functions, achieved by Laplace transforming the crosscorrelation function relating the system output to a white noise input.

(ii) As a means of designing systems for optimum performance, or alternatively calculating the minimum possible error so that the quality of a conventional trial and error design can be assessed. The significance here is that the system is designed with its actual input in mind instead of an artificial input such as a step or ramp, with consideration of the effects of noise and disturbances.

(iii) As an extension of (i) the transfer function, or some other representative function, can be measured while the system is functioning normally, provided the magnitude of the injected white noise is such as not to produce objectionable outputs. Such measured results can be used to alter the system parameters in some manner so as to keep the transfer function constant, as may be necessary with wild

* See, for example, *The Extrapolation, Interpolation and Smoothing of Stationary Time Signals*, N. Wiener, Wiley, 1948; and *Theory of Servomechanisms*, James, Nichols and Phillips, McGraw-Hill, 1947.

changes in the plant, as often occurs in flight and process controls. The process of continuous parameter adjustment as a result of a measured performance index is called 'Adaptive' or 'Self-adaptive' control (ref. 24). It must be realised that the whole object of closed loop control is to correct the system against parameter changes, so that Adaptive control is only necessary when parameter changes are too excessive for the conventional control system to cope with.

This chapter concentrates on the background leading up to a working understanding of (ii) above, often referred to as 'Analytical Design' (see Truxal (ref. 1) and Gibson (ref. 19)), which can be broken down into:

(a) determination of the statistical properties of the signal, noise and disturbances;

(b) formulation of an error criterion on which to base the quality of a system;

(c) design of an optimum system in the light of (a) and (b).

Design can be achieved either by calculating the values of various adjustable parameters (integrator time constants, etc.), or by calculating the required transfer function and then attempting to synthesise this, thus introducing extra problems of physical realisability.

Following this section, Section 10.2 is devoted to definitions of the basic statistical terms necessary to formulate the signals and errors.

Section 10.3 discusses the weighting function as a means of relating, using the convolution integral, the system output and input as functions of time. Analytical design can thus be considered in both the time and frequency domains.

Section 10.4 outlines various possible error criteria necessary if an optimum design is to be achieved.

Section 10.5 defines the correlation functions which are used for time domain calculations.

Section 10.6 defines the power spectral density functions which are used for frequency domain calculations.

10.2 Some Important Statistical Terms

10.2.1 Stationary Time Series

A stationary time series is a continuous function which has the same long-term properties at any time. Since it is necessary for the signal to exist for a sufficiently long time in order to determine the statistical properties, it is the long-term properties that must be constant. The short-term statistical properties are relatively meaningless and will not be constant.

A random 'noise' signal from an electronic device is an example of a stationary time series; conversely, a single transient is an obvious example of a non-stationary time series.

The major principles of statistical design rotate around stationary signals so that a system ideally designed to follow a randomly varying input, such as the target signal input to a radar set, may have a poor transient response, as when positioning after slewing.

10.2.2 Ergodicity

Statistical properties can be determined in two ways:

   (i) by considering a large number of similar signals at any one instant in time, termed an 'ensemble value';
   (ii) by considering one signal at a number of intervals of time, termed a 'time value'.

If these two properties are equal, the function, or set of functions, is said to be 'ergodic'. To be ergodic a function must be a stationary signal, although a stationary process need not be ergodic.

10.2.3 Stochastic (or Random) Signals (or Processes)

A signal, the future value of which cannot be predicted with certainty is called a 'stochastic' signal. If an ensemble of signals is derived, the

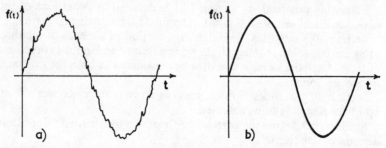

Fig. 10.1 (*a*) Stochastic signal, e.g. filtered noise; (*b*) predictable signal, e.g. a sine wave.

values of the function at some time *t* form a random variable, the probability distribution of which can be defined. Thus the future behaviour of a single stochastic signal is predicted on the basis of a known ensemble probability. If the statistical properties of the random variable at *t* are the same at all other times the signal is called a 'stationary stochastic process'.

10.2.4 Probability and Probability Density Functions

If the variable has only a fixed number of possibilities, i.e. it is only possible to throw 1, 2, ... or 6 on a dice with no in-betweens, then the probability of throwing, say, 3, is simply one in every six throws, provided there are enough throws, i.e. $P(3) = \frac{1}{6}$. If the process is continuous, with the variable able to take up any value, either positive or negative, then the chance of achieving any one particular value is zero, since it theoretically only exists for an infinitesimally small time, i.e. the probability is $1/\infty = 0$. The probability, however, of a function, $f(t)$, lying between f and $f + \delta f$ is finite, but is still dependent upon δf; f is a particular value of $f(t)$. If the increment is small enough that $f(t)$ can be considered constant in this period, then a probability density can be formulated. Thus the first probability density function, $p_1(f)$, is the probability that $f(t)$ lies between f and $f + \delta f$, divided by δf. As $\delta f \rightarrow 0$ $p_1(f)$ takes on a finite value which in general is a function of f and t. For stationary processes, however, the properties are independent of time so that $p_1(f)$ is simply a function of f.

Mathematically stated:

$$P(f < f(t) < f + \delta f) = p_1(f) \cdot \delta f$$

From this definition

$$p_1(f) \geqslant 0$$

and

$$\int p_1(f) \, df = 1$$

where the limits of integration include all possible values of f. It follows that the probability that $f_a < f(t) < f_b$ is

$$P(f_a < f(t) < f_b) = \int_{fa}^{fb} p_1(f) \, df$$

and that the probability of $|f(t)| > A$ is

$$P(|f(t)| > A) = \int_{-\infty}^{-A} p_1(f) \, df + \int_{A}^{\infty} p_1(f) \, df$$

In order to completely define a stochastic process an infinite set of probability density functions is required, but fortunately, in using such criteria as the mean square error, the first two will suffice. The second probability density function, then, is the probability that $f(t)$ lies between f_1 and $f_1 + \delta f$ at time t_1 *and* between f_2 and $f_2 + \delta f$ at t_2. Since the second probability density function is a function of f_1, f_2, t_1 and t_2 it is written

$$p_2(f_1, f_2, t_1, t_2)$$

For a stationary signal, p_2 is independent of t, but must be dependent upon $t_2 - t_1 = \tau$.

Thus the second probability density function for a stationary signal is written $p_2(f_1, f_2, \tau)$.

The first and second probability density functions are, for stationary signals, related thus:

$$p_1(f) = \int_{-\infty}^{\infty} df_2 \cdot p_2(f_1, f_2, \tau)$$

10.2.5 Some Examples of $p_1(f)$

To get some idea of the shape of a graph of $p_1(f)$, consider the following
example shown in Fig. 10.2. Let $f(t)$ be represented by uniformly spaced
beads on wires which are spaced in height by increment $\triangle f$. If the whole
set-up is now inverted so that the beads run to one end they will repre-

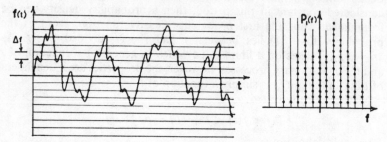

Fig. 10.2 Simple representation of first probability density function.

sent the basic shape of the curve $p_1(f)$ to a base of f. The most important
common form of $p_1(f)$ is called the 'Normal' or 'Gaussian' distribution:

$$p_1(f) = \frac{1}{\sqrt{2\pi}\,\sigma}\,e^{-\frac{f^2}{2\sigma^2}}$$

where σ is the standard deviation, which is a measure of the dispersion
of f about zero.

σ^2 is called the variance, λ, which is the mean value of $f(t)^2$. In this
case σ is the r.m.s. value.

If the mean value of $f(t)$ is not zero but is μ (in Section 10.2.6, $\mu = \overline{f(t)}$),

then $\qquad\qquad \lambda = \sigma^2$ is the mean value of $(f(t) - \mu)^2$

and $\qquad\qquad p_1(f) = \frac{1}{\sqrt{2\pi}\,\sigma}\,e^{-\frac{(f-\mu)^2}{2\sigma^2}}$

σ is no longer the r.m.s. value.

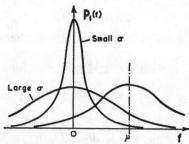

Fig. 10.3 Gaussian distribution. In each case the area under the curve is unity.

10.2.6 Average, Mean or Expected Values

Considering the general case where a large number of long samples of similar signals are available for analysis, the average value can have two meanings:

(a) The ensemble average, written $\widetilde{f(t)}$, i.e. the average value of the various ensemble signals at any one time. From the definition of the first probability density function (E means 'expected' value),

$$\widetilde{f(t)} = \int_{-\infty}^{\infty} p_1(f) \cdot f \, df = E[f(t)]$$

This is also called the first moment of the probability density function.

(b) The time average, written $\overline{f(t)}$, i.e. the average value of any one signal over a large time interval.

$$\overline{f(t)} = \lim_{T \to \infty} \frac{1}{2T} \int_{-T}^{+T} f(t) \, dt$$

and for a stationary signal

$$\overline{f(t)} = \lim_{T \to \infty} \frac{1}{T} \int_{0}^{+T} f(t) \, dt$$

By the ergodic property of the stationary stochastic signal

$$\widetilde{f(t)} = \overline{f(t)}$$

10.2.7 The Mean Square Error

The mean square value of the error is of particular importance since it is used as a common criterion for system quality.

In general

$$\overline{f^2(t)} = \lim_{T \to \infty} \frac{1}{2T} \int_{-T}^{T} f^2(t) \, dt$$

or by the ergodic property

$$\overline{f^2(t)} = \int_{-\infty}^{\infty} p_1(f) \cdot f^2 \, df$$

which is the second moment of the probability density function.

$$\text{If a system output} = f_o(t)$$
$$\text{and the desired output} = f_d(t)$$

the mean square error is defined as

$$\overline{y_e^2(t)} = \lim_{T \to \infty} \frac{1}{2T} \int_{-T}^{+T} [f_d(t) - f_o(t)]^2 \, dt$$

10.2.8 The Integral Square Error

The integral square error is given by

$$I_e = \int_{-\infty}^{\infty} [f_d(t) - f_0(t)]^2 \, dt$$

The solution of integrals of this form is particularly arduous and has merited particular attention. Using Parseval's theorem:

if $x_1(t)$ transforms to $X_1(s)$

and $x_2(t)$ transforms to $X_2(s)$.

Then if $I = \int_{-\infty}^{\infty} dt \; x_1(t) \, x_2(t)$

the equivalent integral in the s domain is

$$I = \frac{1}{2\pi j} \int_{-j\infty}^{j\infty} ds \; X_1(-s) \, X_2(s)$$

If $f_d(t) - f_0(t) = y_e(t)$ and $y_e(t)$ transforms to $Y_e(s)$,

$$I_e = \frac{1}{2\pi j} \int_{-j\infty}^{j\infty} ds \; Y_e(-s) \, Y_e(s)$$

The real significance now comes if $Y_e(s)$ can be expressed as a rational polynomial,

$$Y_e(s) = \frac{C(s)}{D(s)} = \frac{c_0 + c_1 s + c_2 s^2 + \ldots + c_{n-1} s^{n-1}}{d_0 + d_1 s + d_2 s^2 + \ldots + d_{n-1} s^{n-1} + d_n s^n}$$

so that $I_e = \dfrac{1}{2\pi j} \displaystyle\int_{-j\infty}^{j\infty} ds \; \dfrac{C(-s)C(s)}{D(-s)D(s)}$

This integral has been tabulated for various values of n.[*]

This form of integral has a further application in calculating the mean square error in terms of the power spectral density, as will be mentioned later, Section 10.6.3.

10.3 The Weighting Function and the Convolution Integral

The weighting function, denoted $w(t)$ or $g(t)$, is the inverse transform of the transfer function, $G(s)$.

Further, if the input to the system is a unit impulse, $\delta(t)$, then the output in transform notation is

$$F_0(s) = G(s) \cdot 1$$

since $\mathscr{L}[\delta(t)] = 1$

Hence $f_0(t) = \mathscr{L}^{-1}[G(s)]$

$$= w(t)$$

[*] See, for instance, *Analytical Design of Linear Feedback Controls*, Newton, Gould and Kaiser, Wiley, 1964.

Thus the weighting function is alternatively called 'the unit impulse response' of a system; in circuit analysis it is called 'Greens function', with the symbol $h(t)$.

Consider now the general case of Fig. 10.4. Let the input, $f_i(t)$, be considered as comprising an infinite number of pulses of width, $\delta\tau$, and height, $f_i(t)$, so that the continuous function can be represented by a succession of impulses, spaced $\delta\tau$ apart, of weighting, $f_i(t) . \delta\tau$.

The contribution to the output at time t due to the impulse applied τ

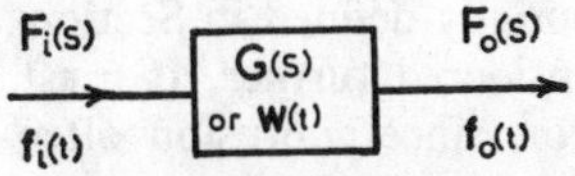

Fig. 10.4]

seconds earlier, i.e. at $t - \tau$, will be the 'weighted' value at τ seconds times the magnitude of the impulse applied at $t - \tau$, viz.:

$$\Delta f_o(t) = w(\tau) . f_i(t - \tau) \, \delta\tau$$

The total output is the sum of the $\Delta f_o(t)$ terms due to impulses applied at $t - \tau$, $t - \tau + \delta\tau$, $t - \tau + 2\delta\tau$, etc., the summation being performed in the limit by an integral giving

$$f_o(t) = \int_0^\infty w(\tau)f_i(t - \tau) \, d\tau$$

Considering only bounded input signals, i.e. $f_i(t) = 0$ for $t < 0$, then $f_i(t - \tau) = 0$ for $\tau > t$ and therefore

$$f_o(t) = \int_0^t w(\tau)f_i(t - \tau) \, d\tau$$

Further, since $w(\tau)$ is zero for $\tau < 0$, the lower limit can be $-\infty$; this can simplify certain manipulations, e.g. in Section 10.5.1.

This is often written

$$f_o(t) = w(t) * f_i(t)$$

compared with $\qquad F_o(s) = G(s) . F_i(s)$

It is this ease of manipulation which makes the Laplace transform so attractive to control system design in general, the direct multiplication of the transfer functions being equivalent to '*convolution*' in the time domain.

The integral for $f_o(t)$ above is called a 'Convolution Integral'; this particular example of convolution, relating f_i and f_o is also called the 'superposition integral'. It can also be shown that

$$f_o(t) = \int_0^t w(t - \tau)f_i(\tau) \, d\tau$$

10.4 Error Criteria

A number of error criteria have been formulated, but two in particular
have gained prominence, their acceptance being mainly on the grounds
that the mathematics involved is simpler and better understood than for
alternative criteria. These two criteria are:

(i) the integral square error, I_e
(ii) the mean square error, $\overline{y_e^2(t)}$.

These terms were initially defined in Sections 10.2.7 and 10.2.8, but
they must now be considered further. It must be clearly understood
what is meant by error, since confusion often arises from the loose
usage of the term error signal to mean actuating signal. This is indicated
diagrammatically in Fig. 10.5.

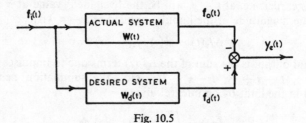

Fig. 10.5

The error $y_e(t) = f_d(t) - f_o(t)$* where $f_o(t)$ is the system output and
$f_d(t)$ the desired output.

It is useful to define the desired transfer function,† i.e. $G_d(s)$ and its
equivalent weighting function, $w_d(t)$.

$$F_d(s) = G_d(s) \cdot F_i(s)$$

and
$$f_d(t) = w_d(t) * f_i(t)$$

$$= \int_0^t w_d(\tau) f_i(t - \tau) \, d\tau$$

Repeating the defining equations of Sections 10.2.7 and 10.2.8:

$$I_e = \int_{-\infty}^{\infty} y_e^2(t) \, dt$$

and
$$\overline{y_e^2(t)} = \lim_{T \to \infty} \frac{1}{2T} \int_{-T}^{T} y_e^2(t) \, dt$$

inspection of these two equations shows that I_e gives a net measure and

* The general symbol for error recommended by the A.I.E.E. is $y_e(t)$; $e(t)$, used in
many articles on statistics, is avoided, since $E(s)$ and $e(t)$ have been used to denote
the actuating signal.

† The desired transfer function is also called the 'ideal transfer function', $G_i(s)$,
and the desired output, the 'ideal value', $I(s)$.

$\overline{y_e^2(t)}$ a mean measure. I_e, then, is used as the criterion for systems subjected to transient inputs, which of course are not stochastic and are only loosely included under the heading of this chapter. $\overline{y_e^2(t)}$ is the common criterion chosen for systems subjected to stochastic inputs.

Because of the square law characteristic these criterion 'weight' against large errors, whether positive or negative, which is to some extent desirable since small errors can usually be tolerated. A more ideal weighting characteristic is shown in Fig. 10.6, in which a small error is allowable and in which errors above a certain level are considered as beyond the scope of the system and will just have to be tolerated, for instance, errors during slewing of an aerial. The mathematical difficulty of using such a characteristic makes it impractical.

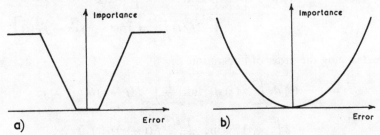

Fig. 10.6 Error-weighting characteristics: (*a*) ideal, (*b*) mean square.

The weighting characteristic of the mean square error is shown for comparison.

Thus, while it is generally accepted that the mean square error is the best practical criterion, for stochastic signals, it does have limitations. This error criterion will produce an optimum system, but will not have any particular significance as to the frequency range of the resulting error, i.e. it will pay little attention to stability requirements. Clearly an optimum design cannot be unstable, but it may have a low damping factor so as to leave inadequate practical tolerance for inherent variations in the plant being controlled.

Other error criteria used are higher orders of $y_e(t)$ than 2 and the mean modulus of error, i.e.

$$\overline{|y_e(t)|} = \lim_{T \to \infty} \frac{1}{2T} \int_{-T}^{T} |y_e(t)| \, dt$$

Another criterion that may be encountered is the 'integral of time times the amplitude of error' (I.T.A.E.) defined as

$$\text{I.T.A.E.} = \int_{0}^{\infty} t |y_e(t)| \, dt$$

10.5 Analysis in the Time Domain

10.5.1 The Mean Square Error in Terms of Correlation Functions

$$\overline{y_e^2(t)} = \lim_{T \to \infty} \frac{1}{2T} \int_{-T}^{T} [f_d(t) - f_o(t)]^2 \, dt$$

$$= \lim_{T \to \infty} \frac{1}{2T} \int_{-T}^{T} [f_d(t) - \int_{-\infty}^{\infty} w(\tau) f_i(t - \tau) \, d\tau]^2 \, dt*$$

In squaring, a term such as $\left[\int x(t) \, dt\right]^2$ occurs. By introducing a dummy time variable, λ, this can be written $\int x(t) \, dt \cdot \int x(\lambda) \, d\lambda$. Thus

$$\overline{y_e^2(t)} = \lim_{T \to \infty} \frac{1}{2T} \int_{-T}^{T} dt \left[\int_{-\infty}^{\infty} w(\tau) f_i(t - \tau) \, d\tau \int_{-\infty}^{\infty} w(\lambda) f_i(t - \lambda) \, d\lambda \right.$$
$$\left. - 2 f_d(t) \int_{-\infty}^{\infty} w(\tau) f_i(t - \tau) \, d\tau + f_d^2(t) \right]$$

rearranging the order of integration

$$\overline{y_e^2(t)} = \int_{-\infty}^{\infty} w(\tau) \, d\tau \int_{-\infty}^{\infty} w(\lambda) \, d\lambda \lim_{T \to \infty} \frac{1}{2T} \int_{-T}^{T} f_i(t - \tau) f_i(t - \lambda) \, dt$$
$$- 2 \int_{-\infty}^{\infty} w(\tau) \, d\tau \lim_{T \to \infty} \frac{1}{2T} \int_{-T}^{T} f_i(t - \tau) f_d(t) \, dt$$
$$+ \lim_{T \to \infty} \frac{1}{2T} \int_{-T}^{T} f_d^2(t) \, dt$$

Now define a CORRELATION FUNCTION as an average of the product of two time functions, such that in general

$$\phi_{ab}(\tau) = \lim_{T \to \infty} \frac{1}{2T} \int_{-T}^{T} f_a(t - \tau) f_b(t) \, dt$$

$$= E \left[f_a(t) f_b(t - \tau) \right]$$

Then $\overline{y_e^2(t)} = \int_{-\infty}^{\infty} w(\tau) \, d\tau \int_{-\infty}^{\infty} w(\lambda) \, d\lambda \cdot \phi_{ii}(\tau - \lambda)$
$$- 2 \int_{-\infty}^{\infty} w(\tau) \, d\tau \cdot \phi_{id}(\tau) + \phi_{dd}(0)$$

While the algebra involved in this derivation is rather messy, the form of the solution is more reasonable since it involves only the system weighting function and the correlation functions.

Thus in the time domain it is sufficient to know only the correlation functions to sufficiently define the signals for a complete knowledge of the mean square error.

* The limits on the convolution integral are here $\pm \infty$ (see Section 10.3).

There are two types of correlation function.

(i) The AUTOCORRELATION FUNCTION, $\phi_{11}(\tau)$, i.e. the autocorrelation of the input $\phi_{ii}(\tau - \lambda)$ and of the desired output $\phi_{dd}(0)$, which is of argument 0 since no shift is involved.

(ii) The CROSSCORRELATION FUNCTION, $\phi_{ab}(\tau)$, i.e. the cross-correlation between the input and the desired output, $\phi_{id}(\tau)$.

10.5.2 Autocorrelation Functions

When considering a single signal it is immaterial whether the shift, τ, is forward or backward, so that

$$\phi_{11}(\tau) = \lim_{T \to \infty} \frac{1}{2T} \int_{-T}^{T} f(t) f(t + \tau)\, dt$$

$$= \lim_{T \to \infty} \frac{1}{2T} \int_{-T}^{T} f(t) f(t - \tau)\, dt$$

In words, the autocorrelation function is the time average of the product of a signal at a specified time and its value at a time shifted by τ seconds.

By ergodic hypothesis, $\phi_{11}(\tau)$ is also the ensemble average

$$\phi_{11}(\tau) = \overline{f(t) . f(t + \tau)}^{*}$$

$$= \int_{-\infty}^{\infty} df_1 \int_{-\infty}^{\infty} df_2 . f_1 . f_2 . p_2(f_1, f_2, \tau)$$

where $p_2(f_1, f_2, \tau)$ is the second probability density function defined in Section 10.2.4.

The basic properties of autocorrelation functions are:

(i) $\phi_{11}(\tau) = \phi_{11}(-\tau)$.

(ii) $\phi_{11}(0)$ is the maximum value.

(iii) $\phi_{11}(0)$ is termed the 'average power' of the signal since it is the r.m.s. value squared. The loose usage of the word power stems from assuming the signal to be a voltage applied to a 1 ohm resistor.

(iv) If $f(t)$ contains periodic components, including d.c., $\phi_{11}(\tau)$ contains the same frequency components, but since time origin is meaningless, no phase information is preserved.

(v) Since in practice many signals are easier measured as derivatives, i.e. velocity instead of position, the autocorrelation function of $f'(t)$, i.e. $\dfrac{df(t)}{dt}$, can be particularly useful, thus:

$$\phi_{1'1'}(\tau) = \lim_{T \to \infty} \frac{1}{2T} \int_{-T}^{T} f'(t) f'(t + \tau)\, dt$$

---

* The autocorrelation function of a non-stationary signal can be defined as

$$\phi_{11}(\tau, t_1) = \overline{f(t_1) . f(t_1 + \tau)}$$

Integration by parts yields

$$\phi_{1'1'}(\tau) = - \frac{d^2}{d\tau^2} \left\{ \lim_{T \to \infty} \frac{1}{2T} \int_{-T}^{T} f(t) f(t + \tau) \, dt \right\}$$

$$= - \frac{d^2}{d\tau^2} \phi_{11}(\tau)$$

(vi) If $f(t)$ contains two components, $f_s(t) + f_n(t)$, e.g. signal plus noise, then

$$\phi_{11}(\tau) = \lim_{T \to \infty} \frac{1}{2T} \int_{-T}^{T} (f_s(t) + f_n(t))(f_s(t + \tau) + f_n(t + \tau)) \, dt$$

$$= \underbrace{\phi_{ss}(\tau) + \phi_{nn}(\tau)}_{\text{autocorrelations}} + \underbrace{\phi_{ns}(\tau) + \phi_{sn}(\tau)}_{\text{crosscorrelations}}$$

In the general analysis of systems subjected to a noisy input signal, it can be assumed that the signal and noise are uncorrelated, i.e. they stem from unrelated sources, so that $\phi_{ns}(\tau) = \phi_{sn}(\tau) = 0$. This is not always true, as, for example, when the noise stems from a mechanical vibration excited by the signal.

Example 1

$$f(t) = E \sin (\omega t + \psi)$$

$$\phi_{11}(\tau) = \lim_{T \to \infty} \frac{1}{2T} \int_{-T}^{T} E \sin (\omega t + \psi) \, . \, E \sin (\omega(t + \tau) + \psi) \, dt$$

Since the integrand is periodic it can be evaluated over one cycle.

$$\therefore \ \phi_{11}(\tau) = \frac{\omega}{2\pi} E^2 \int_{0}^{\frac{2\pi}{\omega}} \sin (\omega t + \psi) \sin (\omega t + \psi + \omega \tau) \, dt$$

$$= \frac{E^2}{2} \cos \omega \tau$$

This confirms (iv) above. It can also be extended to non-sinusoidal waveforms by Fourier series expansion, although the original waveform may not be reproduced due to loss of phase information.

Example 2. If $f(t)$ is a random wave, then $\phi_{11}(\tau)$ will define any one random stochastic signal from another, so that from a practical viewpoint this is a desirable function to measure. However, if certain features of the signal are known, $\phi_{11}(\tau)$ can be calculated.

For the wave of Fig. 10.7 the signal is constant for a time interval, Δt. Values of the amplitude of successive steps are independent and have a normal distribution. If A_n is the amplitude of any step, then

$$P(f < A_n < f + \delta f) = p_1(f) \delta f$$

$$= \frac{1}{\sqrt{2\pi}\sigma} e^{-\frac{f^2}{2\sigma^2}} \, . \, \delta f$$

σ is the standard deviation; $\sigma^2 = \lambda$, the variance, = mean value of $A_n{}^2$; μ, the mean value is assumed zero, i.e. equal possibility of positive or negative values.

Now
$$\phi_{11}(\tau) = \lim_{T \to \infty} \frac{1}{2T} \int_{-T}^{T} f(t)f(t + \tau)\, dt*$$

This integral will be solved by inspection in the following way:

(a) If $\tau = 0$, $\phi_{11}(0)$ = mean value of $f^2(t)$
$$= \lambda, \text{ by definition}$$

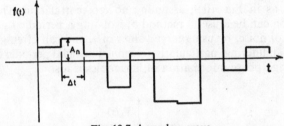

Fig. 10.7 A random wave.

(b) If $\tau > \Delta t$, there is no correlation between $f(t)$ and $f(t + \tau)$ since they cannot be in the same interval, thus
$$\phi_{11}(\tau) = 0$$

(c) If $0 < \tau < \Delta t$ the correlation between $f(t)$ and $f(t + \tau)$ is proportional to the portion of Δt during which $f(t)$ and $f(t + \tau)$ are at the same value.
$$\therefore \phi_{11}(\tau) \propto \frac{\Delta t - \tau}{\Delta t}$$

Since it is immaterial whether τ is positive or negative
$$\phi_{11}(\tau) \propto \frac{\Delta t - |\tau|}{\Delta t} \propto 1 - \frac{|\tau|}{\Delta t}$$

so that
$$\phi_{11}(\tau) = \lambda\left(1 - \frac{|\tau|}{\Delta t}\right) \quad \text{for} \quad |\tau| < \Delta t$$
$$= 0 \quad \text{for} \quad |\tau| > \Delta t$$

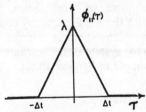

Fig. 10.8 Autocorrelation function of the wave of Fig. 10.7.

* Truxal (ref. 1) solves a similar problem an alternative way using the probability density function definition of $\phi_{11}(\tau)$.

An important feature of this result is that $\phi_{11}(\tau)$ is independent of the amplitude distribution, being dependent only upon the period and the variance. It is interesting, then, that a waveform which can only be $+V$ or $-V$ in the intervals Δt, the choice of plus or minus being random, has the same autocorrelation function as the worked example. The example has particular significance to 'chain codes' as mentioned in the conclusions, Section 10.7.

Example 3. From the solutions of examples 1 and 2 the autocorrelation of a noise signal containing a periodic component is given by simple addition, as in Fig. 10.9, assuming no crosscorrelation. Thus, auto-correlation can be a useful method of isolating a periodic signal in the presence of noise, by using large values of τ, and can often give better results than filtering, particularly for non-sinusoidal periodic functions, even though phase information of harmonics is lost.

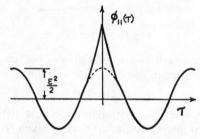

Fig. 10.9 Autocorrelation function of a sinusoid, $E \sin (\omega t + \psi)$, in the presence of noise.

Example 4. White noise. White noise is defined as being a completely random signal. It is a theoretical signal since by its definition it contains peaks of infinite power. Being purely random the autocorrelation function is zero for all τ, but $\phi_{11}(0)$ is defined as the mean power and therefore must be finite.

The autocorrelation function of white noise therefore is an impulse at $\tau = 0$.

Band-limited white noise, or 'pink' noise, which is the practical form of white noise with the high frequency components removed, has an autocorrelation function which is a pulse about $\tau = 0$ of short duration compared to the system time constants. The shape of the pulse is calculated as an example in Section 10.6.1.

10.5.3 Crosscorrelation Functions

The crosscorrelation function is a measure of the correlation existing between two functions. When measured as an ensemble average, the ensemble must be an ensemble of pairs of functions.

Let the two functions to be crosscorrelated be $f_a(t)$ and $f_b(t)$, then

$$\phi_{ab}(\tau) = \lim_{T \to \infty} \frac{1}{2T} \int_{-T}^{T} f_a(t) \cdot f_b(t + \tau) \, dt$$

$$= \lim_{T \to \infty} \frac{1}{2T} \int_{-T}^{T} f_a(t - \tau) f_b(t) \, dt$$

However, shifting $f_a(t)$ forward yields a different function:

$$\phi_{ba}(\tau) = \lim_{T \to \infty} \frac{1}{2T} \int_{-T}^{T} f_a(t + \tau) f_b(t) \, dt$$

$$= \lim_{T \to \infty} \frac{1}{2T} \int_{-T}^{T} f_a(t) f_b(t - \tau) \, dt$$

The principal properties of crosscorrelation functions of stationary signals are:

(a) $\phi_{ba}(\tau) = \phi_{ab}(-\tau)$ but $\phi_{ba}(\tau) \neq \phi_{ba}(-\tau)$, i.e. the crosscorrelation function is not an even function as is the autocorrelation function.

(b) $\phi_{ab}(\tau)$ does not necessarily have a maximum at $\tau = 0$ as does the autocorrelation function.

10.5.4 Measurement of System Functions by Crosscorrelation

In Section 10.5.1 it was shown that by selecting the mean square error as the error criterion the problem could be completely described by the weighting function of the system, the autocorrelation functions of the input and the desired output signals, together with the crosscorrelation function between the input and the desired output. Another interesting crosscorrelation function will be that between the input and the actual output.

$$\phi_{io}(\tau) = \lim_{T \to \infty} \frac{1}{2T} \int_{-T}^{T} f_i(t - \tau) f_o(t) \, dt$$

The output can be expressed as the convolution of the input and the system weighting function; it is necessary to introduce the time variable λ here so as to avoid confusion with the shift τ used in the crosscorrelation. Thus

$$f_o(t) = \int_{-\infty}^{\infty} w(\lambda) f_i(t - \lambda) \, d\lambda$$

Then $\quad \phi_{io}(\tau) = \lim_{T \to \infty} \frac{1}{2T} \int_{-T}^{T} f_i(t - \tau) \, dt \int_{-\infty}^{\infty} w(\lambda) f_i(t - \lambda) \, d\lambda$

$$= \int_{-\infty}^{\infty} w(\lambda) \, d\lambda \left[\lim_{T \to \infty} \frac{1}{2T} \int_{-T}^{T} f_i(t - \tau) f_i(t - \lambda) \, dt \right]$$

The bracketed term is the autocorrelation function of the input signal, of argument $(\tau - \lambda)$.

$$\therefore \; \phi_{io}(\tau) = \int_{-\infty}^{\infty} w(\lambda)\phi_{ii}(\tau - \lambda)\, d\lambda$$

The similarity between this expression and the convolution integral describing $f_o(t)$ can now be put to advantage, i.e.:

If $f_i(t)$ is a unit impulse, then $f_o(t) = w(t)$ from the definition of $w(t)$ as the unit impulse response, Section 10.3.

If $\phi_{ii}(\tau)$ is also a unit impulse, then $\phi_{io}(\tau)$ will be numerically equal to $w(t)$.

It has already been calculated that white noise has an autocorrelation function which is an impulse—Section 10.5.2 example 4.

Thus the system weighting function $w(t)$ can be measured by the cross-correlation of the input and output, provided the input is a white noise signal.

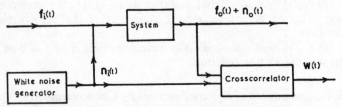

Fig. 10.10 Measurement of system weighting function.

This can be extended to even further practical use, as indicated in Fig. 10.10, since the system can be operating with its normal input without effecting the answer since there is no correlation between the applied noise and the true system output. Clearly the magnitude of the noise signal must be relatively small. Finally, remember that

$$G(s) = \mathscr{L}[w(t)]$$

10.6 Analysis in the Frequency Domain

10.6.1 Power Density Spectra

As with conventional linear theory the Laplace and Fourier transforms can be applied to statistical analysis. Thus the Laplace transform of the autocorrelation function is defined as

$$\Phi_1(s) = \int_{-\infty}^{\infty} \phi_{11}(\tau)e^{-s\tau}\, d\tau$$

and the Fourier transform is

$$\Phi_1(j\omega) = \int_{-\infty}^{\infty} \phi_{11}(\tau)e^{-j\omega\tau}\, d\tau$$

A double subscript notation is not necessary, although commonly used, as no equivalent cross function is used here.

Both these transforms are rather unusual, stemming from the fact that $\phi_{11}(\tau)$ is an even function, i.e. $\phi_{11}(\tau) = \phi_{11}(-\tau)$, so that the function is non-zero for negative time. Thus the Laplace transform is the 'two-sided Laplace transform' having the limits $-\infty$ to $+\infty$. Truxal (ref. 1) includes a description of this transform, the significant feature of which is that poles in the right-half plane can indicate existence for negative time and not necessarily instability. The rules for validity are much more rigorous.

Secondly, the Fourier transform is simplified by substituting for $e^{-j\omega\tau}$, thus

$$\Phi_1(j\omega) = \int_{-\infty}^{\infty} \phi_{11}(\tau)(\cos \omega\tau - j \sin \omega\tau) \, d\tau$$

but since $\phi_{11}(\tau)$ is an even function, the sine term is zero and

$$\Phi_1(j\omega) = \int_{-\infty}^{\infty} \phi_{11}(\tau) \cos \omega\tau \, d\tau$$

Further, this function is wholly real and even, so that

$$\Phi_1(j\omega) = \Phi_1(\omega)$$

and

$$\Phi_1(\omega) = \Phi_1(-\omega)$$

This term, $\Phi_1(\omega)$, is called the POWER SPECTRAL DENSITY, or POWER DENSITY SPECTRUM.*

The inverse transform is

$$\phi_{11}(\tau) = \frac{1}{2\pi} \int_{-\infty}^{\infty} \Phi_1(\omega) e^{j\omega\tau} \, d\omega$$

$$= \frac{1}{2\pi} \int_{-\infty}^{\infty} \Phi_1(\omega) \cos \omega\tau \, d\omega$$

Now for $\tau = 0$,

$$\phi_{11}(0) = \frac{1}{2\pi} \int_{-\infty}^{\infty} \Phi_1(\omega) \, d\omega$$

From the definition of $\phi_{11}(\tau)$ in Section 10.5.2

$$\phi_{11}(0) = \lim_{T \to \infty} \frac{1}{2T} \int_{-T}^{T} f^2(t) \, dt$$

$$= \text{mean square value,} \, \overline{f^2(t)}$$

As with the definition of the decibel, control theory again borrows from the telecommunications field, by assuming that $f(t)$ is a voltage applied

---

* There are inconsistencies in the definition of $\Phi_1(\omega)$; Lee, for example, defines $\Phi_1(\omega)$ as $\frac{1}{2\pi} \times$ the Fourier transform of $\phi_{11}(\tau)$, the reasons for which become clear in the following paragraphs.

across a 1 ohm resistor. The mean square value is then the mean power, thus $\frac{1}{2\pi}\int_{-\infty}^{\infty} \Phi_1(\omega)\, d\omega = \overline{f^2(t)}$, called the mean power. It can be inferred from this (no attempt is made to prove it) that $\Phi_1(\omega)$ is the power density, except that the area under the curve of $\Phi_1(\omega)$ plotted to a base of ω from $-\infty$ to $+\infty$ is, in fact, 2π times the mean power.

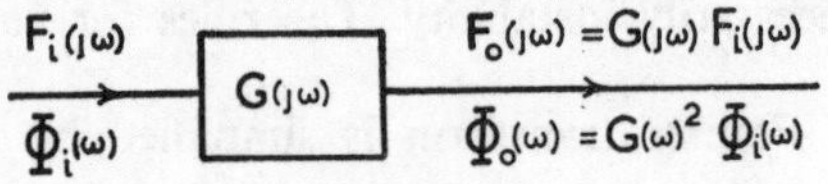

Fig. 10.11

The units of $\Phi_1(\omega)$ are (units of $f(t))^2$ per rad/s.
For the system of Fig. 10.11:

$$F_o(j\omega) = G(j\omega)F_i(j\omega) \ldots \text{(deterministic signals)}$$

and since $\Phi_1(\omega)$ depends upon the amplitude squared:

$$\Phi_o(\omega) = G(\omega)^2 \cdot \Phi_i(\omega) \ldots \text{(random signals)}$$

This can also be written

$$G(\omega)^2 = G(j\omega) \cdot G(j\omega)^* = G(j\omega) \cdot G(-j\omega)$$

Hence $\qquad \Phi_o(s) = G(s) \cdot G(-s) \cdot \Phi_i(s)$

which can help in the evaluation of certain integrals by the theorem of residues.

Example 1. White noise.

White noise is now easily defined as having a uniform content at all frequencies from $-\infty$ to $+\infty$, i.e. $\Phi_1(\omega)$ is a constant. The auto-correlation function given by the inverse transform of a constant is an impulse, as has previously been derived. Note that white noise is only a theoretical function since the mean power represented is infinite due to the range of ω extending to infinity.

Example 2. Band-limited white noise.

The power density spectra is constant for $-\frac{1}{T} < \omega < +\frac{1}{T}$ and zero for higher frequencies.

$$\therefore \ \Phi_1(\omega) = A \quad \text{for} \quad -\frac{1}{T} < \omega < +\frac{1}{T}$$

$$= 0 \quad \text{for} \quad |\omega| > \frac{1}{T}$$

The corresponding autocorrelation function is

$$\phi_{11}(\tau) = \frac{1}{2\pi} \int_{-\infty}^{\infty} \Phi_1(\omega) \cos \omega\tau \, d\omega$$

$$= \frac{A}{\pi} \int_0^{\frac{1}{T}} \cos \omega\tau \, d\omega$$

$$= \frac{A}{\pi} \left[\frac{\sin \omega\tau}{\tau} \right]_0^{\frac{1}{T}}$$

$$= \frac{A}{\pi} \frac{\sin \frac{\tau}{T}}{\tau}$$

for $\tau = 0$ $\phi_{11}(0) = \frac{A}{T\pi}$;

the complete function being plotted in Fig. 10.12.

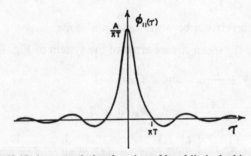

Fig. 10.12 Autocorrelation function of band-limited white noise.

10.6.2 Mean Square Error in Terms of Spectral Density Functions

Referring back to Fig. 10.5 of Section 10.4, the error

$$y_e(t) = f_d(t) - f_o(t)$$

$$\therefore \; Y_e(s) = F_d(s) - F_o(s)$$

$$= G_d(s)F_i(s) - G(s)F_i(s)$$

$$= (G_d(s) - G(s)) \cdot F_i(s)$$

If the power spectral density of the input is $\Phi_i(\omega)$ and of the error $\Phi_e(\omega)$,

$$\Phi_e(\omega) = |G_d(j\omega) - G(j\omega)|^2 \cdot \Phi_i(\omega)$$

N.B. $$|G_d(j\omega) - G(j\omega)| \neq G_d(\omega) - G(\omega)$$

and thus $$\overline{y_e^2(t)} = \frac{1}{2\pi} \int_{-\infty}^{\infty} |G_d(j\omega) - G(j\omega)|^2 \cdot \Phi_i(\omega) \, d\omega$$

10.6.3 Mean Square Error of a Noisy Signal

If the input comprises a signal component of power spectral density $\Phi_s(\omega)$ and a noise component $\Phi_n(\omega)$, then the noise output will all be error.

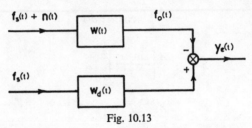

Fig. 10.13

The output power spectral density due to the noise only is

$$\Phi_{on}(\omega) = G(\omega)^2 \,.\, \Phi_n(\omega)$$

$$\therefore \; \overline{y_e^2(t)} = \frac{1}{2\pi} \int_{-\infty}^{\infty} [G(\omega)^2 \Phi_n(\omega) + |G_d(j\omega) - G(j\omega)|^2 \,.\, \Phi_s(\omega)] \, d\omega$$

assuming no correlation between signal and noise.

Example. Find the mean square error of the system of Fig. 10.13, when

$$G(s) = \frac{1}{1 + sT} \quad \text{and} \quad G_d(s) = 1$$

and

$$\Phi_s(\omega) = \frac{A}{a^2 + \omega^2} \quad \text{and} \quad \Phi_n(\omega) = B$$

$$\therefore \; \overline{y_e^2(t)} = \frac{1}{2\pi} \int_{-\infty}^{\infty} \left[B \left| \frac{1}{1 + j\omega T} \right|^2 + \frac{A}{a^2 + \omega^2} \left| 1 - \frac{1}{1 + j\omega T} \right|^2 \right] d\omega$$

This integral can be solved in three ways:

(i) By graphical integration, i.e. finding the area under the curve $G(\omega)^2 \,.\, \Phi_1(\omega)$ plotted to a base of ω.

(ii) By expressing the integrand as a polynomial and using standard tables (see the reference given in Section 10.2.8).

(iii) By expressing as a function of s, replacing $G(\omega)^2$ by $G(s) \,.\, G(-s)$ as previously explained and solving by contour integration, viz.:

$$\overline{y_e^2(t)} = \frac{1}{2\pi j} \int_{-j\infty}^{j\infty} \left[\frac{B}{(1 + sT)(1 - sT)} + \frac{AsT(-sT)}{(a^2 - s^2)(1 + sT)(1 - sT)} \right] ds$$

$$= \frac{1}{2\pi j} \int_{-j\infty}^{j\infty} \left[\frac{B}{T^2} \,.\, \frac{1}{\left(s + \frac{1}{T} \right)\left(-s + \frac{1}{T} \right)} \right.$$

$$\left. - A \,.\, \frac{s^2}{(s + a)(-s + a)\left(s + \frac{1}{T} \right)\left(-s + \frac{1}{T} \right)} \right] ds$$

$$= \Sigma \text{ residues in the L.H. } s\text{-plane}$$

$$= \frac{B}{T^2} \times \text{residue of first term at } s = -\frac{1}{T}$$

$$\quad - A \times \text{residues of second term at } s = -a \text{ and } s = -\frac{1}{T}$$

$$= \frac{B}{T^2} \cdot \frac{1}{\frac{1}{T} + \frac{1}{T}} - A\left[\frac{a^2}{(a + a)\left(-a + \frac{1}{T}\right)\left(a + \frac{1}{T}\right)} \right.$$

$$\left. + \frac{\frac{1}{T^2}}{\left(-\frac{1}{T} + a\right)\left(\frac{1}{T} + a\right)\left(\frac{1}{T} + \frac{1}{T}\right)} \right]$$

$$= \frac{B}{2T} + A\left[\frac{-aT^2}{2(1 + aT)(1 - aT)} + \frac{T}{2(1 + aT)(1 - aT)} \right]$$

$$= \underset{\text{Noise error}}{\underbrace{\frac{B}{2T}}} + \underset{\text{Signal error}}{\underbrace{\frac{AT}{2} \cdot \frac{1}{1 + aT}}}$$

Thus the filter can be designed, by choice of T, to give the best signal-to-noise ratio in the sense of minimising $\overline{y_e^2(t)}$.

For an extremum error

$$\frac{d}{dT}\left[\overline{y_e^2(t)} \right] = -\frac{B}{2T^2} + \frac{A}{2}\left(\frac{(1 + aT) - aT}{(1 + aT)^2} \right) = 0$$

$$\therefore T = \frac{-aB \pm \sqrt{AB}}{a^2 B - A}$$

Only positive T is allowed, and it should be checked that this gives the minimum (not maximum) value.

10.7 Conclusions

The preceding sections have laid the basic rules of manipulation of random signals. It has been shown that if the mean square error is accepted as the error criterion, sufficient information about the input signals can be obtained in one of three ways:

(i) The autocorrelation function.
(ii) The power spectral density.
(iii) The second probability density function.

The correlation functions, for example, can be calculated on a digital computer or by using a tape-recorder with two reading heads, the spacing between which is variable and proportional to the delay τ. The signals from the heads are multiplied and averaged electronically.

Power density spectra are easiest measured by using a variable frequency, narrow band pass filter and measuring the mean power in each band; the narrower the band the smaller the signal but the more accurate the results. The problems involved in practical measurements are numerous, particularly due to the finite length of records. *Measurement of Power Spectra* by Blackman and Tukey, Dover, 1958, is one of the few texts devoted to practical problems. It is, however, a difficult book to read.

Optimum design procedure attempts to minimise the mean square error (or other error criterion) in view of the known statistical properties of the signals and disturbances by adjusting the system parameters, $G(s)$. It must also be shown, in practice, that the desired $G(s)$ is physically realisable; if it isn't the best physically realisable value must be calculated (see Truxal (ref. 1)).

The alternative application of crosscorrelation of the output with a white noise input in order to measure the weighting function, $w(t)$, has the same problem of apparatus, plus the problem that for noise to be truly random it must exist for a long period. It has been shown in the examples of autocorrelation that certain simple waveforms consisting of equal, short intervals with random amplitude distribution have an autocorrelation function approximating to that of white noise. The length of a sufficient sample, dependent upon the time constants of the system to which it is applied, can be made reasonably short. Such signals are called 'chain codes', 'pseudo random binary sequences' (P.R.B.S.) or 'maximum length sequences'. The important characteristic of chain codes is that, while they are of finite duration, the autocorrelation function is exactly known and is repetitive, which is not true of short samples of a random signal, resulting in better approximations to white noise. Further, the binary nature of the signal makes the calculation of the crosscorrelation function much easier. See W. D. T. Davies, *System Identification for Self-Adaptive Control*, Wiley, 1970.

Finally, it must be appreciated that analytical design is most unlikely to replace classical frequency response design, but should be looked on as an extension of existing techniques. It is hoped that the immediate impact of this chapter is to give a greater understanding and foresight of the ordinary problem.

A chapter on the application of analytical design to nonlinear systems, based on Wieners' methods, is included in Mishkin and Braun (ref. 24).

A more rigorous, but readable, treatment of stochastic processes and their application to optimal control systems is included in Meditch (ref. 30).

11
Digital Control Systems

11.1 Sampled Data Systems

A sampled data, or discrete time system is one in which the signal at some point is sampled before being processed. While it is not the only possible sampling system, by far the most common, and the only one considered here, is one with a repetitive, fixed sample rate, i.e. samples equi-spaced in time. The simplest scheme is to close a switch in series with the signal for a short time interval, thus generating a train of fixed

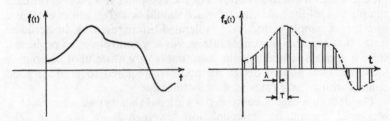

Fig. 11.1 A sampled signal.

width, fixed sample rate pulses, the height of which retains the necessary amplitude information. This is represented in Fig. 11.1. Note that in closed loop systems the sampled signals may be reference, actuating, or feedback signals, the latter pair being inside the loop so that system stability will be affected. Intuitively, such systems will be less stable than continuous ones, due to the interval, T, in which no correcting action can occur.

Sampled data control systems occur in practice when a digital computer is to be used as a controller. Care must be taken to differentiate between the use of a digital computer as an off-line, highly sophisticated, calculating machine, e.g. plotting root-loci, etc., and as an actual element in a control system. The latter prospect has been made possible by the advent of first the minicomputer and now the microprocessor.

One obvious advantage of using a computer as a controller is the increase in sophistication possible, with side advantages such as data-logging, alarm checking, etc. More directly, one computer, which is fundamentally an extremely fast device, can be used on a large number

217

of control loops. Full scale multivariable control with interacting inputs and outputs is possible, but rather complex; an introduction is given in the next chapter. Easier, and therefore more common in practice, is the multiplexing of a number of conventional control loops. An output signal is sampled, processed, and then used to set an electronic 'hold' circuit, which becomes the actuating signal of the loop. A fixed interval in time later, the signal is re-sampled, processed and the hold circuit up-dated. Provided the sample rate is high enough, the resulting 'stepped' actuating signal is a good enough approximation to the usual continuous signal. The process being controlled, however, will be slow acting compared to the computer, so that while the sample rate is high related to the process dynamics, it is slow in terms of the computer. Thus, immediately after sampling the output of one loop, a switch (multiplexor) is incremented so as to sample the output from another loop which is then processed to feed the actuating signal to the appropriate loop. In this way a number of loops can be serviced by one computer on a time-sharing basis. Since each loop is independent of the others, only single loops need be considered here; the number of loops that can be multiplexed is then a question of how fast the computer can process the data sample and set the hold circuit (which should be as fast as possible) and how long a sample period can be tolerated before resetting. It should be noted that a very high sample rate avoids any control design problems, since it appears virtually continuous, but is very wasteful of computing facilities. Thus special design methods are required to produce good control with sample rates as slow as tolerable.

The data in a digital computer is a digital (binary) number which is processed by number manipulation governed by a program. The sequence of logical and arithmetic events which governs the computer operation is called an 'algorithm'. The connections between the computer and the process under control require conversion from the physical variable to digital numbers. Thus the voltages from the transducers are sampled and converted to digital form by an 'analogue-to-digital' convertor (A.D.C.). The resulting binary actuating signal calculated by the computer is converted to a voltage by a 'digital-to-analogue' convertor (D.A.C.); while there are more sophisticated devices, the voltage developed by the D.A.C. will be held constant until a new binary number is 'strobed' into it as shown in Fig. 11.3.

It must be stressed that digital control as described above is sampled data control of a continuous process, i.e. the output from the transducer is continuous, but it is processed as discrete time data. This method of control is called DIRECT DIGITAL CONTROL (D.D.C.). Note that the data processing performed by the computer is equivalent to the compensation networks and summing point in a conventional control system. Reference signals could be continuous signals fed to the computer via an A.D.C., but more often advantage can be taken of the computer to interlink systems, the reference signal thereby being a digital signal in the first place. Similarly the sampled outputs, after conversion to digital form, can be recorded (data-logged) at the same time as being processed for control action.

These principles are shown schematically in Fig. 11.2.

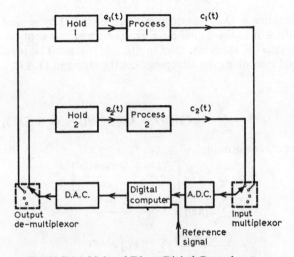

Fig. 11.2 Multiplexed Direct Digital Control system.

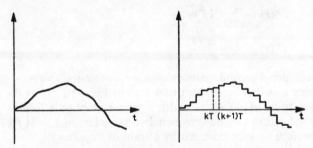

Fig. 11.3 Effect of sampling a continuous function and reforming with a hold circuit.

11.2 Mathematical Representation of the Sampled Signal *

In a digital computer, the continuous signal $f(t)$ will be handled as a series of numbers, equal to the amplitude of the signal, at a series of

* Both Truxal (ref. 1) and Gibson (ref. 19) include good chapters on this topic. Ragazzini and Franklin (ref. 21) is one of the more detailed earlier texts. Savas (ref. 32) is a wider text.

points in time. Here only repetitive sampling is considered, at frequency ω_s rad/s, i.e. sample period $T = \dfrac{2\pi}{\omega_s}$ seconds. Thus $f(t)$ is replaced by the series

$$f(kT), \text{k} = 0, 1, 2, \ldots \text{ etc.}$$

The conventional D.A.C. attempts to recover $f(t)$ by setting a signal level equivalent to the digital number, $f(kT)$, which is kept constant for a sample interval, when it is reset to $f((k + 1)T)$, etc. This is called zero-order hold smoothing, or clamping. Let the clamped D.A.C. output be $f_c(t)$.

Then

$$f_c(t) = \sum_{k=0}^{\infty} f(kT) \left[H(t - kT) - H(t - (k + 1)T) \right]$$

∴ Laplace transforming

$$F_c(s) = \sum_{k=0}^{\infty} f(kT) \left[\frac{e^{-kTs}}{s} - \frac{e^{-(k+1)Ts}}{s} \right]$$

$$= \sum_{k=0}^{\infty} f(kT) \, e^{-kTs} \frac{(1 - e^{-Ts})}{s}$$

$$= \frac{(1 - e^{-Ts})}{s} \sum_{k=0}^{\infty} f(kT) \, e^{-kTs}$$

$$= G_c(s) \, F^*(s)$$

where

$$F^*(s) = \sum_{k=0}^{\infty} f(kT) \, e^{-kTs}$$

and

$$G_c(s) = \frac{1 - e^{-Ts}}{s}$$

Now consider a second problem where a continuous wave is sampled by closing a switch for a short time λ as in Fig. 11.1. Thus the output $f_s(t)$ is a train of pulses of width λ whose amplitude follows $f(t)$. If $\lambda \ll T$, the pulses can be considered rectangular, of height $f(kT)$, and can therefore be approximated by a train of impulses

$$f_s(t) = \lambda f(t) \, \delta_T(t)$$

where $\delta_T(t)$ is the train of impulses, given by

$$\delta_T(t) = \sum_{k=0}^{\infty} \delta(t - kT)$$

$$\therefore \, f_s(t) = \lambda \sum_{k=0}^{\infty} f(kT) \, \delta(t - kT)$$

$$\therefore \, F_s(s) = \lambda \sum_{k=0}^{\infty} f(kT) e^{-kTs}$$

$$= \lambda \, F^*(s)$$

Comparing these equations, it can be seen that the two signals $f_c(t)$ and $f_s(t)$ can be treated as being derived by filtering the signal $f^*(t) = \mathcal{L}^{-1}[F^*(s)]$. The filter transfer functions are $G_c(s)$ or λ respectively, as shown in Fig. 11.4.

Note that

$$f^*(t) = \mathcal{L}^{-1}[F^*(s)] = f(t)\,\delta_T(t) = f(kT)\,\delta_T(t)$$

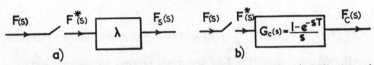

Fig. 11.4 Transfer function diagrams of: (*a*) a simple sampler, and (*b*) a sampler and clamp.

This variable is an artificial function, introduced for mathematical simplicity, and has the dimensions of units of $f(t)$ per second. The nearest physical interpretation of $f^*(t)$ is as the digital number being processed by the computer. In fact the computer can be considered to act as a 'digital filter', filtering $f^*(t)$, with the D.A.C. and hold being mathematically represented by $G_c(s)$. Thus the design of a direct digital controller amounts to designing an algorithm to make the computer first subtract the A.D.C. variable from the reference 'number' and then to filter the resulting $f^*(t)$ in the same way that a phase-lead, phase-lag network filters the actuating signal in a conventional analogue controller (see Section 11.7).

11.3 Analysis of $f^*(t)$

A train of rectangular pulses of height A and width λ, repeated every T seconds, i.e. repetition frequency $\omega_s = 2\pi/T$, can be expressed in the exponential form of the Fourier series (see Cheng (ref. 7)) as

$$f_s(t) = \frac{A}{\pi} \sum_{k=-\infty}^{\infty} \frac{1}{k} \sin\left(k\,\pi\,\frac{\lambda}{T}\right) e^{\pm jk\omega_s(t-\lambda/2)}$$

Thus the train of impulses, $\delta_T(t)$, can be similarly expressed by putting $A = 1/\lambda$ and letting $\lambda \to 0$ (note that $\lim\limits_{\lambda \to 0} \sin \dfrac{k\pi\lambda}{T} = \dfrac{k\pi\lambda}{T}$) giving

$$\delta_T(t) = \sum_{k=0}^{\infty} \delta(t - kT) \qquad \text{by definition}$$

$$= \lim_{\lambda \to 0} \frac{1}{\lambda}\frac{1}{\pi} \sum_{k=-\infty}^{\infty} \frac{1}{k} \cdot \frac{k\pi\lambda}{T}\, e^{-jk\omega_s t}$$

$$= \frac{1}{T} \sum_{k=-\infty}^{\infty} e^{-jk\omega_s t}$$

Now consider the sampled signal representation

$$f^*(t) = f(kT) . \delta_T(t)$$

$$= \sum_{k=0}^{\infty} f(kT) . \delta(t - kT) \dots \text{(time expression)}$$

Laplace transforming, $f(kT)$ can be considered a constant at each value of k and

$$\mathscr{L}[\delta(t - kT)] = e^{-kTs}$$

so that

$$\mathscr{L}[f^*(t)] = F^*(s) = \sum_{k=0}^{\infty} f(kT)e^{-kTs} \dots \text{(direct Laplace transform)}$$

Alternatively, using the Fourier series representation of $\delta_T(t)$

$$f^*(t) = f(t) \frac{1}{T} \sum_{k=-\infty}^{\infty} e^{-jk\omega_s t}$$

Laplace transforming,

$$F^*(s) = \frac{1}{T} \sum_{k=-\infty}^{\infty} \int_0^{\infty} e^{-st} . e^{-jk\omega_s t} . f(t) \, dt$$

$$= \frac{1}{T} \sum_{k=-\infty}^{\infty} F(s + jk\omega_s) \dots \text{(alternative form of Laplace transform)}$$

The frequency response of the impulse represented sampled signal is

$$F^*(j\omega) = \frac{1}{T} \sum_{k=-\infty}^{\infty} F(j\omega + jk\omega_s)$$

Fig. 11.5 shows the cyclic nature of the sampler frequency response,

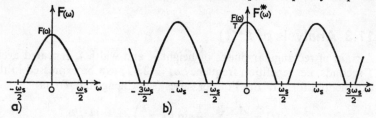

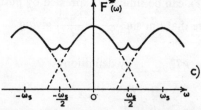

Fig. 11.5 (a) Frequency spectrum of input signal, $f(t)$; (b) frequency spectrum of the sampled signal, $f^*(t)$ [note that since $T \ll 1$ normally, the amplitude of $F^*(\omega)$ appears bigger than the unsampled signal; (c) distorted sampled spectrum when signal frequencies exceed $\omega_s/2$.

diagram (*b*) being the output spectra in response to a typical input as indicated by (*a*).

Inspection of the diagrams will show that if the input contains frequencies higher than half the sampling frequency, the output frequency spectrum will be distorted, as shown in Fig. 11.5 (*c*):

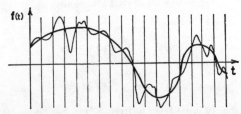

Fig. 11.6 Different continuous functions with the same sampled function.

This is essentially Shannon's sampling theorem, one of the fundamental concepts of digital information theory, which states that information at more than half the sampling frequency cannot be reclaimed, and conversely, if a signal is sampled at a rate twice the maximum frequency content, *all* the information can be reclaimed. The 'aliasing' problem is demonstrated in Fig. 11.6.

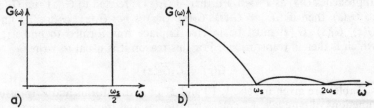

Fig. 11.7 Characteristics of: (*a*) an ideal low pass filter, and (*b*) a clamp circuit. For (*b*) the phase lag, $\phi(\omega) = \omega T/2$ radians.

From Fig. 11.5 it is clear that the sampling action introduces a wide range of harmonics. In order to reclaim the original signal, the sampled signal, $f^*(t)$, must be filtered to remove the harmonics. The ideal filter characteristic is low-pass, with sharp cut-off at $\omega_s/2$ and low frequency gain T, with no phase shift. Such a filter cannot be realised in any circumstances, but the D.A.C. and hold circuit previously discussed is an effective approximation. The transfer function, $G_c(s)$, was shown to be

$$G_c(s) = \frac{1 - e^{-Ts}}{s}.$$

the frequency response of which is shown in Fig. 11.7 (*b*).

These results indicate that, while a sample rate of twice the highest frequency component of the signal is theoretically sufficient, a factor of 5 to 10 times higher is more practical.

11.4 The Z Transform

The Z transform is a particular form of Laplace transform specially useful in sampled data systems. Cheng (ref. 7) gives a detailed treatment.

The transform is given by the simple substitution

$$z = e^{sT} \quad \text{(so that } s = \frac{1}{T} \log_e z)$$

Thus

$$F^*(s) = \sum_{k=0}^{\infty} f(kT) e^{-kTs}$$

$$\therefore \; F(z) = \sum_{k=0}^{\infty} f(kT) z^{-k}$$

$F(z)$ could probably have been written $F^*(z)$, but the asterisk is deemed superfluous since $F(z)$ is so directly related to sampled data.

A Z transfer function can also be defined as the Z transform of the impulse response of the system, i.e. if $\mathscr{L}^{-1}[G(s)] = w(t)$, then

$$G(z) = Z[w(t)] = \frac{F_o(z)}{F_i(z)}$$

Note that $G(z)$ is not $G(s)$ with e^{sT} replaced by z; in fact $TG(z)\Big|_{z \,=\, e^{sT}}$

approaches $G(s)$ as $T \to 0$. Further, if $G_1(z)$ is related to $G_1(s)$ and $G_2(z)$ to $G_2(s)$, then if $G(s) = G_1(s) \cdot G_2(s)$, $G(z)$ is *not* $G_1(z) \cdot G_2(z)$. To find $G(z)$, $G_1(s) \cdot G_2(s)$ must be inverse Laplace transformed to find $w(t)$, which is then Z transformed. For this reason it is usual to write

$$G(z) = \overline{G_1 G_2}(z)$$

Tables are given in refs. 1, 10 and 21 of Z transforms and Z transfer functions of equivalent Laplace transforms and transfer functions.

	$f(t),\ t = kT$	$F(z)$
1	$f(t + mT)$	$z^m F(z) - z^m f(0) - \ldots - zf(m-1)$
2	$f(t - mT)\, H(t - mT)$	$z^{-m}\, F(z)$
3	$H(t)$	$\dfrac{z}{z-1} \text{ or } \dfrac{1}{1 - z^{-1}}$
4	t	$\dfrac{Tz}{(z-1)^2}$
5	e^{at}	$\dfrac{z}{z - e^{aT}}$
6	$e^{at} \sin \omega t$	$\dfrac{z e^{aT} \sin \omega T}{z^2 - 2z e^{aT} \cos \omega T + e^{2aT}}$
7	$e^{at} \cos \omega t$	$\dfrac{z(z - e^{aT} \cos \omega T)}{z^2 - 2z\, e^{aT} \cos \omega T + e^{2aT}}$

Fig. 11.8 Table of basic Z transforms.

Note that an inverse Z transform is related to the time instant kT and not continuous time t, i.e. the solution attained by using Z transforms applies only to the sample instants.

The most important property of the Z transform is

$$Z[f(k - m)T] = z^{-m}F(z)$$

where $F(z) = Z[f(kT)]$, since e^{sT} is the Laplace transform of the unit delay. Thus multiplying $F(z)$ by z^{-1} is equivalent to delaying $f(kT)$ by one sample period. This property is the direct parallel to multiplying $F(s)$ by s to give the transform of the first derivative of $f(t)$.

As an example, let

$$\frac{F_o(z)}{F_i(z)} = G(z) = \frac{1 + 2z^{-1}}{2 + 3z^{-1} - 5z^{-2}}$$

Then

$$2F_o(z) + 3z^{-1}F_o(z) - 5z^{-2} F_o(z) = F_i(z) + 2z^{-1} F_i(z)$$

and hence

$$2f_o(k) + 3f_o(k - 1) - 5f_o(k - 2) = f_i(k) + 2f_i(k - 1)$$
$$\therefore f_o(k) = -1 \cdot 5f_o(k - 1) + 2 \cdot 5f_o(k - 2) + 0 \cdot 5 f_i(k) + f_i(k - 1)$$

The notation $f_i(k)$ means $f_i(kT)$, etc.

The rather awkward form of $G(z)$, in terms of negative powers of z, stems from the original choice of definition. A positive power of z corresponds to an increment forward in time.

11.5 Manipulation of Transfer Functions in Sampled Systems

Compare the three circuits of Fig. 11.9:

(a) $C(s) = G(s) \cdot R^*(s)$

(b) $C(s) = M^*(s) = (G(s) \cdot R(s))^*$

(c) $C(s) = M^*(s) = (G(s) \cdot R^*(s))^* = G^*(s) \cdot R^*(s)$ since $R^*(s)$ sampled again is unaltered, i.e. $R^{**}(s) = R^*(s)$

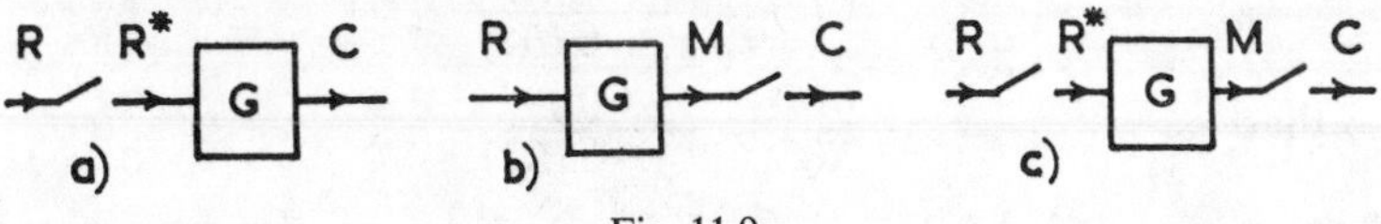

Fig. 11.9

Remember that $G^*(s)R^*(s)$ is a different function to $(G(s) \cdot R(s))^*$, the notation used to denote the second type of term being $GR^*(s)$, or $\overline{GR}^*(s)$, indicating that the product must be made before sampling.

When applying the Z transform, if $R(z)$ is the Z transform equivalent to $R(s)$ or $R^*(s)$ † and $G(z)$ is the Z transform equivalent to $G(s)$ or $G^*(s)$, then $G^*(s)R^*(s)$ transforms to $G(z)R(z)$, but $GR^*(s)$ transforms to $GR(z)$, which is *not* $G(z)R(z)$. The product $GR(s)$ must first be calculated and then transformed.

The equivalent Z transforms of the circuits of Fig. 11.9 are:

$$(a) \quad C(z) = G(z)R(z)$$
$$(b) \quad C(z) = GR(z)$$
$$(c) \quad C(z) = G(z)R(z)$$

N.B. while (a) and (c) apparently give the same answer, this is only true at the sampling instants and it can be seen from the circuits that case (a) is a continuous output, while (c) is zero between samples.

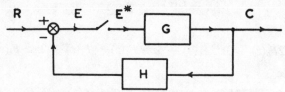

Fig. 11.10 A closed loop involving a sampled signal.

Similar consideration can now be given to closed loops, as in Fig. 11.10.

Here $E(s) = R(s) - C(s)H(s)$

and $C(s) = G(s)E^*(s)$

Thus $E(s) = R(s) - G(s)H(s)E^*(s)$

sampling this expression,

$$E^*(s) = R^*(s) - GH^*(s)E^*(s)$$

$$\therefore \ E^*(s) = \frac{R^*(s)}{1 + GH^*(s)}$$

and $$C(s) = \frac{G(s)R^*(s)}{1 + GH^*(s)}$$

Note that a simple overall transfer function $\dfrac{C(s)}{R(s)}$ is not derived. However, the equivalent Z transform is

$$C(z) = \frac{G(z)R(z)}{1 + GH(z)}$$

so that $$\frac{C(z)}{R(z)} = \frac{G(z)}{1 + GH(z)}$$

Ragazzini and Zadeh tabulated $C(s)$ and $C(z)$ for a number of single loop systems, sampled at different points, which are reproduced in the references given.

† Since $R(z)$ is only defined at the sampling instant, $R(s)$ and $R^*(s)$ have the same $R(z)$. Inverse transforming $R(z)$, however, can only give $R^*(s)$ or $r^*(t)$ with no information between samples.

Example. For the system of Fig. 11.10, let G be a hold circuit feeding a linear system with the transfer function $4/(s + 1)$; the feedback is direct, $H = 1$. The sample period, T, is 0.1 second. Find the output at $t = 0, 0.1, 0.2$ and 0.3 second in response to a unit ramp input, if the system is initially dead.

$$G(s) = \frac{1 - e^{-sT}}{s} \cdot \frac{4}{s + 1} = (1 - e^{-sT})\frac{4}{s(s + 1)}$$

Multiplying $F(s)$ by e^{sT} produces the Laplace transform of the time function delayed by T seconds. Thus using transform 2 in Fig. 11.8, $(1 - e^{sT})G(s)$ is equivalent to $(1 - z^{-1})G(z)$. It is thus required to find the equivalent Z transfer function to $G'(s) = 4/s(s + 1)$. The inverse Laplace transform (impulse response) is

$$w(t) = 4(1 - e^{-t})$$

$$\therefore \ G'(z) = 4\left(\frac{z}{z - 1} - \frac{z}{z - e^{-T}}\right) = \frac{4z(1 - e^{-T})}{(z - 1)(z - e^{-T})}$$

Hence
$$G(z) = (1 - z^{-1})\frac{4z(1 - e^{-T})}{(z - 1)(z - e^{-T})}$$

$$= \frac{4(1 - e^{-T})}{z - e^{-T}}$$

Since $H(s) = 1$, $GH(z) = G(z)$ and so

$$\frac{C(z)}{R(z)} = \frac{4(1 - e^{-T})/(z - e^{-T})}{1 + 4(1 - e^{-T})/(z - e^{-T})} = \frac{4(1 - e^{-T})}{z + 4 - 5e^{-T}}$$

$$= \frac{0.381}{z - 0.525} = \frac{0.381z^{-1}}{1 - 0.525z^{-1}}$$

$$\therefore \ (1 - 0.525z^{-1})C(z) = 0.381z^{-1}R(z)$$

$$\therefore \ c(k) - 0.525c(k - 1) = 0.381r(k - 1)$$

The times specified in the question correspond to $k = 0, 1, 2,$ and 3; for a ramp input $r(k) = kT = 0.1k$; the initial condition gives $c(0) = 0$.

$$\therefore \ c(1) = 0.381 \times 0 + 0.525 \times 0 = 0.0$$
$$c(2) = 0.381 \times 0 + 0.525 \times 0.1 = 0.0525$$
and
$$c(3) = 0.381 \times 0.0525 + 0.525 \times 0.2 = 0.125$$

11.6 Stability Analysis of Sampled Systems

Many of the techniques used for continuous systems can be adapted for discrete systems.

(a) The z-plane
Since $z = e^{sT} = e^{(\sigma + j\omega)T} = e^{\sigma T}(\cos \omega T + j \sin \omega T)$, it can be seen that the right-half of the s-plane 'maps' into the outside of the unit circle on

the z-plane, i.e. for positive σ, the modulus of z is greater than 1. The cyclic properties of the sampled signal shown in Fig. 11.5 can also be seen in that the section of the imaginary axis of the s-plane ($\sigma = 0$) from $-\pi/T$ to $+\pi/T$ maps into the unit circle on the z-plane; further displacement along the s-plane imaginary axis causes further encirclements of the same z-plane unit circle. Thus the sampled system is stable if all the poles of the closed loop $G(z)$ lie *inside* the unit circle (the mapping of the left-half s-plane).

Example. For the previous example, determine the maximum value of gain constant to keep the system just stable.

$$G(z) = \frac{K(1 - e^{-T})}{z + K - (1 + K)e^{-T}}$$

$\therefore$ the closed loop z-plane pole is

$$z = (1 + K)e^{-T} - K$$

$\therefore$ the maximum K is given by $|z| = 1$, i.e.

$$(1 + K)e^{-T} - K = \pm 1$$

$$\therefore K(e^{-T} - 1) = \pm 1 - e^{-T}$$

$\therefore$ for K positive and $T = 0.1$,

$$K = \frac{1 + e^{-T}}{1 - e^{-T}} = 20.02$$

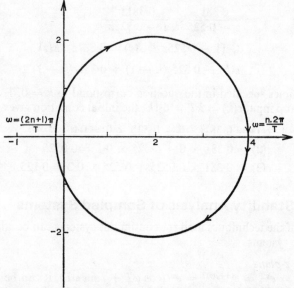

Fig. 11.11 Nyquist diagram first-order system plus sample and hold.

Note that even though the linear part of the system is only first order, the closed loop system (including the sample and hold) can be unstable for high K.

(b) Frequency response methods

The frequency response of a sampled system can readily be determined from the Z transfer function by the substitution $s = j\omega$, giving

$$z = e^{j\omega T} = \cos \omega T + j \sin \omega T$$

Thus for the running problem

$$GH(z) = \frac{4(1 - e^{-T})}{z - e^{-T}} = \frac{0.381}{z - 0.905}$$

$$\therefore \quad GH^*(j\omega) = \frac{0.381}{\cos (0.1\,\omega) - 0.905 + j \sin (0.1\,\omega)}$$

The Nyquist diagram for this system is shown in Fig. 11.11, from which it can be seen that the closed loop will be stable. Note that the locus of $GH^*(j\omega)$ continues to circle as ω increases, repeating after $2\pi/T$.

11.7 Direct Digital Control

In the foregoing, only the problem of sampling the continuous signal and smoothing (reforming) it has been considered. However, the justification for conversion to digital form is to take advantage of the processing capability of the digital computer, prior to reforming. D.D.C. then uses the computer to perform a similar function to the 'three-term controller' discussed in Chapter 8. The computer takes the digital number corresponding to the sample taken from the output transducer and subtracts it from the reference number (set point) so as to create the (digital) actuating signal. This number is then processed so as to be equivalent to the action of the continuous $P + D + I$ control, i.e.

$$M(s) = \left(1 + sT_d + \frac{1}{sT_i}\right)E(s)$$

or

$$m(t) = e(t) + T_d\dot{e}(t) + \frac{1}{T_i}\int e(t)dt$$

The use of digital signal processing is in general called 'Digital or Numerical Filtering'; for more details see Munroe (ref. 36).

A simple example can be given here to indicate the concept involved:

(i) the derivative can be approximated by the slope between two consecutive samples,

$$\therefore \quad \dot{e}(t)\Big|_{t = kT} \simeq \frac{e(k) - e(k - 1)}{T}$$

(ii) the integrand can be approximated by the mid-ordinate rule, thus

$$\int e(t)dt\Big|_{t=kT} \simeq T(0\cdot5\, e(0) + e(1) + \ldots + e(k-1) + 0\cdot5\, e(k))$$

The sum of the past values of (ek) can be saved (accumulated) as a separate variable, $y(k)$, i.e.

$$y(k) = y(k-1) + e(k),\ \text{with}\ y(0) = 0\cdot5\, e(0)$$

thus

$$\int e(t)dt\Big|_{t=kT} \simeq T(y(k-1) + 0\cdot5\, e(k))$$

The three-term control law can thus be approximated by

$$m(k) = e(k) + \frac{T_d}{T}(e(k) - e(k-1)) + \frac{T}{T_i}(y(k-1) + 0\cdot5\, e(k))$$

$$= \left(1 + \frac{T_d}{T} + \frac{T}{2T_i}\right)e(k) - \frac{T_d}{T}e(k-1) + \frac{T}{T_i}y(k-1)$$

Note that $e(k)$ must be added to $y(k-1)$ *and* saved so as to create new values of $y(k-1)$ and $e(k-1)$ for use at the next sample.

In practice more sophisticated 'algorithms' may be used, based on more accurate approximations to the derivative and integral terms.

12
Multivariable Systems—
State Variables and Matrices

12.1 Introduction

Analysis in the time domain by solution of the nth-order differential equation of the system was introduced in Chapter 3 under the general title of transient analysis. Solution of the equation by Laplace transformation was suggested, but, for practical reasons, methods of analysis and design in terms of the transformed equations, i.e. the s-plane and frequency response techniques, were preferred without recourse to the inverse transform. Thus the approach used through Chapters 4 to 8 is the accepted method for engineering design. However, as the complexity of life increases, so does the complexity of systems under control, so that methods of design for such complex systems need reconsidering.

The following points are particularly relevant:

(i) Multivariable systems—systems with more than one interrelated input and output, e.g. in complex process plants, flight controls and systems engineering.*

(ii) Nonlinear systems—Chapter 9 has already introduced state variables and the state space as the simplified, graphical phase-plane method. Nonlinearity is a general headache for simple as well as complex systems, so that any promising technique is welcomed.

(iii) Design for better dynamic performance than achievable by rule of thumb phase-lead, phase-lag, etc., compensation. The desirable aim is an 'engineering' method of optimum design.

In parallel with the demand for more complex systems came the development of the digital computer, so that a fourth requirement of the method of analysis of complex systems can reasonably be a suitability to solution by digital computer.

While it is not immediately obvious, the upshot of the above is that time domain analysis becomes preferable to s-plane or frequency response methods; not, however, by straight-forward solution of the

---

* Systems engineering (see Chestnut (ref. 26)) is a term coined to describe, for example, a factory workshop, relating stores, machining, handling, orders, costs, etc.; in effect, control processes involving human operators.

*n*th-order differential equation (simultaneous equations for a multi-variable system) but by a method known as 'state space analysis', employing matrix methods. The use of matrices can simplify the manipulation to be performed by the designer, the tedious final solution being performed by the computer.

12.2 Matrix Algebra

Relatively little knowledge of matrices is required and the simple rules following will suffice for this introduction. De Russo, Roy and Close (ref. 25) will provide any extra requirements.

Matrices

A set of linear equations can be represented by a single matrix equation, thus:

$$a_{11}x_1 + a_{12}x_2 + \ldots + a_{1n}x_n = y_1$$
$$a_{21}x_1 + a_{22}x_2 + \ldots + a_{2n}x_n = y_2$$
$$\cdot \quad \cdot \quad \cdot \quad \cdot \quad \cdot$$
$$a_{m1}x_1 + a_{m2}x_2 + \ldots + a_{mn}x_n = y_m$$

is written

$$\mathbf{A}\,\mathbf{x} = \mathbf{y}$$

or

$$[A][x] = [y]$$

where $\mathbf{A}$ is a *coefficient matrix* of order $m \times n$ (m rows by n columns) and represents the constant parameters of the system,

$$\mathbf{A} = [A] = \begin{bmatrix} a_{11} & a_{12} \ldots a_{1j} \ldots a_{1n} \\ a_{21} & a_{22} \ldots a_{2j} \ldots a_{2n} \\ \vdots & \\ a_{i1} & a_{i2} \ldots a_{ij} \ldots a_{in} \\ \vdots & \vdots \\ a_{m1} & a_{m2} \ldots \quad\quad a_{mn} \end{bmatrix}$$

The general term, a_{ij}, is in the *i*th row, *j*th column.

$\mathbf{x}$ and $\mathbf{y}$ are *column matrices* of order $n \times 1$ and $m \times 1$ respectively which represent the system variables, also called *column vectors* or simply *vectors*. While the *n*-dimension case cannot be physically represented, an idea of the vector notation can be got by considering a three-dimensional example

$$\mathbf{x} = [x] = \begin{bmatrix} x_1 \\ x_2 \\ x_3 \end{bmatrix} = \begin{bmatrix} 3 \\ 5 \\ 1 \end{bmatrix}$$

If a three-dimensional space is visualised with axes x_1, x_2 and x_3 mutually at right angles, then the vector $\mathbf{x}$ is the line joining the origin to the point $x_1 = 3$, $x_2 = 5$, $x_3 = 1$.

A *row matrix* or *row vector* is a $1 \times n$ matrix, i.e.:

$$\mathbf{v} = [v] = [v_1 \; v_2 \ldots v_n]$$

A *transpose matrix* is a matrix formed from a given matrix by interchanging the rows and columns, i.e. the transpose of the previous matrix, $\mathbf{A}$, is

$$\mathbf{A}' = [A]' = \begin{bmatrix} a_{11} & a_{21} & \ldots & a_{m1} \\ a_{12} & a_{22} & \ldots & a_{m2} \\ \vdots & & & \\ a_{1n} & a_{2n} & \ldots & a_{mn} \end{bmatrix}$$

The transpose of an $n \times m$ matrix is an $m \times n$ matrix. The transpose is sometimes used to simplify the printing of column vectors, e.g. the column matrix $\mathbf{x}$ in the original example could be written

$$\mathbf{x} = [x_1 \; x_2 \ldots x_n]'$$

If $m = n$ the matrix is called a *square matrix*.

A matrix with all elements zero except those on the principal diagonal is called a *diagonal matrix*, e.g.:

$$\begin{bmatrix} 3 & 0 & 0 \\ 0 & -7 & 0 \\ 0 & 0 & 1 \end{bmatrix}$$

For a *symmetrical matrix* the element $a_{ij} = a_{ji}$ or $\mathbf{A} = \mathbf{A}'$, e.g.:

$$\begin{bmatrix} 1 & 3 & 7 \\ 3 & 6 & -1 \\ 7 & -1 & 1 \end{bmatrix}$$

The diagonal matrix with all the diagonal elements unity is called an *identity matrix*, $\mathbf{I}$:

$$\mathbf{I} = \begin{bmatrix} 1 & 0 & 0 \\ 0 & 1 & 0 \\ 0 & 0 & 1 \end{bmatrix}$$

A *null matrix*, $\mathbf{0}$, has all elements zero.

Summation of matrices (only applicable to matrices of the same order); the corresponding coefficients of the matrices are added, e.g.:

$$\begin{bmatrix} 3 & 2 & 1 & 1 \\ 2 & -1 & 3 & -1 \\ 1 & 0 & 0 & -1 \end{bmatrix} + \begin{bmatrix} 1 & -1 & 2 & 3 \\ 0 & 1 & -1 & 0 \\ 1 & 1 & -3 & 2 \end{bmatrix} = \begin{bmatrix} 4 & 1 & 3 & 4 \\ 2 & 0 & 2 & -1 \\ 2 & 1 & -3 & 1 \end{bmatrix}$$

Multiplication by a scalar multiplies all coefficients by the scalar, e.g.:

$$-2\begin{bmatrix} 1 & 2 \\ -1 & 3 \end{bmatrix} = \begin{bmatrix} -2 & -4 \\ 2 & -6 \end{bmatrix}$$

Multiplication of matrices. The matrix product $\mathbf{AB}$ is defined only if the number of columns of $\mathbf{A}$ equals the number of rows of $\mathbf{B}$; $\mathbf{BA}$ is defined only if the number of columns of $\mathbf{B}$ equals the number of rows of **A**, e.g. if **A** is of order $m \times n$ and **B** $n \times p$ the product **A . B** is of order $m \times p$. The product **B . A** does not exist unless $p = m$, i.e. **A** and **B**′ are of the same order, and then **B . A** is not normally equal to **A . B**. Thus matrices are said, in general, not to *commute*, unless **A . B = B . A**, e.g. the identity matrix, **I**, commutes with any matrix of the same order.

While the order of multiplication cannot be changed, the sequence can, e.g.:

$$\mathbf{A . B . C} = \mathbf{(A . B) . C} = \mathbf{A . (B . C)}$$

A is of order $m \times n$; **B**, $n \times p$ and **C**, $p \times q$, so that **ABC** is $m \times q$.

Rules for multiplication. If **P = A . B**, then the term in the ith row and jth column of **P**, p_{ij} = sum of the ith row terms of **A** multiplied sequentially by the jth column terms of $\mathbf{B} = \sum_{r=1}^{n} a_{ir} b_{rj}$, e.g.:

$$\begin{bmatrix} 3 & -1 \\ 2 & 1 \\ 1 & 0 \end{bmatrix} \cdot \begin{bmatrix} 2 & 1 \\ -2 & 3 \end{bmatrix} = \begin{bmatrix} 3.2 + (-1)(-2) & 3.1 + (-1).3 \\ 2.2 + 1(-2) & 2.1 + 1.3 \\ 1.2 + 0(-2) & 1.1 + 0.3 \end{bmatrix} = \begin{bmatrix} 8 & 0 \\ 2 & 5 \\ 2 & 1 \end{bmatrix}$$

Since it is essential to define the order of multiplication, **P** is said to be **B** *premultiplied* by **A**, or **A** *postmultiplied* by **B**. Note that a column vector of order $m \times 1$ premultiplied by a row vector of order $1 \times m$ is a matrix of order 1×1, which is a scalar. The same column vector post-multiplied by the row vector gives a square matrix of order m, e.g.:

$$[1 \quad 3 \quad -2] \begin{bmatrix} 2 \\ -1 \\ -2 \end{bmatrix} = [1.2 + 3.(-1) + (-2)(-2)] = 3$$

while $\begin{bmatrix} 2 \\ -1 \\ -2 \end{bmatrix} . [1 \quad 3 \quad -2] = \begin{bmatrix} 2 & 6 & -4 \\ -1 & -3 & 2 \\ -2 & -6 & 4 \end{bmatrix}$

Check now the rules of multiplication by reconstructing the set of linear equations from the matrix equation **A . x = y** used at the beginning of this section.

Matrix inversion. If **A . x = y**, where **A** is a *square matrix*, then **x = A⁻¹ . y** where **A⁻¹** is the *inverse matrix* of **A** which is not simply the reciprocal of the coefficients. See Section 12.5 for further details of matrix inversion.

12.3 State Space Analysis

The primary object of this chapter is to provide the basic understanding of state space analysis so that a simple linear servo will be used as an example for ease of comparison with earlier chapters. The reader must be quite clear, however, that this method is proposed for more complex systems and must not expect the worked examples used here to show any improvement over other methods.

The preferred text on the subject is Schultz and Melsa (ref. 27), with De Russo, Roy and Close (ref. 25), an extensive reference work.

Consider a single input, single output system,* represented by a nth-order differential equation relating the input variable (forcing function) to *one* system variable, conventionally the output and occasionally the actuating signal. By introducing additional variables such as the rate of change of output the same system can be represented by n first-order equations involving n system variables.

The future performance of the system from some time t_o, here taken as zero, can be determined from these equations for any forcing function provided that the values of the n system variables are known at time t_o, irrespective of how they achieved this state. For this reason these n variables are called the *state variables of the system*.

Consider a general case of a linear system with a transfer function†

$$\frac{Z(s)}{U(s)} = \frac{b_0 + b_1 s + \ldots + b_m s^m}{a_0 + a_1 s + \ldots + a_n s^n} \quad \text{with} \quad n > m$$

Create a function with unity numerator by putting

$$X(s) = \frac{Z(s)}{b_0 + b_1 s + \ldots + b_m s^m} \tag{12.1}$$

so that

$$\frac{X(s)}{U(s)} = \frac{1}{a_0 + a_1 s + \ldots + a_n s^n} \tag{12.2}$$

The equivalent differential equations of (12.2) and (12.1) respectively are

$$a_0 x(t) + a_1 \frac{dx(t)}{dt} + \ldots + a_n \frac{d^n x(t)}{dt^n} = u(t) \tag{12.3}$$

and

$$z(t) = b_0 x(t) + b_1 \frac{dx(t)}{dt} + \ldots + b_m \frac{d^m x(t)}{dt^m} \tag{12.4}$$

Put

$$
\left.
\begin{aligned}
x(t) &= x_1 \\
\frac{dx(t)}{dt} &= \frac{dx_1}{dt} = x_2 \\
\frac{d^2 x(t)}{dt^2} &= \frac{dx_2}{dt} = x_3, \quad \text{etc.} \\
\frac{d^{n-1} x(t)}{dt^{n-1}} &= \frac{dx_{n-1}}{dt} = x_n
\end{aligned}
\right\} \tag{12.5}
$$

and

Substituting in equation 12.3 gives

$$a_0 x_1 + a_1 x_2 + \ldots + a_{n-1} x_n + a_n \frac{dx_n}{dt} = u(t) \tag{12.6}$$

---

* The use of matrices to aid conventional transfer function analysis of multivariable systems was suggested in Section 4.10.

† The notation used here is fairly standard in state space analysis. It is most unfortunately different from that used in classical control theory.

From equations 12.5 and 12.6 a 'set' of first-order equations can be written:

$$\frac{dx_1}{dt} = x_2$$

$$\frac{dx_2}{dt} = x_3, \quad \text{etc.}$$

$$\frac{dx_{n-1}}{dt} = x_n$$

and

$$\frac{dx_n}{dt} = -\frac{a_0}{a_n}x_1 - \frac{a_1}{a_n}x_2 - \ldots - \frac{a_{n-1}}{a_n}x_n + \frac{1}{a_n}u(t)$$

Expressing in matrix form,

$$\frac{d}{dt}\mathbf{x}(t) = \mathbf{A} \cdot \mathbf{x}(t) + \mathbf{B} \cdot u(t) \tag{12.7}$$

where

$$\mathbf{x}(t) = \begin{bmatrix} x_1 \\ x_2 \\ \vdots \\ x_n \end{bmatrix}; \mathbf{A} = \begin{bmatrix} 0 & 1 & 0 \ldots 0 \\ 0 & 0 & 1 \ldots 0 \\ \vdots & & \\ 0 & 0 & \ldots 1 \\ -\dfrac{a_0}{a_n} & -\dfrac{a_1}{a_n} & -\dfrac{a_{n-1}}{a_n} \end{bmatrix} \text{ and } \mathbf{B} = \begin{bmatrix} 0 \\ 0 \\ \vdots \\ \dfrac{1}{a_n} \end{bmatrix}$$

which is called the *state vector differential equation*.

The output, $z(t)$, can then be derived in matrix form from equation 12.4 as

$$z(t) = [b_0 \ b_1 \ldots b_m \ 0 \ 0 \ldots 0][x] \tag{12.8}$$

The general state equations for a multivariable system with r inputs and m outputs can be deduced from equations 12.7 and 12.8 as

$$\frac{d}{dt}\mathbf{x}(t) = \mathbf{A}\mathbf{x}(t) + \mathbf{B}\mathbf{u}(t)$$

and

$$\mathbf{z}(t) = \mathbf{C}\mathbf{x}(t) + \mathbf{D}\mathbf{u}(t)$$

where the first equation is an expansion of equation 12.7 to include the m inputs and the second equation includes the $\mathbf{D}\mathbf{u}(t)$ term because each output may interact with the other inputs:

$\mathbf{x}(t)$ is the state vector (n elements)
$\mathbf{z}(t)$ is the output vector (m elements)
$\mathbf{u}(t)$ is the input vector (r elements)
$\mathbf{A}$ is the coefficient matrix ($n \times n$)
$\mathbf{B}$ is the driving matrix ($n \times r$)
$\mathbf{C}$ is the output matrix ($m \times n$)
$\mathbf{D}$ is the transmission matrix ($m \times r$)

The variables $x_1, x_2, \ldots, x_n$ are called the *state variables** and $\mathbf{x}(t)$ is the *state vector* in the n-dimension *state space*. It has been pointed out that this has little physical significance since the general n-dimension space cannot be visualised. The state space for a second-order system, however, has already been considered graphically as the phase plane, Section 9.8, plotting $\dot{x}$ against x, i.e. the phase-plane is a graphical representation of the state equations of a second-order system.

Example 1. Find the matrix equations for a constant field d.c. motor with net inertia J, viscous friction coefficient F, torque constant K_T, e.m.f. constant K_e, armature resistance R_a and inductance L_a, with an input to the armature $v(t)$, in terms of the state variables position, θ_0, velocity, $\dot{\theta}_0$ and acceleration $\ddot{\theta}_0$ (see Section 15.8 for details of dimensions, etc.).

Put $$x_1 = \theta_0, \quad x_2 = \dot{\theta}_0 \quad \text{and} \quad x_3 = \ddot{\theta}_0$$

then $$K_T i_a = J x_3 + F x_2$$

and $$v(t) - K_e x_2 = R_a i_a + L_a \frac{di_a}{dt}$$

$$= \frac{R_a}{K_T}(J x_3 + F x_2) + \frac{L_a}{K_T}\left(J \frac{dx_3}{dt} + F x_3 \right)$$

$$= \frac{F R_a}{K_T} x_2 + \left(\frac{J R_a}{K_T} + \frac{F L_a}{K_T} \right) x_3 + \frac{L_a J}{K_T} \frac{dx_3}{dt}$$

$$\therefore \frac{dx_1}{dt} = x_2, \quad \frac{dx_2}{dt} = x_3$$

and $$\frac{dx_3}{dt} = -\left(K_e + \frac{F R_a}{K_T} \right) \frac{K_T}{L_a J} x_2 - \left(\frac{J R_a + F L_a}{L_a J} \right) x_3 + \frac{K_T}{L_a J} v(t)$$

$$\therefore \frac{d}{dt} \begin{bmatrix} x_1 \\ x_2 \\ x_3 \end{bmatrix}$$

$$= \begin{bmatrix} 0 & 1 & 0 \\ 0 & 0 & 1 \\ 0 & -\left(\dfrac{K_e K_T + F R_a}{L_a J} \right) & -\left(\dfrac{J R_a + F L_a}{L_a J} \right) \end{bmatrix} \begin{bmatrix} x_1 \\ x_2 \\ x_3 \end{bmatrix} + \begin{bmatrix} 0 \\ 0 \\ \dfrac{K_T}{L_a J} \end{bmatrix} v(t)$$

and $\theta_0(t) = x_1 = [1 \ 0 \ 0]x$

Example 2. Repeat the above example in terms of the state variables θ_0, $\dot{\theta}_0$ and i_a.

---

* There is more than one possible set of state variables, the following example expressing equation 12.7 in two alternative sets. The set of variables used to date is often called the 'phase variables' and the resulting A matrix of equation 12·7 is said to be 'phase canonical'.

Manipulating the above equations with $\theta_0 = x_1$, $\dot{\theta}_0 = x_2$ and $i_a = x_4$

$$\frac{dx_1}{dt} = x_2$$

$$K_T x_4 = J\frac{dx_2}{dt} + F x_2$$

$$\therefore \frac{dx_2}{dt} = -\frac{F}{J} x_2 + \frac{K_T}{J} x_4$$

Also

$$v(t) - K_e x_2 = R_a x_4 + L_a \frac{dx_4}{dt}$$

$$\therefore \frac{dx_4}{dt} = -\frac{K_e}{L_a} x_2 - \frac{R_a}{L_a} x_4 + \frac{1}{L_a} v(t)$$

Hence

$$\frac{d}{dt}\begin{bmatrix} x_1 \\ x_2 \\ x_4 \end{bmatrix} = \begin{bmatrix} 0 & 1 & 0 \\ 0 & -F/J & K_T/J \\ 0 & -\dfrac{K_e}{L_a} & -R_a/L_a \end{bmatrix}\begin{bmatrix} x_1 \\ x_2 \\ x_4 \end{bmatrix} + \begin{bmatrix} 0 \\ 0 \\ \dfrac{1}{L_a} \end{bmatrix} v(t)$$

$$\theta_0(t) = x_1$$

Note that the state variables are associated with energy storage, i.e. current through inductance, voltage across capacitance, extension of springs, acceleration and velocity of inertial loads. The first form is phase canonical, the second is not.

12.4 Solution of the State Vector Differential Equation

The general state vector differential equation is

$$\dot{\mathbf{x}}(t) = \mathbf{A}\mathbf{x}(t) + \mathbf{B}\mathbf{u}(t)$$

This differential equation can be solved by Laplace transforming,

$$s\mathbf{X}(s) - \mathbf{x}(o) = \mathbf{A}\mathbf{X}(s) + \mathbf{B}\mathbf{U}(s)$$

where $\mathbf{x}(o)$ is the vector of initial conditions (states) of $x(t)$. s is not a matrix, so that in order to manipulate the transform equation $s\mathbf{X}(s)$ is replaced by $s\mathbf{I} \cdot \mathbf{X}(s)$, where $\mathbf{I}$ is an identity matrix.

$$\therefore [s\mathbf{I} - \mathbf{A}]\mathbf{X}(s) = \mathbf{x}(o) + \mathbf{B}\mathbf{U}(s)$$

$$\therefore \mathbf{X}(s) = [s\mathbf{I} - \mathbf{A}]^{-1}\mathbf{x}(o) + [s\mathbf{I} - \mathbf{A}]^{-1}\mathbf{B}\mathbf{U}(s)$$

Inverse transforming

$$\mathbf{x}(t) = \varphi(t)\mathbf{x}(o) + \int_0^t \varphi(t - \tau)\mathbf{B}\mathbf{u}(\tau)\, d\tau$$

where $\varphi(t) = \mathscr{L}^{-1}[s\mathbf{I} - \mathbf{A}]^{-1}$ which is called the *transition matrix*. The integral term on the R.H.S. is a matrix convolution integral, cf.

$\mathscr{L}^{-1}[F_1(s) . F_2(s)] = \displaystyle\int_0^t f_1(t-\tau) . f_2(\tau)\, d\tau$ (Section 10.3, Cheng (ref. 7) or

Healey, (ref. 10), and Schultz and Melsa (ref. 27), for an extension to matrices). The term $\varphi(t)\mathbf{x}(o)$ is called the *initial condition response* and the integral term the *forced response*, for self explanatory reasons.*

The initial condition solution can also be found by solving the equation

$$\dot{\mathbf{x}}(t) = \mathbf{A}\mathbf{x}(t)$$

by assuming a Taylor series expansion

$$\mathbf{x}(t) = \mathbf{E}_0 + \mathbf{E}_1 t + \mathbf{E}_2 t^2 + \dots$$

where $\mathbf{E}_k$ is a column vector of the coefficients of the t^k terms.

$$\therefore \quad \text{when} \quad t = 0, \quad \mathbf{E}_0 = \mathbf{x}(o)$$

Differentiating and putting $t = 0$,

$$\frac{d}{dt}\mathbf{x}(o) = \mathbf{E}_1 = \mathbf{A}\mathbf{x}(o) \quad \text{from the defining equation}$$

$$\therefore \ \mathbf{E}_1 = \mathbf{A}\mathbf{x}(o)$$

Further differentiation gives

$$\mathbf{x}(t) = \left[\mathbf{I} + \mathbf{A}t + \frac{\mathbf{A}^2}{2}t^2 + \frac{\mathbf{A}^3}{3!}t^3 + \dots\right]\mathbf{x}(o)$$
$$= e^{\mathbf{A}t}\mathbf{x}(o)$$

The transition matrix can therefore be alternatively defined as

$$\varphi(t) = e^{\mathbf{A}t}$$

Note that $\dot{\varphi}(t) = \mathbf{A}\varphi(t)$ with $\varphi(o) = \mathbf{I}$.

12.5 The Transition Matrix

It has now been shown that the performance of a system can be determined in terms of the input function and the initial state of the system. The system, in transform analysis described by a transfer function, is described by the transition matrix $\varphi(t)$ and the matrices $\mathbf{B}$, $\mathbf{C}$ and $\mathbf{D}$. The last three are constant matrices, but the transition matrix involves a matrix inversion, viz.:

$$\varphi(t) = \mathscr{L}^{-1}[s\mathbf{I} - \mathbf{A}]^{-1}$$

* These terms are commonly called the complementary solution and the particular integral in the literature, which is most undesirable since they are not analogous with the scalar differential equation solutions, for which the complementary function is the transient and the particular integral the steady state solution. Both steady state and some transient terms are in the forced response above. Note also that simple transfer function theory, which assumes zero initial conditions, leads only to the forced solution.

$\varphi(t)$ is here calculated by matrix inversion. There are superior numerical methods available for more complex systems.

$$\mathscr{L}[\varphi(t)] = [s\mathbf{I} - \mathbf{A}]^{-1}$$

$$= \frac{\text{Adj }[s\mathbf{I} - \mathbf{A}]}{|s\mathbf{I} - \mathbf{A}|} \quad \text{(see DeRusso, Roy and Close (ref. 25))}$$

where Adj $[s\mathbf{I} - \mathbf{A}]$ is the adjoint of the $n \times n$ matrix

$$= \begin{bmatrix} \alpha_{11} & \alpha_{21} \ldots \alpha_{n1} \\ \alpha_{12} & \cdots \\ \vdots & \\ \vdots & \\ \alpha_{1n} & \alpha_{2n} \ldots \alpha_{nn} \end{bmatrix}$$

where α_{ij} is the cofactor of the jith term of $[s\mathbf{I} - \mathbf{A}]$. Note the transpose effect, i.e. α_{21} is in the 1st row 2nd column! The cofactor, α_{ij} is $(-1)^{i+j}$ $\times$ determinant formed from the coefficients of the matrix less the ith row and jth column. $|s\mathbf{I} - \mathbf{A}|$ is the determinant of the matrix.

$$|s\mathbf{I} - \mathbf{A}| = 0$$

is called the *characteristic equation*; see also Section 6.2.1. The roots of the equation, the system poles, are also called the characteristic values or *eigenvalues*, λ_i, $i = 1, 2 \ldots n$, of the matrix $\mathbf{A}$. The related vectors $\mathbf{w}_i$, derived from the equation

$$[\lambda_i\mathbf{I} - \mathbf{A}]\mathbf{w}_i = \mathbf{0}$$

are called *eigenvectors*.

Example. A second-order r.p.c. servo with inertia, J, viscous damping, F, and torque per unit error, K, has $\theta_0 = 1$ and $\dot{\theta}_0 = -1$ when $t = 0$. Find the response when θ_i is abruptly changed to 2 at $t = 0$.

The state variables are $x_1 = \theta_0$ and $x_2 = \dot{\theta}_0$

$$\therefore J\frac{d\theta_0}{dt} + F\dot{\theta}_0 = K(\theta_i - \theta_0)$$

$$\therefore J\frac{dx_2}{dt} + Fx_2 = K\theta_i - Kx_1$$

Hence $$\frac{dx_1}{dt} = x_2$$

and $$\frac{dx_2}{dt} = -\frac{K}{J}x_1 - \frac{F}{J}x_2 + \frac{K}{J}\theta_i$$

Let $\frac{K}{J} = 6$ and $\frac{F}{J} = 5$, so that

$$\frac{d}{dt}\begin{bmatrix} x_1 \\ x_2 \end{bmatrix} = \begin{bmatrix} 0 & 1 \\ -6 & -5 \end{bmatrix}\begin{bmatrix} x_1 \\ x_2 \end{bmatrix} + \begin{bmatrix} 0 \\ 6 \end{bmatrix}\theta_i$$

and $$\theta_0(t) = x_1$$

$$\therefore \mathbf{A} = \begin{bmatrix} 0 & 1 \\ -6 & -5 \end{bmatrix}; \quad \mathbf{B} = \begin{bmatrix} 0 \\ 6 \end{bmatrix}; \quad \mathbf{C} = [1\,0] \quad \text{and} \quad \mathbf{D} = 0$$

$$[s\mathbf{I} - \mathbf{A}] = \left[\begin{bmatrix} s & 0 \\ 0 & s \end{bmatrix} - \begin{bmatrix} 0 & 1 \\ -6 & -5 \end{bmatrix}\right] = \begin{bmatrix} s & -1 \\ 6 & s+5 \end{bmatrix}$$

$$\therefore \text{ the characteristic equation is } \begin{vmatrix} s & -1 \\ 6 & s+5 \end{vmatrix} = 0$$

$$\therefore \ s(s+5) - (-1) \,.\, 6 = 0$$

$$\therefore \ (s+2)(s+3) = 0$$

The eigenvalues are $\lambda_1 = -2$ and $\lambda_2 = -3$.

Thus

$$\mathcal{L}[\varphi(t)] = [s\mathbf{I} - \mathbf{A}]^{-1} = \frac{\text{Adj } [s\mathbf{I} - \mathbf{A}]}{|s\mathbf{I} - \mathbf{A}|}$$

$$= \frac{1}{(s+2)(s+3)} \begin{bmatrix} s+5 & 1 \\ -6 & s \end{bmatrix}$$

$$= \begin{bmatrix} \dfrac{s+5}{(s+2)(s+3)} & \dfrac{1}{(s+2)(s+3)} \\ \dfrac{-6}{(s+2)(s+3)} & \dfrac{s}{(s+2)(s+3)} \end{bmatrix}$$

Inverse transforming gives

$$\varphi(t) = \begin{bmatrix} 3e^{-2t} - 2e^{-3t} & e^{-2t} - e^{-3t} \\ -6e^{-2t} + 6e^{-3t} & -2e^{-2t} + 3e^{-3t} \end{bmatrix}$$

N.B. Check that $\varphi(o) = \mathbf{I}$. Thus

$$\mathbf{x}(t) = \varphi(t)\mathbf{x}(o) + \int_0^t \varphi(t - \tau)\mathbf{B}\mathbf{u}(\tau) \, d\tau$$

$$= \begin{bmatrix} \phi_{11} & \phi_{12} \\ \phi_{21} & \phi_{22} \end{bmatrix} \begin{bmatrix} 1 \\ -1 \end{bmatrix} + \int_0^t \begin{bmatrix} \phi_{11}(t-\tau) & \phi_{12}(t-\tau) \\ \phi_{21}(t-\tau) & \phi_{22}(t-\tau) \end{bmatrix} \begin{bmatrix} 0 \\ 6 \end{bmatrix} . \, 2 \,.\, d\tau$$

$$\therefore \ x_1 = \phi_{11} - \phi_{12} + \int_0^t \phi_{12}(t-\tau) \,.\, 6 \,.\, 2 \, d\tau$$

and

$$x_2 = \phi_{21} - \phi_{22} + \int_0^t \phi_{22}(t-\tau) \,.\, 6 \,.\, 2 \, d\tau$$

From this, the output $\theta_0(t) = x_1$ is

$$\theta_0(t) = 3e^{-2t} - 2e^{-3t} - e^{-2t} + e^{-3t} + 12\int_0^t (e^{-2(t-\tau)} - e^{-3(t-\tau)}) \, d\tau$$

$$= 2e^{-2t} - e^{-3t} + 12\left(\frac{1}{2} - \frac{1}{3} - \frac{e^{-2t}}{2} + \frac{e^{-3t}}{3}\right)$$

$$= -4e^{-2t} + 3e^{-3t} + 2$$

12.6 State Variable Diagrams *

The state vector differential equations can be shown diagrammatically in a manner similar to the analogue computer diagram (see Section 13.2), using integrators, summing points and constant multipliers. The phase

* 'Signal flow diagrams' can be used in a similar manner.

reversal inherent in the analogue computer integrator is not necessary since the state variable diagram has no physical counterpart.

If
$$\frac{dx_1}{dt} = x_2$$

then
$$sX_1(s) - x_1(o) = X_2(s)$$

and
$$X_1(s) = \frac{1}{s} X_2(s) + \frac{1}{s} x_1(o)$$

which is represented diagrammatically by Fig. 12.1 (*a*). The alternative forms of Fig. 12.1 (*b*) and (*c*) are commonly used; (*b*) is misleading and

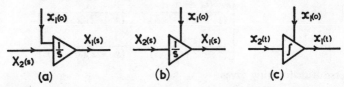

Fig. 12.1 Alternative ways of drawing the state variable diagram integrator.

should be avoided; (*c*) is a real time diagram where $x_1(o)$ must be added at the output.

Diagrams can more readily be constructed by first ignoring initial conditions, as in using transfer functions, and then inserting them at the appropriate points.

For
$$\frac{Z(s)}{U(s)} = \frac{b}{s + a}$$
$$sZ(s) + aZ(s) = bU(s)$$
$$\therefore \ sZ(s) = bU(s) - aZ(s)$$

This is a first-order equation and requires only one integrator, the output from which is $Z(s)$ and the input to which is $sZ(s)$. This input

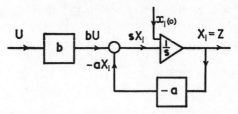

Fig. 12.2 State variable diagram for $\dfrac{Z(s)}{U(s)} = \dfrac{b}{s + a}$.

can be constructed at a summing point according to the above equation by subtracting a times the output from b times the input. All inputs to summing points on state variable diagrams are drawn positive so that the output multiplier in Fig. 12.2 is labelled $-a$.

Similarly, Fig. 12.3 is the construction procedure for the state variable diagram for the nth-order equation, involving n integrators:

$$\frac{Z(s)}{U(s)} = \frac{1}{s^n + a_{n-1}s^{n-1} + \ldots + a_1s + a_0}$$

$$\therefore \ s^n Z(s) = U(s) - a_{n-1}s^{n-1}Z(s) - \ldots - a_1sZ(s) - a_0Z(s)$$

Functions with s terms in the numerator can be drawn by introducing another state variable, x, as in Section 12.3 and then summing the

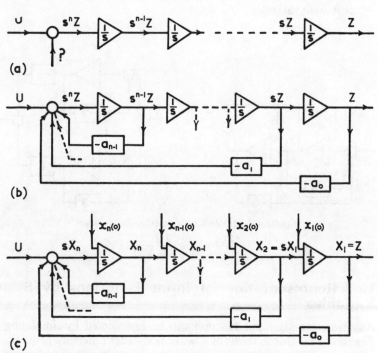

Fig. 12.3 Construction rules for a nth order system: (a) draw the n integrators; (b) complete the diagram by generating the queried input to the summing point according to the defining equation; (c) substitute the state variables $X_1(s) = Z(s), \ldots,$

$$X_n(s) = \mathcal{L}\left[\frac{d^{n-1}}{dt^{n-1}} z(t)\right] = s^{n-1}Z(s) \text{ and add the initial conditions.}$$

variables multiplied by the appropriate coefficients to create $Z(s)$. Further, it has been shown in the worked examples in Section 12.3 that more than one set of state variables can be used for one system. Both these points are brought out in the following example:

$$\frac{Z(s)}{U(s)} = \frac{s+1}{s^2 + 5s + 6}$$

put $\hspace{2em} Z(s) = (s+1)X(s)$ } This pair of equations are con-

so that $\hspace{1.5em} \dfrac{X(s)}{U(s)} = \dfrac{1}{s^2 + 5s + 6}$ } structed in Fig. 12.4 (a) with state variables $X_1 = X(s)$ and $X_2 = sX_1$

Alternatively $\dfrac{Z(s)}{U(s)} = -\dfrac{1}{s+2} + \dfrac{2}{s+3}$ by partial fraction expansion of the defining equation.

Put new state variables

$$X_3 = \frac{1}{s+2}\,U(s) \quad \text{and} \quad X_4 = \frac{1}{s+3}\,U(s)$$

Then Fig. 12.4 (b) is the equivalent of Fig. 12.4 (a), but in terms of different state variables.

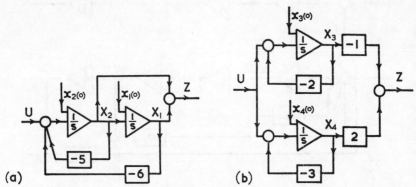

(a)   (b)

Fig. 12.4 Alternative state variable diagrams for:

$$\frac{Z(s)}{U(s)} = \frac{s+1}{(s+2)(s+3)}$$

12.7 Representation of Input Functions by State Variables

Any time-varying input function can be represented by simulating a function generator in terms of a set of first-order differential equations with appropriate initial conditions. Some examples are given below:

(a) *Polynomial Input*

$$u(t) = u_0 + u_1 t + u_2 t^2 + \ldots + u_{n-1}\,t^{n-1} \quad \text{for} \quad t \geqslant 0$$

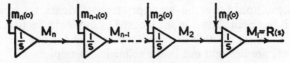

Fig. 12.5 State variable diagram of the function:

$$u(t) = u_0 + u_1 t + u_2 t^2 + \ldots + u_{n-1} t^{n-1},$$

where $\qquad m_k(o) = (k-1)!\,u_{k-1}$

The symbol m will be used as the state of an input signal to differentiate from the system states, x.

Let $\qquad u(t) = m_1$

differentiating $\quad \dfrac{du(t)}{dt} = u_1 + 2u_2t + \ldots + (n-1)u_{n-1}\,t^{n-2}$

$$= m_2, \text{ etc.}$$

$$\frac{d^{n-1}u(t)}{dt^{n-1}} = \frac{dm_{n-1}}{dt} = (n-1)!\,u_{n-1} = m_n$$

This is shown in the state variable diagram of Fig. 12.5.

(b) *Step Input*

From the previous example, put $m_1(o) = u_0$ and $m_2(o) = m_3(o) =$ etc. $= 0$ (Fig. 12.6 (a)). Then $u(t) = m_1 = u_0 \cdot H(t)$.

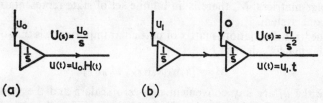

(a)   **(b)**

Fig .12.6 State variable diagrams of: (a) step, and (b) ramp functions.

(c) *Ramp Input*

Put $m_2(o) = u_1$ and $m_1(o) = m_3(o) =$ etc. $= 0$ (Fig. 12.6 (b)). Then

$$u(t) = m_1 = u_1 \cdot t$$

(d) *Sinusoidal Input*

Consider the equation of an undamped second-order system, $\ddot{u} + \omega^2 u = 0$.

Put $\qquad\qquad u(t) = m_1 \text{ and } \dot{m}_1 = \omega m_2$

then $\qquad\qquad \dot{m}_1 = \omega \dot{m}_2 = -\omega^2 m_1$ from defining equation

$\therefore \quad \dot{m}_2 = -\omega m_1$

The solution of this equation is $\quad u(t) = A \sin \omega t + B \cos \omega t$
$$= m_2(o) \sin \omega t + m_1(o) \cos \omega t$$
by substituting initial conditions of $u(t)$ and $\dot{u}(t)$.

Fig. 12.7 is the state variable diagram of this function.

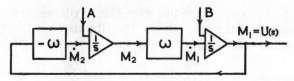

Fig. 12.7 State variable diagram for a sinusoidal function.

12.8 Transformation of the State Vector

Let $\mathbf{x}(t) = \mathbf{M}\mathbf{y}(t)$, where $\mathbf{M}$ is a constant non-singular (invertable) $n \times n$ matrix.

Substituting into the state equations gives

$$\mathbf{M}\dot{\mathbf{y}}(t) = \mathbf{A}\mathbf{M}\mathbf{y}(t) + \mathbf{B}\mathbf{u}(t)$$

$$\therefore\ \dot{\mathbf{y}}(t) = \mathbf{M}^{-1}\mathbf{A}\mathbf{M}\mathbf{y}(t) + \mathbf{M}^{-1}\mathbf{B}\mathbf{u}(t)$$

also $$\mathbf{z}(t) = \mathbf{C}\mathbf{M}\mathbf{y}(t)$$

Thus the system originally defined by the state vector $\mathbf{x}(t)$ is now defined by the state vector $\mathbf{y}(t)$, with appropriate modifications to the system matrices. It follows that since there is an infinite number of non-singular matrices, $\mathbf{M}$, there is an infinite set of state representations of the same system.

One 'transformation' matrix of particular importance is the so called 'Modal matrix', defined by

$$\mathbf{M} = [k_1\mathbf{w}_1\ k_2\mathbf{w}_2 \ldots k_n\mathbf{w}_n]$$

where the k_i are any convenient non-zero scalars and the $\mathbf{w}_i$ are the eigenvectors of $\mathbf{A}$ (see Section 12.5). Note that $\mathbf{M}$ will have complex elements corresponding to complex eigenvalues. With this particular $\mathbf{M}$ matrix, the modified matrix becomes

$$\mathbf{\Lambda} = \mathbf{M}^{-1}\mathbf{A}\mathbf{M}$$

where $\mathbf{\Lambda}$ is a diagonal matrix, the diagonal elements being the eigenvalues, λ_i, of $\mathbf{A}$. This diagonal form of the system matrix is called the 'canonical form'.*

Thus

$$\dot{\mathbf{y}}(t) = \begin{bmatrix} \lambda_1 & & & \\ & \lambda_2 & & 0 \\ & & \cdot & \\ & & & \cdot \\ 0 & & & \cdot \\ & & & & \lambda_n \end{bmatrix} \mathbf{y}(t) + \mathbf{M}^{-1}\mathbf{B}\mathbf{u}(t)$$

and expanding

$$y_1(t) = \lambda_1 y_1(t) + (\mathbf{M}^{-1}\mathbf{B})_1\mathbf{u}(t) \text{ etc.}$$

where $(\mathbf{M}^{-1}\mathbf{B})_i$ is the ith row of $\mathbf{M}^{-1}\mathbf{B}$.

It can be seen that $\dot{y}_i(t)$ will be a function of $y_i(t)$ and the input, but is independent of the other states. Thus the states in this canonical representation correspond to the 'natural modes of vibration' of the system.

* Compare this with the phase-canonical form used previously and realise that 'canonical' means 'simple'.

12.9 Observability and Controllability

From the previous section, it can be seen that the output vector is

$$\mathbf{z}(t) = \mathbf{CM}\mathbf{y}(t)$$

If any column of the matrix $\mathbf{CM}$ is all zero, then the measurements, $\mathbf{z}(t)$, contain no information about the corresponding elements of $\mathbf{y}(t)$, e.g.

$$z_1(t) = [1\ 2\ 0\ -1] \begin{bmatrix} y_1(t) \\ y_2(t) \\ y_3(t) \\ y_4(t) \end{bmatrix}$$

$$\therefore\ z_1(t) = y_1(t) + 2y_2(t) - y_4(t).$$

Since the state variables, $\mathbf{y}(t)$, are the natural modes of the system, the particular mode, y_3 in the example, is *unobserved*. Note that columns of $\mathbf{C}$ may be zero and the system still be observed. If any mode is unobserved by the measurements, the system is said to be *unobservable*.

Similarly, since

$$\dot{\mathbf{y}}(t) = \mathbf{\Lambda}\mathbf{y}(t) + \mathbf{M}^{-1}\mathbf{B}\mathbf{u}(t).$$

if any row of the matrix $\mathbf{M}^{-1}\mathbf{B}$ is all zero, then that particular mode is unaffected by changes in $\mathbf{u}(t)$ and the system is said to be *uncontrollable*. If such a mode is unstable, no control law can be devised to stabilise the system.

Example

Let

$$\frac{Z(s)}{U(s)} = \frac{4(1-s)(1+s)}{s(1-s)(1+2s)}$$

which, by inspection, has a unstable mode, which could be disguised by pole-zero cancellation, which of course is incorrect. Applying the method of Section 12.3, the state equations, in phase-canonical form, are

$$\dot{\mathbf{x}}(t) = \begin{bmatrix} 0 & 1 & 0 \\ 0 & 0 & 1 \\ 0 & \frac{1}{2} & \frac{1}{2} \end{bmatrix} \mathbf{x}(t) + \begin{bmatrix} 0 \\ 0 \\ -\frac{1}{2} \end{bmatrix} u(t)$$

and

$$z(t) = [4\ \ 0\ \ -4]\mathbf{x}(t)$$

The eigenvalues are given by the roots of

$$|s\mathbf{I} - \mathbf{A}| = \left| \begin{bmatrix} s & 0 & 0 \\ 0 & s & 0 \\ 0 & 0 & s \end{bmatrix} - \begin{bmatrix} 0 & 1 & 0 \\ 0 & 0 & 1 \\ 0 & \frac{1}{2} & \frac{1}{2} \end{bmatrix} \right|$$

$$= \begin{vmatrix} s & -1 & 0 \\ 0 & s & -1 \\ 0 & -\frac{1}{2} & s-\frac{1}{2} \end{vmatrix} = 0$$

$$\therefore\ s(s(s-\tfrac{1}{2}) - \tfrac{1}{2}) = s(s^2 - \tfrac{1}{2}s - \tfrac{1}{2}) = 0$$

$\therefore \lambda = 0$, 1 and $-\frac{1}{2}$, which can be seen in this simple example directly from the transfer function.

The eigenvector for $\lambda = 1$ is given by

$$(\lambda_i \mathbf{I} - \mathbf{A})\mathbf{w}_i = 0$$

$$\therefore \left[\begin{bmatrix} 1 & 0 & 0 \\ 0 & 1 & 0 \\ 0 & 0 & 1 \end{bmatrix} - \begin{bmatrix} 0 & 1 & 0 \\ 0 & 0 & 1 \\ 0 & \frac{1}{2} & \frac{1}{2} \end{bmatrix} \right] \begin{bmatrix} w_1 \\ w_2 \\ w_3 \end{bmatrix} = 0$$

$$\therefore \begin{bmatrix} 1 & -1 & 0 \\ 0 & 1 & -1 \\ 0 & -\frac{1}{2} & \frac{1}{2} \end{bmatrix} \begin{bmatrix} w_1 \\ w_2 \\ w_3 \end{bmatrix} = 0$$

Expanding,
$$w_1 - w_2 = 0$$
$$w_2 - w_3 = 0$$
and
$$-\tfrac{1}{2}w_2 + \tfrac{1}{2}w_3 = 0$$

The third equation is $-\frac{1}{2}$ times the second, so that only two equations are available to solve for three unknowns. Thus one element of the eigenvector is given an arbitrary non-zero value and the other elements are solved relative to this element. In this sense there is no 'length' associated with an eigenvector, only a direction; i.e. for any 'starting' value, the values of the remaining elements will be such as to lie along a line in the n-dimensional state space. The eigenvalue can be considered as a 'length' in a 'direction' defined by the eigenvector.

Thus let $w_1 = 1$, so that in this case $w_2 = 1$ and $w_3 = 1$. By finding the eigenvectors corresponding to the other two eigenvalues, the modal matrix becomes*

$$\mathbf{M} = \begin{bmatrix} 1 & 1 & 4 \\ 0 & 1 & -2 \\ 0 & 1 & 1 \end{bmatrix} \quad \therefore \mathbf{M}^{-1} = \begin{bmatrix} 1 & 1 & -2 \\ 0 & \frac{1}{3} & \frac{2}{3} \\ 0 & -\frac{1}{3} & \frac{1}{3} \end{bmatrix}$$

It can now be shown that, as predicted,

$$\mathbf{\Lambda} = \mathbf{M}^{-1}\mathbf{A}\mathbf{M} = \begin{bmatrix} 0 & 0 & 0 \\ 0 & 1 & 0 \\ 0 & 0 & -\frac{1}{2} \end{bmatrix}$$

$$\mathbf{CM} = [4 \quad 0 \quad 12]$$

and
$$\mathbf{M}^{-1}\mathbf{B} = \begin{bmatrix} 1 \\ -\frac{1}{3} \\ -\frac{1}{6} \end{bmatrix}$$

From these results it can be seen that one mode is unobserved, but all the modes are controllable. An uncontrollable mode also leads to pole/zero cancellation in the transfer function.

---

* As explained in the previous section, each column of M could be multiplied by any non-zero scalar. This should be evident from the derivation of the eigenvectors.

12.10 Transfer Function Matrices

Laplace transforming the state vector differential equation and assuming zero initial conditions,

$$sX(s) = AX(s) + BU(s)$$

and

$$Z(s) = CX(s)$$

$$\therefore (sI - A)X(s) = BU(s)$$

and

$$X(s) = (sI - A)^{-1}BU(s)$$

$$\therefore Z(s) = C(sI - A)^{-1}BU(s)$$

$$= G(s)U(s)$$

Where $G(s)$ is an $m \times r$ matrix transfer function. The element in the ith row, j^{th} column of $G(s)$ is a normal polynomial transfer function relating the ith output to the jth input.

Multivariable frequency response methods follow from this formulation, an introduction to which is given in the paper 'Return-difference and return-ratio matrices and their use in analysis and design of multivariable feedback control systems' by A. G. J. MacFarlane, *Proc. I.E.E.*, Vol. 117, No. 10, Oct. 1970, pp. 2037–49. This paper in fact generalises Bode work to the multivariable case.

12.11 State Variable Equations for Digital Systems

The concept of a digital control system, introduced in Chapter 11, can be extended to cover multivariable systems using state variables. The effect of a digital control on a continuous system is that the control variable, $u(t)$, is the output from a 'hold' circuit, so that it is constant over the sample period, T. Thus:

$$\mathbf{u}(t) = \mathbf{u}(kT) \text{ for } kT \leqslant t < (k + 1)T$$

Now the solution of the state vector differential equation was given in Section 12.4, where the initial time is implied as zero. More generally, the solution given in Section 12.4 can be written as

$$\mathbf{x}(t) = \varphi(t - t_0)\mathbf{x}(t_0) + \int_{t_0}^{t} \varphi(t - \tau)\mathbf{B}\mathbf{u}(\tau)d\tau$$

where t_0 is the initial time.

Thus the state at the end of a sample period can be determined given the state at the beginning of the interval and the value of the control, which is constant during the interval, by solving the equation for $t = (k + 1)T$ with $t_0 = kT$.

$$\mathbf{x}((k + 1)T) = \varphi(T)\mathbf{x}(kT) + \left[\int_{kT}^{(k+1)T} \varphi((k + 1)T - \tau)d\tau \right] \mathbf{B}\mathbf{u}(kT)$$

Since the system is time-invariant (autonomous), $\mathbf{A}$, $\mathbf{B}$ and $\varphi(T)$ are constants. Thus it can be seen from Fig. 12.8 that

$$\int_{kT}^{(k+1)T} \varphi((k+1)T - \tau)d\tau = \int_0^T \varphi(\tau)d\tau = \theta(T)$$

Hence

$$\mathbf{x}((k+1)T) = \varphi(T)\mathbf{x}(kT) + \theta(T)\mathbf{B}\mathbf{u}(kT)$$

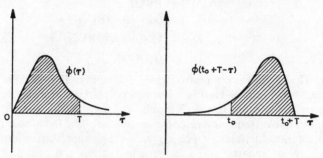

Fig. 12.8 To show that

$$\int_{t_0}^{t+T_0} \varphi(t_0 + T - \tau)d\tau = \int_0^T \varphi(\tau)d\tau$$

Now put $\theta(T)\mathbf{B} = \psi(T)$ and simplify the notation by inferring the constant sample period, T, so that

$$\mathbf{x}(k+1) = \varphi\mathbf{x}(k) + \psi\mathbf{u}(k)$$

This is called the state vector difference equation. Note that the state of the process is a continuous variable $\mathbf{x}(t)$; the state vector difference equation gives only the value of $\mathbf{x}(t)$ at the sampling instants, with $\mathbf{u}(t)$ held constant during the interval.

Example. Consider the system of Fig. 12.9

$$\frac{Z(s)}{U(s)} = \frac{1}{s+a}$$

Fig. 12.9 First-order system with sample and hold; ramp input.

Let

$$x_1(t) = z(t)$$

then

$$\dot{x}_1(t) = -ax_1(t) + u(t)$$
$$\therefore \ A = -a \text{ and } B = 1$$
$$\therefore \ \varphi(T) = e^{-aT}$$

and

$$\psi(T) = \int_0^T e^{-a\tau} d\tau \, . \, B = \frac{1}{a}(1 - e^{-aT})$$

Now since $u_r(t) = t$, $u(k) = kT$

$$\therefore \ x(k + 1) = e^{-aT}x(k) + \frac{1}{a}(1 - e^{-aT})kT$$

and $$z(k) = x(k)$$

12.12 Conclusions

This chapter has given an introduction to the analysis of multivariable systems. Design of controllers for multivariable systems is a varied and complex subject, related in current research to optimisation problems. Rosenbrock and Storey (ref. 31), Noton (ref. 28) and Prime (ref. 35) are good introductions to the problem, which is a major topic of current control theory research.

Part Four

PRACTICAL ASPECTS

13
Methods of Measuring, Computing and Simulating Systems

13.1 Measurement of Transfer Functions

13.1.1 Some General Comments

The transfer function of a system can be found in one of four ways, viz.:

(i) Direct measurement, i.e. $G(s)$ is found as a rational polynomial in s.

(ii) Harmonic measurement, i.e. the frequency response function $G(j\omega)$ is found by measurement with sinusoidal signals and $G(s)$ found by replacing $j\omega$ by s. It has already been pointed out that the bulk of conventional design is carried out in the frequency domain so that $G(j\omega)$, as measured, tends to be more useful than $G(s)$.

(iii) Transient measurement, i.e. the interpretation of the system output to various deterministic inputs.

(iv) Statistical evaluation by finding the crosscorrelation function between the output and input with a 'white' noise input. The cross-correlation function measured is the 'weighting function' or response to a unit impulse.

The transfer function of any dynamic system can be measured whether it is open loop or closed loop. With control systems it is most practical to measure the open loop transfer function so that a suitable design can be achieved, with measurement of the closed loop system as a final check. Since the output and input are not often in the same form, e.g. a position output–voltage input, transducers for the output are required. All practical measurements are then performed on electrical signals, i.e. voltage input and voltage from the output transducer. The output transducer used in the closed loop can normally be utilised, but the transfer function of any transducers must be known, or measured separately.

In measuring open loop transfer functions it is often advisable, because of large gains, to measure the transfer function in separate pieces, i.e. $G = G_1 . G_2$—measure G_1 and G_2 separately.

Pure integration terms in the loop should be avoided because of the effects of drifting, i.e. a position control system has the open loop transfer function

$$G(s) H(s) = \frac{K \cdot H}{s(1 + sT_m)}$$

H is the transfer function of the transducer, assumed here to be frequency independent.

It is advisable to measure the output velocity rather than the position, i.e. measure $\dfrac{K}{1 + sT_m}$ and then multiply by $\dfrac{H}{s}$. In this way the large output swings at low frequency and drifting in the d.c. level of the output, both of which can cause excursions beyond the range of the transducer, can be avoided.

Care should always be taken to try and maintain all applied measuring signals at amplitudes similar to those encountered in practice, so that any possible excessive nonlinearities can be spotted. An oscilloscope is a very powerful tool in this respect!

The remaining problem is random noise in the output due to valve noise, gear chatter, stray friction, transducer noise, etc. Such noise may well swamp the desired output signal and make readings difficult. With a sinusoidal input signal a variable frequency band-pass filter tuned to the input frequency will reject most of the noise and 'clean up' the output signal. Wattmeter type instruments as used in some transfer function analysers have a natural noise rejection of about 15 db. In extreme cases a correlator is the most desirable; such techniques are inherently designed into modern Transfer Function Analysers (T.F.A.).

13.1.2 Measurement of $G(s)$ as a Polynomial

Let the transfer function to be measured be

$$G(s) = \frac{b_0 + b_1 s + b_2 s^2 + \ldots + b_m s^m}{a_0 + a_1 s + a_2 s^2 + \ldots + a_n s^n} = \frac{f(b)}{f(a)}$$

A circuit for measuring $G(s)$ in this way is shown in Fig. 13.1 which is basically the method used by Wayne-Kerr.

$f(a)$ and $f(b)$ can be generated by a chain of differentiators and coefficient 'pots' as in Fig. 13.2.

In practice, differentiators are troubled by noise problems so that, purely for practical reasons, it is desirable to use integrators. A value must be set for n (or m), the maximum order of equation liable to be encountered, which is normally limited to 3, 4 or 5, when a modified circuit can be used.

The coefficient 'pots' are adjusted until the Lissajous figure indicates that the two inputs to the C.R.O. are similar in amplitude and phase; when $f(a)$ and $f(b)$ can be read off. In theory, any form of input must give the same correct answer. In practice, the higher order terms have little effect at low frequency, so by using a sinusoidal low frequency I, balance can be achieved of the a_0 and b_0 pots; increasing the frequency,

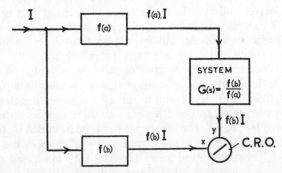

Fig. 13.1 Method of measuring $G(s)$ as a polynomial. The symbol I is used for an input applied for measuring purposes only and O for the corresponding output.

balance can be achieved using only the a_1, b_1 pots, the a_0, b_0 pots needing little or no resetting, etc.

This technique is a powerful teaching tool because of the elementary form of the answer. It has not been widely used in industry, mainly because of the preference for frequency response ideas for design, this

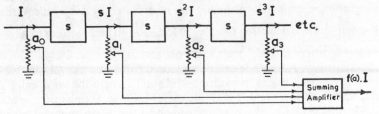

Fig. 13.2 System for simulating a polynomial.

being more suited to root loci design. An electronic root solver to find the poles and zeros of $G(s)$ is an essential if this technique is to gain ground in industrial application.

13.1.3 Frequency Response Measurements

Frequency response, or harmonic response, is the most commonly used form of system analysis. The system is excited by a sinusoidal signal and the resulting steady state sinusoidal output is compared in phase and amplitude with the input. A transient associated with the initial application of the input must be allowed to die away, and any nonlinearities giving non-sinusoidal outputs must be given special considerations, i.e. a describing function can be measured. It is always wise to use an output display so that many obvious nonlinearities can be spotted. The output display can be on a long-persistence C.R.O., a pen recorder or a storage oscilloscope. Due to the low frequencies involved, a conventional C.R.O. is very limited.

The frequency of the input signal will need to range as low as 0·01 Hz for most servos, and even lower for process controls, and up to about 200 Hz. The signal will in nearly all cases be an electrical signal, although special equipment will be required to generate mechanical (e.g. swash plate angle of a pump-controlled hydraulic servo) or pneumatic (pneumatic process controllers) input signals.

In the past, electro-mechanical systems were used because of their cheapness, made up, for example, of a slab potentiometer, a sine–cosine pot or a synchro (with a phase-sensitive rectifier), driven by a speed-controlled motor. The amplitude can then be varied by the magnitude of the excitation, and the frequency by the speed of the motor. It is, however, possible nowadays to produce L.F. electronic oscillators at competitive prices which are appreciably simpler and easier to use, to such an extent that all other forms of generator can be considered obsolete.

To obtain the maximum information about the frequency response function the output should be measured in amplitude and phase compared to the input. It has already been shown, however, that for minimum phase systems (most practical cases) there is a fixed relationship between amplitude and phase which are mathematically related by Bode's theorems. Thus for many practical systems, particularly the simpler ones, enough information can be acquired by simply measuring the amplitude, calculating $G(\omega) = 20 \log_{10} \dfrac{|O|}{|I|}$ db and plotting to a base of log ω, where O and I are the output and input respectively and ω is the frequency. Amplitude can be measured either as r.m.s., mean, peak or peak-to-peak values, provided both O and I are in the same units. A d.c. valve voltmeter or a C.R.O. are useful L.F. measuring instruments.

More accurate methods of measurement are often required which give amplitude and phase results, viz.:

(i) **Double beam oscilloscope or a two channel pen recorder used to display the input and output waveforms to a common time base** (Fig. 13.3). It is then possible by inspection to estimate the phase shift and the amplitudes. This is a useful, but inaccurate technique.

The inherent 180° phase shift in the open loop characteristic necessary for negative feed back must be considered when measuring phase shift.

In both (*a*) and (*b*) of Fig. 13.4 the open loop again is basically G.H. The major difference is that the summing point has been included in (*b*) but not in (*a*). More accurately, then:

$$(a)\ \frac{O}{I} = \text{G.H, and in } (b)\ \frac{O}{I} = -\text{G.H}$$

It is felt that provided this point is fully appreciated it can cause no practical problems which common sense cannot overcome. It is case (*a*) which is generally considered since Nyquist's stability criterion is based on G.H, not −G.H.

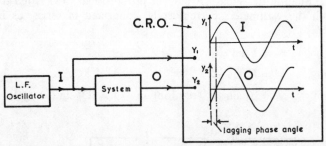

Fig. 13.3 Measurement of frequency response using a double-beam C.R.O.

Finally, when making open loop measurements, impedance mismatching due to disconnecting part of the loop should be allowed for.

(ii) **Variable phase oscillator and C.R.O.** Oscillators are now marketed* at reasonable prices which comprise a basic variable low

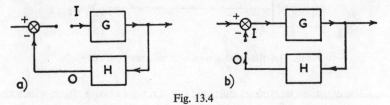

Fig. 13.4

frequency oscillator and a 'slave' oscillator ganged in frequency but 0–360° variable in phase. The system output can then be compared with the phase-shifted oscillator output and the variable phase shift adjusted until it is in phase with the output as indicated by a Lissajous

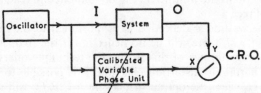

Fig. 13.5 Measurement of phase angle by comparison with a variable phase signal.

figure on a C.R.O. The system phase shift can then be read off as equal to the phase shift introduced in the variable phase shift oscillator. The amplitude of the output signal can also be read off the C.R.O. to a usually acceptable degree of accuracy. Phase accuracy of about 2–3° can be achieved. The circuit is shown in Fig. 13.5.

* FEEDBACK and SERVOMEX.

(iii) **Resolved component indicator using a two-phase oscillator.** Two wattmeters are used, one excited from the same signal as the system input (the reference signal) and the other from an oscillator of the same frequency as, but with 90° phase shift to, the reference input (the quadrature signal). A current proportional to O is fed through both the wattmeters current coils, connected in series, as in Fig. 13.6.

If $\qquad v_r = V \sin \omega t$

and $\qquad v_q = V \cos \omega t$

then $\qquad I = |I| \sin \omega t$ resulting in $O = |O| \sin (\omega t + \phi)$

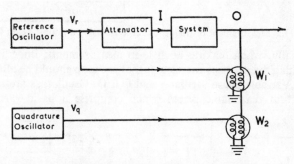

Fig. 13.6 Resolved component measurement of $G(j\omega)$.

The common wattmeter current, $i \propto |O| \sin (\omega t + \phi)$ from which the wattmeter readings are

$$W_1 \propto V|O| \cos \phi$$

and $\qquad W_2 \propto V|O| \sin \phi$

If V is constant the wattmeters can be calibrated to read $|O| \cos \phi$ and $|O| \sin \phi$ direct.

Since the wattmeter deflection is proportional to the mean value of the product $v \cdot i$, there will be no deflection if the frequencies are different. The significance of this is that the technique offers an inherent noise and harmonic distortion rejection. It is possible to get a usable reading even when the output display on a C.R.O. will not show the sinusoidal content as distinct from the noise.

The conventional dynamometer wattmeter is not sensitive enough and therefore not accurate enough, so that thermocouple instruments are used in commercial transfer function analysers.*

Such instruments are expensive and are only justified in use where the high accuracy and inherent noise rejection are needed. They are also

* SOLARTRON. Digital versions are also available.

direct reading, no adjustments having to be made such as setting the phase shift oscillator as in (ii).

13.1.4 Statistical Measurements

The technique of crosscorrelating the system output with a white noise input as described in Chapter 10 has gained some practical ground of recent years. This is due to improved equipment, particularly in the use of pseudo-random noise signals. Commercial correlators are available.

13.1.5 Transient Measurements

Measurement of the transient response of f_o to various forms of input functions, f_i, are important since specifications are often written in this form. Measurement presents no particular problems since the output can be displayed direct on a C.R.O. or pen recorder; the real problems stem from interpretation of the results. A long persistence C.R.O. is usually used, although if the input function is repeated at a rate slow enough to allow all transients to decay before the next step is applied, then a conventional C.R.O. can be synchronised to the repetitive input, giving a consistent display of the transient.

13.2 Computation and Simulation

13.2.1 The Field of Application

Digital computers are of little immediate significance to the analysis of closed loop control systems. They are, however, of extreme significance in the data processing necessary in most overall systems involving closed loops, i.e. processing of information in machine tools, on-line computation of process controls, telemetering, etc. Their possible use for finding the roots of characteristic equations and calculating correlation functions should not be overlooked. New methods of analysis using state variables and matrices, Chapter 12, has led to great use of digital computers for the design of complex systems.

The analogue computer, on the other hand, is an invaluable tool for design problems. The analogue computer uses electronic integrators to set up the differential equation of a system, e.g. if a voltage proportional to velocity is fed into an integrator, then the output from the integrator is proportional to position. Two distinct methods of application are used:

(*a*) Analogue computation—the overall differential equations of the system are set up and solved.

(*b*) Simulation—the analogy of the physical system is set up on the computer bit by bit, with no overall simplifications.

In the following sections the basic principles are outlined. For any reader interested in using an analogue computer, reference should first be made to a specialised text, for instance Jackson (ref. 23).

13.2.2 The Basic Elements of the Analogue Computer

(i) The computing amplifier—often called a virtual earth amplifier. This is a precision, d.c.-coupled, very high-gain amplifier. The frequency response should be d.c. to 20 kHz and the gain bigger than 10^5 and preferably 10^8, achieved with an odd number of stages so that the input and output are in antiphase, i.e. the gain is $-A$. The amplifier must be as free from drift as possible and have high input impedance.

(ii) An input function generator.

(iii) An output display unit.

(iv) Multiplier units.

(v) Squarers and square-root units.

(vi) Various nonlinear units.

(vii) Stabilised d.c. supplies and a number of potentiometers. The stabilised supply is used for setting the potentiometer coefficients and for initial conditions.

(viii) A patch panel is used to bring all the connections from the amplifiers and potentiometers and various fixed resistors and capacitors to a central point to make easy the various connections.

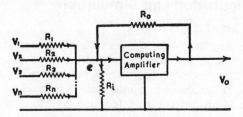

Fig. 13.7 A summing amplifier.

The amplifier has two basic modes of connection, viz.:

(*a*) *Summing Amplifier* (Fig. 13.7)

Let

$\qquad e =$ actuating signal of the amplifier

$\qquad A =$ open loop amplifier gain

$\qquad R_i =$ input resistance of the amplifier

$\qquad R_o =$ feedback resistance

$R_1, R_2,$ etc. $=$ input resistances

Then $\qquad v_o = -Ae$; negative due to phase reversal and

$$\frac{v_1 - e}{R_1} + \frac{v_2 - e}{R_2} + \ldots + \frac{v_n - e}{R_n} + \frac{v_o - e}{R_o} = \frac{e}{R_i}$$

$$\therefore \frac{v_1}{R_1} + \frac{v_2}{R_2} + \ldots + \frac{v_n}{R_n} = -\frac{v_o}{R_o} + e\left[\frac{1}{R_i} + \frac{1}{R_o} + \frac{1}{R_1} + \ldots \frac{1}{R_n}\right]$$

Since A is very large, $e \ll v_0$, the second term on the R.H.S. is negligible.

$$\therefore v_0 \simeq - \left[\frac{R_o}{R_1} v_1 + \frac{R_o}{R_2} v_2 + \ldots + \frac{R_o}{R_n} v_n \right]$$

Note that if one of the inputs or its resistance is varied it will not affect the relation between v_0 and the other inputs, and that the gain can be varied simply by changing R_n. Also since the principle of the computing amplifier is that e is always very small the input to the amplifier is a 'virtual earth'.

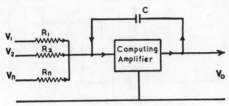

Fig. 13.8 An integrating amplifier.

(b) *Integrator* (Fig. 13.8)

V_0 can be found from the above reasoning by replacing R_o by $\frac{1}{sC}$

Note that R_i is still 'shorted' by the virtual earth and has been omitted.

Then

$$v_0 = - \left[\frac{1}{sCR_1} v_1 + \frac{1}{sCR_2} v_2 + \ldots + \frac{1}{sCR_n} v_n \right]$$

If $\qquad R_1 = R_2 = \ldots = R_n = R$

$$v_0 = - \frac{1}{CR} \left[\frac{1}{s} v_1 + \frac{1}{s} v_2 + \ldots + \frac{1}{s} v_n \right]$$

The output, then, is the sum of the integrands of the inputs scaled in amplitude by $\frac{1}{CR}$ and reversed in phase. The practical circuit also has facilities for applying initial conditions and for holding the outputs.

13.2.3 Analogue Computation

Consider a system for which the transfer function has been found. It is now desired to investigate the response of this system to various inputs. One way of doing this is to make the physical system and test it accordingly. This, however, can be very expensive if the system proves to be unsatisfactory, involving redesign. The analogue computer is used to find the system performance by drawing electrical analogues to the system equations so that any necessary design can more easily be achieved.

Let the transfer function of a simple position control system be

$$\frac{\theta_o(s)}{\theta_i(s)} = \frac{1}{1 + bs + as^2}$$

Thus
$$a\frac{d^2\theta_o}{dt^2} + b\frac{d\theta_o}{dt} + \theta_o = \theta_i$$

and
$$\therefore \; a\ddot\theta_o = \theta_i - b\dot\theta_o - \theta_o$$

This equation is then set up on the computer as in Fig. 13.9.

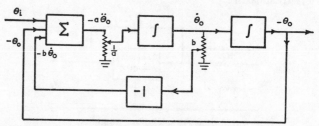

Fig. 13.9 Analogue computer set up for a second-order system.

The phase reversal units are amplifiers arranged for unity gain. In practice, the problem must be scaled:

(a) in amplitude to avoid signals θ_o, $\dot\theta_o$ and $\ddot\theta_o$ which either saturate or are too small; and

(b) in time, if desired, so that the problem can be solved in a shorter time, i.e. a process control problem which takes $\frac{1}{2}$ hour to reach steady state can be time-scaled to give an answer from the computer in, say, one minute.

The waveforms of the 'state variables', θ_i, θ_o, $\dot\theta_o$ and $\ddot\theta_o$ can readily be displayed as required.

13.2.4 Simulation

The limitation for the control system designer in straightforward analogue computation arises from the fact that the effect of changes in any one parameter cannot be investigated. In the previous example, the effect of changing the coefficient b could be observed, but in practice b could comprise, say, the velocity feedback coefficient and the viscous friction. Worse still, change in system inertia would effect a and b. The technique of simulation, then, is to set up the analogue for all parts of the system with no intermediate mathematical steps.

14
System Components

14.1 General Considerations

The choice of components and the general layout of practical servo-mechanisms is an engineering design problem in which experience is essential. Many times the final choice of components is made, not for the best servomechanism but because of other factors, primarily cost, and also environment and standardisation. For example, in an aircraft, weight and varying temperatures and pressures are more important than cost; for certain transportable radar units, hydraulic fluids are unacceptable; and so on; for each individual design certain components are acceptable and others aren't.

Each system can be broken down into a number of pigeon-holes:

(a) the actuator;
(b) the transducer or monitor;
(c) the controller;
(d) data processing of the input to the servo;
(e) the physical layout of the system.

It must be realised that the components of a servo are subjected to severe loading, due to the constant rapid correcting for errors and the high peak demands during changes in the commands. For this reason cheap gearboxes and motors, e.g. a hydraulic gear motor compared to a piston motor, are unsuitable and have very short lives in servos.

It cannot be overstressed that the best start to any problem is a clear specification of what is required of the servo. Obvious though this may seem, the desired information is not always forthcoming, so that guessing an initial specification is all too commonplace.

The servos in the following sections are typed by the actuator as d.c., a.c. or hydraulic. Such categories are not watertight, so that, say, any transducer can be used on any type of system, e.g. the synchro, which is an a.c. device, is commonly used on d.c. and hydraulic servos, so that the controller must convert the signal into a suitable form. It is possible to design all-mechanical servos, e.g. an hydraulic profile-copying attachment, but most systems use electrical devices for monitoring and controlling.

Pneumatics, which are of importance in process control, are relatively insignificant in servos, although fluid logic devices, due to their cheapness and reliability, should find application in some data processing applications where the response required is less severe.

14.2 Comparison of D.C., A.C. and Hydraulic Servomechanisms

The d.c. servo, which uses a d.c. commutator type rotary motor, is by far the most common in use, and can be considered as a general-purpose system. In comparison with the other systems, the a.c. servo, using rotary induction motors, is a much lower performance system, which, however, being a.c. coupled, is a modulated carrier system and does not suffer from drift. Hydraulic servos, using hydraulic motors controlled by electro-hydraulic devices, are capable of higher performance at a higher cost. The all-mechanical servo is rugged and cheap, but has a limited field of application and accuracy.

The a.c. servo, which has the advantage of cheapness, is at present used only for very low power applications such as remote position indicators, potentiometer setting, etc., with ratings up to a maximum of 200 W. The desirability of the cheap, reliable squirrel-cage induction motor as the power source, however, has led to development of various thyristor switching circuits which utilise higher power motors.

Though the d.c. motor is capable of higher *controlled* power and thus higher response in servo applications than the a.c. motor, its performance is limited by the maximum torque–inertia ratio achievable, a limit set by heating due to armature current and magnetic saturation of iron paths. The hydraulic rotary motor is less limited by such considerations, so that considerably higher torque–inertia ratios can be achieved, resulting in higher performance. Such higher performance servos are more costly and there is some inherent resistance to the use of oil on the grounds of bursts and filtering.

In positioning systems the motor speed must reduce to zero while still under control. Thus the low-speed performance of the motor has particular significance, in which respect hydraulic motors can be superior to other motors.

A further application of hydraulics lies in using linear actuators, i.e. jacks or rams, which produce cheaper systems with reduced performance.

Of all-mechanical servos, the hydraulic profile-copying system and elevator-type control systems are in common, satisfactory use, since they are cheap and reliable with an acceptable accuracy. They are unversatile and are limited to a few specific applications.

No one system rises above the others on all counts, so that each individual application must be assessed in its own light. The following is a general guide and it must be recalled that outside factors must often have an appreciable effect on the final choice, e.g. standardisation, replacements, servicing, availability, etc.

Type of system	Type of application
D.c.—field control	Position and velocity control systems up to about 1·5 kW. Capable of quite high performance.
D.c.—armature control	General purpose systems above 750 W, including some very high power applications of hundreds of kilowatts.
A.c.	As yet limited to low power, up to 200 W, position systems. Relatively low performance, but cheap. Not troubled by drift, making them ideal small position servos. Used with thyristor frequency convertors, high power induction and synchronous motors can be used for speed controlled drives.
Hydraulic—rotary motor	Expensive, high performance systems with ratings from a few watts up to about 100 kW. Particularly good low-speed performance.
Hydraulic—rams	Cheaper, medium performance system, limited to about 40 cm stroke. Longer strokes can be used, but with lower performance.
Variable couplings and brakes	Usually eddy current types, suitable for speed controls of 1–50 kW. Unsuitable for position control since they will only drive in one direction. Good low-speed performance, comparatively cheap, but with low power efficiency (see Section 15.6).
Relay or on–off	Discontinuous type of control for which it is possible to use any type of actuator, most commonly a d.c. motor (see Section 15.7).
Pneumatic	Systems using air instead of oil are of little significance in servos, but are used extensively for process controls.

14.3 Servomotors and Drives

The term servomotor is coined simply to indicate that a motor has characteristics suitable for use in servos. These characteristics are controllability, good linearity and high sensitivity, reliability and high max. torque–inertia ratio. The basic principles and construction of such motors are no different from conventional motors, and in fact conventional 'off the shelf' motors are commonly used.

The servomotor is far better thought of as a torque device rather than a power device, with speed as more of a limiting factor which has some effect on the torque characteristic and power rating.

Except for hydraulic rams all servomotors are rotary. There are linear d.c., induction and hydraulic piston motors under development, but none are particularly suitable for servo application as yet. Thus, for linear displacement systems, the rotary torque must be converted to linear force by using either a leadscrew and nut or a rack and pinion. The recirculating ball and nut leadscrew is by far the best for control systems, since backlash and friction are kept down to a minimum. In most cases the design of the servo will require reduction gearing between the motor and the leadscrew, or in the case of a rotary final output, the shaft itself.

One of the biggest problems is the mechanical assembly of the motor and drive, since friction and backlash must be avoided. Bearings and structures must be of high accuracy and extremely stiff.

14.4 Referred Inertia and Friction

It was shown in Section 3.2.1 that inertia and viscous friction could be referred from the motor shaft to the output and vice versa.

Thus if

$$J_m = \text{motor inertia, kg m}^2$$
$$J_L = \text{load inertia, kg m}^2$$
$$F_m = \text{motor viscous friction coefficient, N m s}$$
$$F_L = \text{load viscous friction coefficient, N m s}$$
$$n:1 = \text{reduction gear ratio.}$$

Then   Total inertia referred to the output $= J_L + n^2 J_m$

and   Total viscous friction referred to the output $= F_L + n^2 F_m$

The   Total inertia referred to the motor $= \dfrac{1}{n^2} J_L + J_m$

and   Total viscous friction referred to the motor $= \dfrac{1}{n^2} F_L + F_m$

For a leadscrew drive, if

$$M = \text{load mass, kg}$$
$$f = \text{load viscous friction force, N per m/s}$$
$$P_L = \text{load force, N}$$
$$x = \text{leadscrew pitch, m/rev}$$

Then Total inertia referred to leadscrew input $= J + M\left(\dfrac{x}{2\pi}\right)^2$

Total viscous friction referred to leadscrew input $= F + f\left(\dfrac{x}{2\pi}\right)^2$

Load force referred as effective torque at leadscrew $= P_L \cdot \dfrac{x}{2\pi}\,\mathrm{N\,m}$

14.5 Gearing

The loop gain of a system is increased by using reduction gearing, of ratio $n:1$, between the motor and the output, thus utilising a smaller motor capable of developing $1/n$ times the output torque required. The motor must, of course, be capable of running at n times the output speed and of providing the losses in the gearing itself. Spur gearing is normally used, although worm and wheel drives are sometimes employed, when single reductions of hundreds to one can be achieved. About $10:1$ is the maximum for a single pair of spur gears. Pulleys and belts can sometimes be used instead of gears.

Two factors are of extreme importance in servomechanism gearing; backlash and friction. These two factors are conflicting since 'tight' gears to eliminate backlash introduce friction and vice versa. Servo gearboxes, then, are made costly by the stringent requirements of accurate manufacture. Many ingenious tricks are used to improve gearboxes, the simplest being to split the gear in two halves at right angles to the axis of rotation and to spring load the two halves torsionally, so that when meshed with a simple gear-wheel all backlash is taken up. The problem with such devices is power rating and possible additional time constants added into the system by the springs.

The choice of gear ratio is by no means straightforward and is a most important factor of experience in design.

Consider first the ideal situation as follows:

Acceleration torque $\qquad T_a = (n^2 J_m + J_L)\ddot{\theta}_o$

For a given motor and load, one value of n will give maximum acceleration, thus:

Load acceleration $\qquad \ddot{\theta}_o = \dfrac{T_a}{n^2 J_m + J_L}$

$$= \dfrac{n T_m}{n^2 J_m + J_L}$$

where T_m is the torque produced by the motor. Differentiating $\ddot{\theta}_o$ w.r.t. n and equating to 0 to find a maximum,

$$\frac{d\ddot{\theta}_o}{dn} = \frac{(n^2 J_m + J_L)\cdot 1 - n\cdot 2n\cdot J_m}{(n^2 J_m + J_L)^2}\cdot T_m = 0$$

giving $\quad n^2 J_m + J_L = 2n^2 J_m$

$\therefore\ n^2 J_m = J_L\quad$ for maximum acceleration

In this condition the load and motor are said to be matched.

Clearly, the bigger the ratio torque/inertia of the motor the higher the acceleration possible for any value of n.

The above is a theoretical design consideration which leaves no choice of motor. More often it is the load acceleration and other forces which are specified, with the load inertia known as well as a maximum speed requirement. The design is now a matter of trial and error along the following lines:

Calculate the total load torque, T_o, as outlined in Section 14.6.

Assume that the motor torque is, say, $2 \cdot \dfrac{1}{n} \cdot T_o$.

Maximum motor speed $= n \times$ maximum output speed.

This now sets two requirements of the motor; the maximum torque and the maximum speed. The factor of 2 in the torque equation is empirical and must be checked with the calculated values of n and J_m for the selected motor, i.e. is the factor of 2 sufficient to allow for the $J_m \ddot{\theta}_m$ term? Thus, a trial and error selection will reveal a suitable motor.

With electric motors, gearing is common because the low-speed performance is erratic, the gear ratio often being chosen simply as the ratio of the maximum allowable motor speed to the maximum desired output speed. However, large hydraulic motors are being developed that will deliver the full required torque without gearing and yet, due to the very high torque–inertia ratios, can produce the desired acceleration (note the *desired* acceleration, and not the maximum possible). Further, as they must with no gearing, these motors have superior low-speed characteristics. This boils down to using a larger, and therefore more expensive motor than necessary but saving on the cost and problems of gearboxes.

In general, reduction gearing will increase the resonant frequency of the open loop transfer function, particularly so with hydraulic servos.

14.6 Motor Rating

In rating a motor three factors must be considered:

    (*a*) the torque;
    (*b*) the speed and speed range;
    (*c*) the power.

The motor must be capable of developing the required torque over the whole speed range. Some motors are capable of delivering very high, short duration peaks of torque, so that the duration of any peak demand torques can be important. This is more so in considering the power rating of a motor, since such a rating is set by heating due to internal power dissipation, not the useful output torque. It is found in practice, however, that a motor rated to provide the desired torque over the rated speed range will in fact be conservatively rated powerwise, since the mean working conditions are low compared to the peaks.

In specifying speed it is important to note the minimum as well as the maximum speed. With positioning systems, for which the minimum speed is zero, it is advisable to define the maximum angular accuracy required of the motor itself, a factor which may well be decisive in choosing the gear ratio. About 5° is usually the smallest controllable angle for an electric motor and about 1° for an hydraulic motor, figures which are vague and depend very much upon the particular motor.

The major problem lies in defining the motor torque requirements. The motor must provide $1/n \times$ the output torque plus say 10 or 20 per cent gearing inefficiency and, in addition, any accelerating and viscous torques of its own. The problem of choosing the gear ratio, n has already been discussed, so that the important factor here is to calculate the output torque required for a particular specification.

The output torque, $T_0 = J_L \ddot{\theta}_0 + F \dot{\theta}_0 + T_L$ where T_L includes external loads and such inherent loads as static friction.

Example. One axis of a milling machine has a total mass of 120 kg and a viscous friction coefficient of $1 \cdot 9 \times 10^{-4}$ N per m/s. Static friction is equivalent to a force of 250 N and reaction forces on the table, due to cutting, of 380 N are encountered. The machine is to have a maximum traverse rate of $2 \cdot 5$ cm/s and is to reach this velocity in 1 second. The worst cutting conditions are to be equivalent to the cutter centre moving around a $0 \cdot 5$ cm diameter circle at a constant tangential velocity of 1 cm/s, the servo errors not exceeding $\pm 0 \cdot 0005$ cm. The motor drives through a leadscrew of $0 \cdot 5$ cm pitch for which $J = 0 \cdot 04$ kg m^2 and $F = 0 \cdot 002$ N m s.

Design of the second axis will be similar except that the mass, etc., will be different.

Referring everything to the leadscrew as in Section 14.4:

$$\text{Total inertia} = 0 \cdot 04 + 120 \left(\frac{0 \cdot 5 \times 10^{-2}}{2\pi} \right)^2$$

$$= 0 \cdot 04 + 0 \cdot 75 \cdot 10^{-4} \simeq 0 \cdot 04 \text{ kg m}^2$$

$$\text{Total viscous friction coeff.} = 0 \cdot 002 + 1 \cdot 9 \times 10^{-4} \left(\frac{0 \cdot 5 \times 10^{-2}}{2\pi} \right)^2$$

$$= 0 \cdot 014 \text{ N m s}$$

$$\text{Cutting + Static torque} = (250 + 380) \frac{0 \cdot 5 \times 10^{-2}}{2\pi} = 0 \cdot 5 \text{ N m}$$

$$\text{Traverse velocity} = 2 \cdot 5 \times 10^{-2} \times \frac{2\pi}{0 \cdot 5 \times 10^{-2}} = 31 \cdot 4 \text{ rad/s}$$

There are two distinct requirements, the traverse and the cut. The motor must be rated for the larger of the two.

Note the important part the leadscrew pays in system requirements, the load inertia being largely due to the leadscrew, the table mass contributing a negligible amount.

(*a*) *Traverse*

Viscous torque at constant speed $= 0\cdot014 \ . \ 31\cdot4 = 0\cdot44$ N m

Torque to accelerate the table to 2·5 cm/s in 1 second

$$= 0\cdot04 \ . \ \frac{31\cdot4}{1} = 1\cdot26 \, \text{N m}$$

A detailed calculation from the differential equation could be made allowing for the inertia and viscous torques together, but an adequate approximation can easily be made.

Thus let the maximum traverse torque $= 1\cdot5$ N m.

(*b*) *Cutting*

The accuracy requirements specified will not effect the torque requirements but are necessary in designing the loop gain, as will be the details of the input velocity and acceleration. The following piece of work, then, has additional importance in writing specifications, as detailed in Chapter 7.

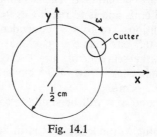

Fig. 14.1

From Fig. 14.1 the component velocities are $\dot{x}$ and $\dot{y}$ where

$$\sqrt{\dot{x}^2 + \dot{y}^2} = 1\cdot0 \text{ cm/s}$$

The angular velocity $\quad \omega = \dfrac{1}{0\cdot5} = 2$ rad/s

With constant angular velocity, x and y are sinusoidal.

Thus $\quad x = 0\cdot5 \sin 2t$ cm

$\therefore \quad \dot{x} = \cos 2t$ cm/s

and $\quad \ddot{x} = -2 \sin 2t$ cm/s^2

With a 0·5 cm pitch (4π rad/cm), the equivalent leadscrew displacement, velocity and acceleration are:

$$\theta = 4\pi \,.\, 0 \cdot 5 \sin 2t$$
$$= 2\pi \sin 2t \text{ rads}$$
$$\dot\theta = 4\pi \cos 2t \text{ rad/s}$$
$$\ddot\theta = -8\pi \sin 2t \text{ rad/s}^2$$
$$\therefore \text{ the torque } T = -J \,.\, 8\pi \sin 2t + F \,.\, 4\pi \cos 2t + \text{static loads}$$

The cutting reaction and static loads are, in fact, not constant, but it is conservative to the design to simply add these torques as a constant.

$$\therefore \ T = -0 \cdot 04 \,.\, 8\pi \sin 2t + 0 \cdot 014 \,.\, 4\pi \cos 2t + 0 \cdot 5$$
$$= -\sin 2t + 0 \cdot 18 \cos 2t + 0 \cdot 5 \text{ N m}$$

This function is plotted in Fig. 14.2.

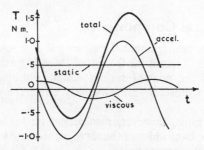

Fig. 14.2 Torque cycle for the example of Fig. 14.1.

From Fig. 14.2, the maximum torque $= 1 \cdot 5$ N m

and the maximum velocity $= 4\pi = 12 \cdot 5$ rad/s

Note, however, from Fig. 14.2 that the velocity, which has the same shape as the viscous torque curve, and the total torque are roughly 90° out of phase. Thus maximum speed occurs at low torque and maximum torque at zero speed. A motor rated to give the required speed and torque will therefore be very conservatively rated powerwise.

(c) *Summary*

The maximum load requirements are:
Traversing—

Peak torque $= 1 \cdot 5$ N m

Maximum speed $= 31 \cdot 4$ rad/s

Continuous torque at maximum speed $= 0 \cdot 44$ N m

Cutting—

Maximum torque $= 1 \cdot 5$ N m

Maximum speed $= 12 \cdot 5$ rad/s

Finally, let the motor and gearing have an inertia of J_m and equivalent friction coefficient, F_m, both referred to the motor, then,

$$\text{Motor torque} = nJ_m\ddot{\theta}_{max} + \frac{1\cdot2}{n}\cdot1\cdot5$$

$$= nJ_m\cdot8\pi + \frac{1\cdot8}{n}\,\text{Nm}$$

The 1·2 factor is included to allow for gearing inefficiency. A check should also be made to see that $nF_m\dot{\theta}_{max}$, which has been neglected since it occurs when the above torques are a minimum, is not significant.

$$\text{Maximum motor speed} = n\cdot31\cdot4\,\text{rad/s}$$

Thus a trial and error selection of various motors must be undertaken to find one which fits these requirements.

14.7 Structural Components

It has so far been assumed that the structural components of a system are perfect, which in fact is far from true. Mostly these factors are outside the field of the control system design, although, for example, compliance of a leadscrew can adversely affect the closed loop performance.

First and foremost, however, mechanical resonances in structures must be above the bandwidth of the servo, so that the servo in operation does not excite these resonances.

The geometry of structures can be very important, e.g. in a machine-tool control system the co-ordinates x and y are located; if the machine slides are not truly at right angles, then an error uncontrollable by the servo is introduced.

Stiffness of structures and particularly bearings is of similar importance, e.g. if a gun is directed at a target by angular control of the base, then any bending of the gun barrel due to external loads produces errors outside the closed loop.

Twisting of shafts inside a control loop introduces extra time constants, thereby reducing stability, e.g. in one particular hydraulic design having a bandwidth of 5 Hz it was calculated that introduction of a 15:1 gearbox would increase the bandwidth to 24 Hz. On fitting the gearbox the measured bandwidth was 4 Hz being due entirely to the frequency response of the particular gearbox! The remedy is simple—fit a better gearbox!; but the point is made of care in selection of the dynamic properties of components. A further example comes from using rubber couplings for mounting tacho's, measured results showing a marked resonance peak due to the coupling and not the system under test.

Example 1. Twisting of drive shafts and gears must be considered, since these often affect the system transfer function.

Fig. 14.3 represents a motor driving a load through a shaft or lead-screw, which is represented by a torsion spring. Gearing ratios are neglected.

The stiffness of the shaft is K Nm/rad twist.

Let the motor output torque $= T$, and the load torque $= T_L$, other symbols as previously defined. Then

$$T = J_m \ddot{\theta}_m + F_m \dot{\theta}_m + K(\theta_m - \theta_o)$$

and for the load

$$K(\theta_m - \theta_o) = J_L \ddot{\theta}_o + F_L \dot{\theta}_o + T_L$$

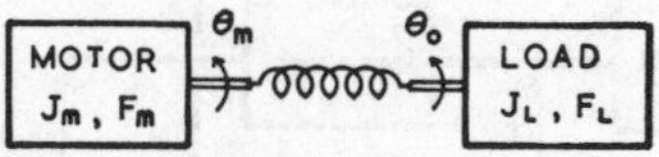

Fig. 14.3

Eliminating θ_m and writing in operational form,

$$T_L(s) + (s^2 J_L + sF_L + K)\theta_o(s) = K \cdot \frac{T(s) + K\theta_o(s)}{s^2 J_m + sF_m + K}$$

$$\therefore \ (s^2 J_m + sF_m + K)T_L(s)$$
$$+ [(s^2 J_m + sF_m + K)(s^2 J_L + sF_L + K) - K^2]\theta_o(s) = KT(s)$$

Note that if $T_L = 0$, T and θ_o are related by a fourth-order equation whereas with $K = \infty$, i.e. neglecting twisting of the shaft, the relationship was second order.

Example 2. In using a tachometer to measure velocity a similar problem arises, except that it is necessary to solve the equations to find $\dot{\theta}_T$, the tacho velocity, as in the Fig. 14.4.

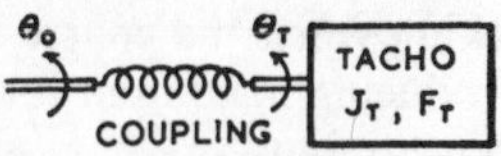

Fig. 14.4

Then
$$K(\theta_o - \theta_T) = J_T \ddot{\theta}_T + F_T \dot{\theta}_T$$
where $K =$ coupling stiffness

$$\therefore \ \frac{\dot{\theta}_T(s)}{\dot{\theta}_o(s)} = \frac{\theta_T(s)}{\theta_o(s)} = \frac{1}{1 + \dfrac{F_T}{K}s + \dfrac{J_T}{K}s^2}$$

14.8 Coarse and Fine Systems

In order to obtain the desired accuracy and resolution of a system the transducer range and the system gain are such that any large change in the input demand will completely saturate the system. This is accentuated by the fact that many precision transducers are cyclic, having a high resolution and accuracy but with an absolute range of, say, 0·1 cm, so that 2·072 cm and 6·872 cm will give the same signal. Such transducers are described in Section 14.11.

For these reasons two (or possibly three) transducers are used, geared together so that one transducer has an unambiguous reading over the full range and a resolution such that when correct position is indicated

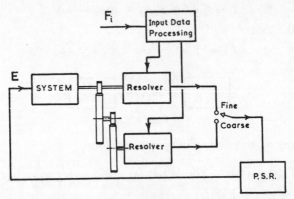

Fig. 14.5 A coarse–fine system. The reduction gear ratio is chosen to give one cycle of the coarse transducer for the full stroke of the system.

it is within the range of the more accurate transducer, i.e. for the above system the 'coarse' transducer must have an accuracy better than 0·1 cm. For long strokes this accuracy may not be forthcoming, so that an intermediate, 'medium' transducer is used before finally switching to the fine transducer. A typical arrangement is shown in Fig. 14.5.

The switching signal from coarse to fine is derived from the error signal of the active transducer. It is convenient, in practice, not to bother about linear working when the coarse transducer is active, the system saturating and producing maximum torque over the bulk of the travel with linear final positioning.

This technique can be extended to use the switching signal to alter the system characteristics.

Example. A radar aerial using a coarse–fine transducer system is to be slewed through 90°. The large input ensures that the coarse transducer

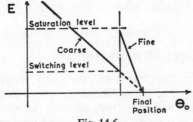

Fig. 14.6

is in operation, under which condition the system is only lightly damped. The system saturates and produces maximum error during most of the slew, ensuring maximum torque and minimum slewing time. When the system comes out of saturation it, being under-damped, will rapidly approach final position about which it would oscillate wildly. However, when the error reduces to the range of the fine transducer, switching occurs which as well as changing to the fine transducer, greatly increases the damping, e.g. the velocity feedback loop is not connected until the

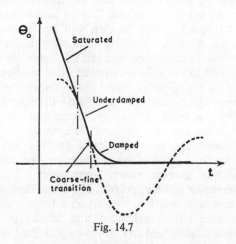

Fig. 14.7

fine transducer switches in. The approach to final position is now achieved without overshoot. Damping of this magnitude during coarse positioning would have greatly increased the response time. This is shown in Figs. 14.6 and 14.7.

14.9 Assisted Braking

While the completely linear continuous system is normally desired, the power rating of the drive to brake down under servo control from a high traverse rate may be many times in excess of that required for more normal operation. It is possible therefore, if the deceleration torque demanded by the error signal is in excess of the motor rating, to sense this demand and actuate assisted braking which can be disconnected when the error signal has reduced sufficiently. Braking can be achieved by an electromagnetic brake or by regenerative braking, i.e. reconnecting the motor as a generator to a resistance load.

Certain types of system are only active for positive error signals, e.g. a single ended thyristor amplifier, or a simple electromagnetic coupling, so that no positive braking can be applied. This is satisfactory in many speed controls where the slow down rates are slower than the normal 'switched-off' slow-down rate, thus producing an actuating signal that never goes negative. Use of switched braking, actuated by any negative error signal can greatly enhance the field of application of such systems, being much cheaper than the extra amplifiers, etc., necessary for complete linear control.

14.10 Transducers

In order to take advantage of the ease of electrical computation and compensation, the controlled output, C, must be converted to a proportional electrical signal, which in turn will be subtracted from the reference voltage, R, at the summing point. Certain transducers, however, e.g. the synchro, act as their own error detector, the output voltage being proportional to the difference in the shaft position and the position demanded by the reference signal fed to the synchro stator.

Since a general theory has been introduced it is plausible that C may be any quantity, e.g. position, velocity, temperature, pressure, etc., so that specialist transducers are required for each system. The fallibility of this text concerning process control systems has already been admitted, so that for a study of temperature, pressure, etc., transducers and other specialist process control components reference should be made to Coughanowr and Koppel (ref. 22). Here, then, a short study is made of the common position, velocity and force transducers. More specialist transducers for acceleration and relative position, such as gyros, are omitted. Mansfield (ref. 17) is a specialised text.

Position can be measured with either linear or rotary transducers. Since most linear drives are actuated by a rotary motor with a leadscrew and nut, linear position is often measured as an equivalent leadscrew rotation.* For higher accuracy direct linear measurement or rotary measurement using a separate instrument leadscrew can be used. Transducers suitable for long strokes as used on numerically controlled machine tools have been described in an article by Healey.† Short-stroke transducers (of the order of 0.1 mm) have been developed, using variable capacitance and inductance effects suitable for profile copying similar to the hydraulic device described in Section 17.4.3. For example, a conducting plate is held close to the template such that the capacitance between the plate and earth increases as the gap gets smaller. This capacitance is used to vary the frequency of a tuned oscillator which is then frequency detected to give a d.c. voltage proportional to the gap.

The synchro is the most common form of position transducer, since it is cheap, accurate, small and convenient. For this reason it is the only position transducer enlarged upon here.

---

* If leadscrew angle is monitored, the directly controlled variable, C, is leadscrew position. The indirectly controlled variable, F_o, the linear position, is related by
$$Z = \frac{\text{linear displacement}}{\text{angular displacement}}.$$
† *Instrument and Control Engineering*, Mar. and May 1966.

Velocity transducers can also be linear or rotary. Linear velocity transducers, using a fixed coil coupling with a moving magnet, are not common, due to low outputs and poor linearity; it is better, in practice, to use a rotary d.c. permanent magnet tachometer. Note that in using such a device to provide velocity feedback damping of position servos, brush noise and ripple only are of importance, but when used as an output transducer for a velocity control, linearity must also be good.

Force or torque transducers work on the principle of allowing the force to deflect a structure and then measuring the deflection. This is commonly achieved by attaching strain gauges to a specially constructed structure and measuring the change in resistance, using a Wheatstone bridge which can give a d.c. voltage proportional to force; or by a swinging arm working against a spring with a resistance potentiometer to measure the deflection proportional to force. To measure torque, twisting of the drive shaft is measured, but this is an extremely difficult problem.

14.11 Synchros and Resolvers

There are a variety of synchros in size, frequency and details of the windings, but basically the synchro has a three-phase type of stator winding and the resolver two-phase.* The physical construction is

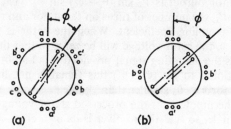

Fig. 14.8 (*a*) a synchro, and (*b*) a resolver.

similar to a motor, having a fixed cylindrical outer casing called the stator, carrying a winding of three coils (synchro) or two coils (resolver) as in Fig. 14.8. The basic rotor is single phase in each case.

Some confusion existed over terminology, but military specifications have led to standardisation of sizes, voltages, frequencies, etc., so that older units such as Selsyns and Magslips are similar in principle but different in size. The resolver, which is alternatively called a synchro-resolver or a sine–cosine resolver, has extended application to computers where extra attention is paid to high frequency performance.

All position control is of rotor angle relative to stator so that datum setting can be achieved by alignment of the stator.

* The Muirhead handbook on synchros is the best complete piece of literature available.

Let θ_o = actual rotor angle (relative to the stator)

      θ = desired rotor angle defined by the input, F_i

and ω_c = frequency of the excitation voltage

$$f_c = \frac{\omega_c}{2\pi} \text{ lies normally between 50 Hz and 1000 Hz}$$

There are two modes of operation using either alternating or rotating fields. An alternating field is a magnetic field of fixed direction with a sinusoidally varying amplitude. A rotating field is of constant amplitude rotating at a constant speed equal to the excitation frequency (cf. a permanent magnet rotating at 'synchronous' speed).*

(i) Single phase–amplitude-detected output–alternating field.

To produce an alternating field, single phase voltages only are required. The two resolver stator windings are fed with a.c. voltages of *amplitude* proportional to $V_1 \sin \theta$ (reference) (i.e. $v_1 = V_1 \sin \theta \sin \omega_c t$); and $V_1 \cos \theta$ (quadrature). This produces a field at an angle $\frac{\pi}{2} + \theta$ to the axis of the reference winding so that when the rotor angle θ_o is equal to $\frac{\pi}{2} + \theta$ there will be maximum coupling with the field; in general, the rotor output is $V_2 \sin (\theta - \theta_o) \sin \omega_c t$. The amplitude V_2 depends upon V_1, the number of turns on the stator and rotor windings, and magnetic coupling coefficient. When the rotor is in the desired position, $\theta_o = \theta$, the rotor voltage will be zero, so that the rotor output can be used as the actuating signal provided it is amplitude detected.† Being proportional to $\sin (\theta - \theta_o)$ this signal is not linear, but it is sufficiently so for $\theta - \theta_o$ smaller than $\pm 15°$.

Note that the input, R, is the processed stator voltages, the output, C, the rotor angle, so that the resolver is its own error detector. As an example, if the reference winding voltage is 50 volts and the quadrature voltage -20 volts, $\theta = \tan^{-1} \frac{50}{-20} = 111° \, 48'$. If the rotor position is 180° out of phase a false null can occur, so that a coarse transducer must be used to locate the correct half cycle.

The amplitude detector used to extract the d.c. actuating signal from the amplitude of the modulated carrier rotor voltage is a 'phase sensitive rectifier', used so that the output polarity can be retained. A negative value of $\theta - \theta_o$ will cause a phase reversal of the carrier; thus, by comparing the phase of the rotor voltage with the reference signal, the

---

* Care should be taken not to confuse sinusoidal variations of *amplitude* of signals as functions of position, i.e. $\sin \theta$, with sinusoidal variations with time, i.e. $\sin \omega t$. Hence the voltage $V \cos (\theta - \theta_o) \sin \omega t$ is a sinusoidal voltage of frequency ω, whose amplitude is dependent upon the difference angle $\theta - \theta_o$, and for example will be of zero amplitude when θ_o differs from θ by 90°.

† The rotor voltage can, of course, be used directly in an a.c. servo.

actuating signal can be detected positive or negative as shown in Fig. 14.9.

With a synchro, three stator voltages of *amplitude* $V_1 \sin \theta$, $V_1 \sin \left(\theta + \frac{2\pi}{3} \right)$ and $V_1 \sin \left(\theta - \frac{2\pi}{3} \right)$ are required, otherwise the operation is as for a resolver.

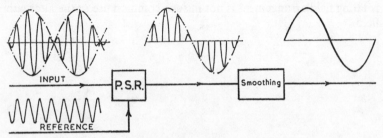

Fig. 14.9 Schematic diagram of a phase-sensitive rectifier. The waveforms correspond to a constant speed rotation of the resolver rotor, with the envelope proportional to $\sin (\theta - \theta_0)$.

Since the command voltages are all a.c. of the same phase, the data processing can be performed by selective switching of tapped transformers. The maximum voltage V_1 is not critical provided it is the same for each winding.

(ii) Polyphase–phase-detected–rotating fields.

To produce a rotating field, polyphase voltages are required, three-phase 120° spaced electrically for a synchro and two-phase 90° spaced electrically for a resolver. Since the rotating field produced is constant in amplitude and speed of rotation, the rotor voltage will be sinusoidal,

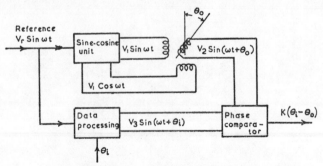

Fig. 14.10 A phase-detected synchro system.

of constant peak amplitude, but of phase dependent upon the position, θ_0. Thus, the resolver must be excited by equal amplitude, quadrature phase signals $V_1 \sin \omega t$ and $V_1 \cos \omega t$ and the induced rotor voltage will be $V_2 \sin (\omega t + \theta_0)$. A signal $V_3 \sin (\omega t + \theta)$ is generated by the input data processing and the actuating signal is developed by comparing this voltage in *phase* with the rotor signal as in Fig. 14.10. The phase detector will give a d.c. voltage proportional to the phase difference between the two a.c. signals.

Since the processing of variable phase signals, $V_3 \sin (\omega t + \theta)$, is much more complex than variable amplitude single phase signals, the rotating field arrangement is not in such common use as the alternating field.

15

D.C. Servomechanisms

15.1 Introduction

The components to be considered in this chapter are the d.c. servomotor and the means of controlling the motor torque, involving therefore some form of power amplification to convert the actuating signal to the full required output power. Various types of power amplifier are considered, ranging from valve amplifiers to motor-generator sets; it is an interesting viewpoint of 'systems' analysis that a d.c. generator can be classed with a transistor circuit as a power amplifier, the output power being controlled by the much lower power field current.

Most systems are of the continuous type where the actuating signal is directly amplified, producing a proportional motor torque. Systems which work on an 'on–off' principle must, however, also be considered.

15.2 D.C. Servomotors

The d.c. servomotor is the same as a conventional d.c. motor in principle, but with a few special design features. For many less demanding

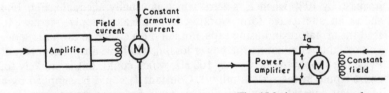

Fig. 15.1 Field control.   Fig. 15.2 Armature control.

systems a conventional motor can quite adequately be used, having the prime advantage of cheapness.

The special features desired of a servomotor are high torque and low inertia; the torque–inertia ratio is made high by reducing the armature diameter and increasing the length compared to conventional design. The motor should be compensated to stand transient changes in armature current or flux, and the commutation must be good over a wide range of speeds, exacting requirements since space does not always permit interpoles.

Two methods of controlling the motor torque are used:

(i) Constant armature current, variable field current—for use on low power (up to 1·5 kW) systems (Fig. 15.1).
(ii) Constant field current, variable armature voltage (Fig. 15.2). Also used with permanent magnet motors.

15.3 Field Control

The advantage of field control is that the input control power is small; the initial calculations are also simplified, since the torque is proportional to the control signal. The disadvantages are the power wastage in providing constant armature current and longer time constants, as shown in Section 15.8.

Constant I_a can be approximately achieved by connecting a large ballast resistor in series with the armature, fed at constant voltage, as in Fig. 15.3. If R is chosen, so that I_aR is, say, ten times bigger than the

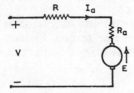

Fig. 15.3 Ballast-resistance constant-current source.

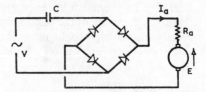

Fig. 15.4 Constant-current source using an a.c. supply and capacitors.

maximum value of the induced armature e.m.f., E, then I_a will only increase by 10% when E is zero, which is usually adequate. Fig. 15.4 shows an alternative form working from an a.c. supply, rectified to feed the armature and using capacitors in the a.c. side to drop the series voltage, thus eliminating the power loss. I_a will have a ripple content at twice the supply frequency, say 100 Hz, which could, but is unlikely to, cause ripple in the system output. Constant I_a could be supplied by a metadyne generator, but a similar set could better be used as armature control.

Since such systems are limited to small power applications ($< 1·5$ kW) the field power will be within the rating of an electronic amplifier. For this reason the specialist, small, (up to 200 W) d.c. servomotors are wound with a 'split-field', i.e. the field is centre-tapped, so that the field windings can be used as the collector (or anode) loads of push-pull amplifiers.

Most of these smaller systems are damped by means of velocity feedback, so that the motor and tachogenerator are marketed as a unit mounted on a common shaft and called a 'Velodyne'.

15.4 Armature Control

The basic system has been outlined in Fig. 15.2. In this case the field supply raises no particular problems, but the controlling amplifier must provide the full power requirements of the motor. The following section, then, is devoted to various forms of power amplifier.

15.5 Power Amplifiers

15.5.1 Ward–Leonard Control

Ward–Leonard control, Fig. 15.5, uses a d.c. generator as the power amplifier; the generator is driven at constant speed, usually by an induction motor, and the e.m.f. varied proportional to the generator field current; the a.c. supply to the induction motor provides the power,

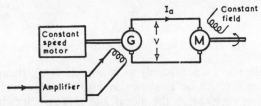

Fig. 15.5 A Ward–Leonard system.

the generator controls it. If the system is a high power one, then the generator field current will be at such a power level as to need more than an electronic amplifier, e.g. an amplidyne, or an auxiliary generator, although modern thyristor (S.C.R.) amplifiers can normally cope.

This is probably the cheapest method for high power applications, but suffers from long time constants.

15.5.2 The Amplidyne Generator

The amplidyne is a cross-field generator, which from the servo viewpoint can be considered as a d.c. generator which has a very much smaller field current than a similarly rated conventional machine. This is achieved by developing a small e.m.f. in the armature, due to the applied field as in a conventional machine, and then shorting the brushes together so that a relatively large current flows, termed the quadrature current. This current flowing in the armature sets up a large armature reaction field which induces an e.m.f., at right angles to the quadrature current, which is used as the output, taken from a second pair of brushes. The technique of using a small applied field to produce a larger main field leads to the amplidyne being often called a 'rotary amplifier'. A variety of auxiliary field windings are used to give a wide variety of load characteristics to the general-purpose amplidyne, but these are of no consequence in this application.

The advantage of the amplidyne in place of a conventional generator in a Ward–Leonard system is that electronic amplifiers can be more readily used. Reference 13 covers the theory of cross-field generators.

15.5.3 Controlled Rectifiers

A conventional rectifier is used for converting a.c. to d.c. The rectifier characteristics are such that with the 'anode' positive with respect to the 'cathode' the resistance is low; with a reversed polarity the resistance becomes very high. If, then, a d.c. motor is connected in series

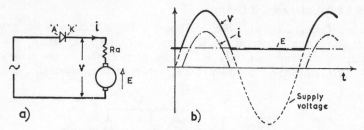

Fig. 15.6 Half-wave rectifier circuit and waveforms.

with a rectifier to an a.c. supply as in Fig. 15.6 (a), then the voltage across the motor will be as shown in Fig. 15.6 (b) which has a d.c. component as required by the motor, but also an undesirable 'ripple' of large amplitude at supply frequency. The rectifier 'cuts off' when the supply voltage is less than the motor e.m.f., E; the diagram is drawn for a constant value of E, which in fact will vary as the motor changes speed.

The use of full wave rectification as in Fig. 15.7 doubles the d.c. component and reduces the ripple, which is now at twice supply frequency.

Use of a three-phase supply improves the d.c. component and reduces

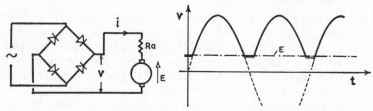

Fig. 15.7 Full-wave rectification.

the ripple further, while still further improvement could be obtained with full-wave rectification of a three-phase supply.

The higher the ripple frequency, the less likely is the motor to follow, so that the ripple should be of minimum amplitude and maximum frequency. For most servos the 50 Hz mains is adequate, but 400 Hz has also been used, particularly for small motors.

To vary the d.c. component of the rectified voltage with any of the conventional rectifier systems shown above, the a.c. supply voltage must be varied. This is not a very practical form of control, although motorised 'Variacs' have been used.

Controlled rectifiers are three-terminal devices. They are a form of rectifier which, as convential rectifiers, are non-conducting with negative anode–cathode voltage, but are also non-conducting with positive anode–cathode voltage until switched on by a control signal applied to the third 'gate' terminal. Once 'struck', controlled rectifiers behave as normal rectifiers with a small anode–cathode volt drop; they cannot be turned off by gate signals but only by interrupting the anode current. They are ideally suited to use with a.c. supplies since they are automatically switched off each cycle as the supply goes negative. Their use with d.c. supplies, of little consequence to d.c. servos, is complicated by the requirement of negative pulses applied to the anode to switch the rectifier off.

The various forms of controlled rectifier used are:

(i) The Thyratron—a hot cathode, gas-filled triode for low powers.

(ii) The Ignitron—a cold cathode, gas-filled valve for powers up to 1 kW.

(iii) The Mercury-arc rectifier, grid controlled—for very high power applications.

(iv) The Thyristor also called a Silicon Controlled Rectifier (S.C.R.). This is the semi-conductor equivalent of the thyratron. It is suitable for all powers and is still being developed to such an extent as to make the other types of controlled rectifier obsolete. See reference 14 for details of ratings, circuitry, etc.

The control signal is a voltage for (i) (ii) and (iii) but a current for the thyristor. Typical thyratron characteristics are shown in Fig. 15.8; it is

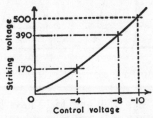

Fig. 15.8 Control characteristic for a type of controlled rectifier (Thyratron).

not practical to show typical thyristor characteristics, but they will not conduct with zero gate current for anode–cathode voltages less than 500–1500 volts. The supply voltage is kept below this level and the thyristor can then be switched on by a current pulse, of sufficient

duration and voltage to supply the necessary minimum gate energy
condition, for all positive anode–cathode voltages.

Control is sometimes possible by d.c. levels of gate signal as shown
in Fig. 15·9 for the thyratron of Fig. 15.8, but this is impractical, i.e.
$v_c > 10$ volts inhibits conduction completely so that control is only
possible over the first quarter cycle. Full control can be achieved by

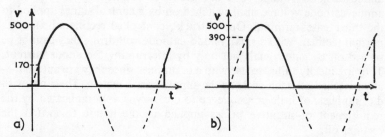

Fig. 15.9 Output voltage waveforms for a controlled rectifier system similar to
Fig. 15.6, with $E = 0$, for d.c. control voltages of: (*a*) −4 volts and (*b*) −8 volts.

variable phase a.c. control signals, commonly used on mercury-arc
rectifiers, but for servos by using pulse switching. The rectifier is non-
conducting (this will require a negative d.c. bias control voltage except
for the thyristor) until switched on by a pulse of sufficient amplitude at
the appropriate instant in time (phase). Two examples are shown in
Fig. 15.10. A trigger unit is used which generates a train of pulses of
supply frequency, the 'phase angle' of which is varied proportional to a
d.c. input voltage (the actuating signal) between 180° ($\alpha = 0$) for zero
and 0 ($\alpha = 180°$) for maximum input.

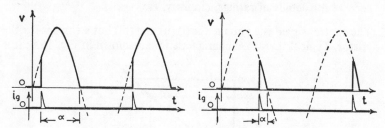

Fig. 15.10 Pulse control of a controlled rectifier.

The waveforms in Figs. 15.9 and 15.10 have been drawn for single
phase rectification with $E = 0$, but as previously explained full-wave,
three-phase rectification has decided advantages, and non-zero E will
affect the conduction period.

The controlled rectifier is not reversible, so that the d.c. output is only
of a fixed polarity. This, in fact, may be adequate for some systems
such as a velocity regulator, in which the actuating signal is always

positive. For position control systems, however, parallel amplifiers must be employed as in Fig. 15.11. Problems arise in the accurate timing and synchronising of the pulses required; also if both positive and negative rectifiers are switched on simultaneously, short circuiting occurs, so that the polarity element must incorporate a delay of one pulse before changing its output line, ensuring that one rectifier is always switched off before the other is switched on.

Regeneration requires additional rather complex circuitry since for deceleration, V_{mean} demanded by the actuating signal is less than E (the 'motor' is required to generate) the rectifiers cut off, so that the maximum rate of deceleration is the unassisted run-down rate of the machine. With the parallel amplifier system, regeneration will occur

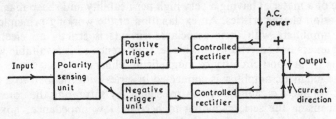

Fig. 15.11 Schematic diagram of a 'positive and negative' controlled rectifier amplifier.

when the negative amplifier is switched on, but this will only occur for negative V_{mean}, while regeneration is often required for a positive V_{mean}, less than E. The controlled rectifier amplifier is thus markedly nonlinear and difficult to analyse. All the same, they are very satisfactory for the bulk of velocity control systems, in which the natural run-down rate is not exceeded, and it is anticipated that future detail developments will make them equally suitable for more complex systems.

Current limiting is essential—consider the severe example of a parallel amplifier unit running in the steady state with $V_{mean} = +50$ volts $E = +45$ volts and $R_a = 0.5$ ohms; the steady current fed from the positive amplifier is $\frac{V - E}{R_a} = 10$ amperes. The input is now reversed, switching to the negative amplifier with $V_{mean} = -50$ volts. E is unaltered (until the speed changes) so that the current fed from the negative amplifier is $\frac{-50 - 45}{0.5} = -180$ amperes. Current limiting is achieved practically by measuring the p.d. across a series resistor and feeding this into the control circuits so as to switch off the amplifier when the p.d. exceeds a pre-set limit—note yet another discontinuity in the amplifier characteristics.

Finally, a diode is normally connected across the output in reverse polarity, i.e. it is open circuit for the normal amplifier output. Termed a 'flywheel' diode, its purpose is to short out any reverse polarity pulses which are fed back from the load, due to inductive effects which could otherwise be of such magnitude that even when the supply voltage is maximum negative it is still positive with respect to the pulse voltage on the cathode of the rectifier, which is therefore not switched off and is not controlled for the next cycle.

15.5.4 Magnetic Amplifiers

Another form of power amplifier working on not dissimilar lines to a controlled rectifier, although by no means the same principles, is the magnetic amplifier, based on the saturable reactor or transductor. These devices are similar in construction to a transformer, but the core is made of a material having a very high permeability and sharp magnetic saturation characteristics. An explanation of the working principles of such amplifiers with their associated circuits is strictly an electrical topic and is not entertained in this text; reference 15 is a suitable work on the theory and circuitry of magnetic amplifiers.

The magnetic amplifier is connected in series with the load to a constant voltage a.c. supply. A control signal, by saturating the core and thus causing the series winding to become low impedance, governs the portion of the supply period during which current flows, so that the average load current is proportional to the control signal; hence the similarity to a controlled rectifier. The output must be rectified, because of which, and the necessity of an internal feedback connection in the circuitry, the output is of one polarity. The units are thus used in pairs for reversible types of drives.

Magnetic amplifiers have on the whole been replaced in new designs by the thyristor but they are still reliable and cheap. In fact, certain commercial thyristor amplifiers utilise magnetic amplifiers for the trigger circuits of the output stages.

The amplifier may be regarded as having one time constant; it is in fact more a finite delay, but this varies as the input signal varies.

15.6 Other Forms of Continuous Control

15.6.1 Variable Couplings

A coupling which is driven at constant speed with the output torque proportional to an electric control signal can also be used for servo control. The advantage of this system is that a simple induction motor can be used for the drive motor and that the power level of the control signal is low.

Fig. 15.12 shows a sectionalised diagram of a commercial unit. The coupling is an eddy current device with a constant speed outer rotor and a fixed concentric coil magnetic system; the torque produced by the inner rotor, which is the output, is proportional to the d.c. control field

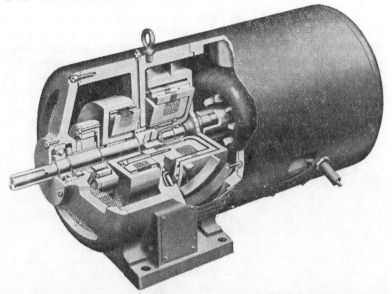

Fig. 15.12 An eddy current coupling drive unit (courtesy of Newbridge Engineers Ltd., Bury).

current (note that reversal of current does not reverse the torque, this being a function of the direction of rotation of the prime mover). Correct design of the magnetic circuit results in a nearly linear torque–speed curve, shown in Fig. 15.13. The speed range and power rating is limited by heating of the rotors.

Since current reversal does not produce negative torque (it in fact

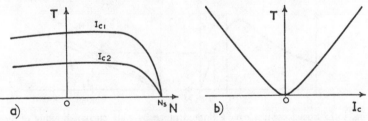

Fig. 15.13 Eddy current coupling characteristics: (*a*) torque–speed, $I_{c1} > I_{c2}$ N_s = synchronous speed, (*b*) torque–control current, N constant.

causes positive feedback in a loop and should be eliminated by a rectifier in series with the coil) there is no inherent regeneration as previously discussed at the end of Section 15.5.3. This is overcome by using a second coupling on the output shaft producing a torque against a fixed 'rotor', thus producing braking; such a device is integral, as is the drive motor, in the unit of Fig. 15.12. The amplifier feeding the coils is effectively a push–pull type, feeding the drive unit for positive input signals and the brake unit for negative signals, resulting in a linear characteristic as shown in Fig. 15.14. Note that this driving–braking characteristic is only achievable provided the output is not required to drive in the opposite direction to the prime mover—reverse drive can never be achieved, so that such devices are used exclusively for velocity and tension controls.

One of the major advantages is good low-speed performance resulting

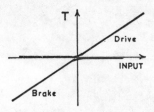

Fig. 15.14 Characteristic of an eddy current coupling with a second coupling for braking.

in speed ranges of 1000 : 1 compared to 50 : 1 with more conventional thyristor–d.c. motor controls.

15.6.2 Variable Brakes

For some speed control drives where the range of output speed is limited, it is possible to use a motor which has a linear torque–speed characteristic over the desired speed range and to apply control by means of a brake, the torque of which is proportional to an electrical control signal. For this purpose an induction motor, the speed-torque

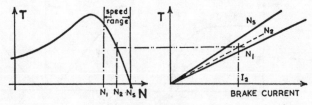

Fig. 15.15 Speed control over a limited range by eddy current braking of an induction motor.

characteristic of which is shown in Fig. 15.15, can be used and the brake can be of the eddy current type, since the speed range is small. All the braking power must be supplied by the brake which will have a low efficiency, so that this method of control is only suitable for low power, constant mean speed applications.

Disc brakes can be used for a limited range of tension controls when back tension only is required—obviously no drive can be achieved.

15.7 Relay Servomechanisms

15.7.1 The General On–Off Type of Control

All the systems considered previously have been continuous in nature, if not necessarily linear. A type of servo is possible in which the motor torque (or other power source, e.g. room heating is an obvious example) is either full on or off, or possibly reversed. A positive error will switch the drive full on in the correct direction and a negative error will cause full reversal. A refinement is a null centre position giving no output for small errors. The basic system applied to a position control is shown in

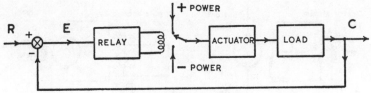

Fig. 15.16 A relay system.

Fig. 15.16. These systems are known variously as 'Bang–bang', 'on-off' or 'discontinuous' as well as relay servos.

Considering relay systems in general:

Advantages—

(i) The system is basically simple, requiring no control signal amplifier and no precision components.

(ii) Because of (i) they are cheap.

(iii) Since maximum correcting torque is produced for any error, and not a torque proportional to the magnitude of the error, very rapid response can be achieved.

Disadvantages—

(i) The relay which is required to perform a large number of interruptions of large currents in inductive circuits is prone to unreliability.

(ii) The stability compared to continuous control is reduced as was shown in the describing function analysis of Section 9.4.5.

(iii) Because of (ii) the application of relay control must be limited to less demanding applications. This is not helped by the difficulty of analysis.

Consider now the components of the system of Fig. 15.16 which are special to relay control.

(*a*) The load, monitor, error detector, etc., will be the same as for continuous control, although the error detector and any possible amplifier required to drive the relay coil and provide the loop gain will not have any stringent requirements.

(*b*) The motor will be subjected to a maximum forward torque and then, as the output is driven past the desired position, thus reversing the actuating signal, being subjected to full reverse torque, a process referred to as 'plugging'. The d.c. motor is commonly used for this application, although good compensation to avoid brush flash over will be necessary, since reversal of the control signal will produce full reversal of torque, resulting in good 'braking'. This is not true of the a.c. motor since it may be working at a large 'slip', thereby producing only a small braking torque when reversed.

(*c*) The relay unit required to switch the power source. The relay

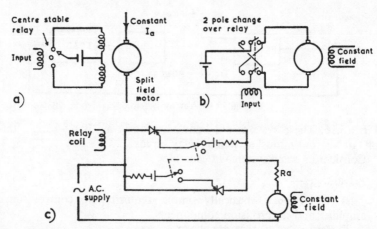

Fig. 15.17 Some simple relay servos: (*a*) field control, (*b*) armature control, (*c*) thyristor assisted switching—(*b*) and (*c*) in practice require additional current limiting.

contacts can be connected direct into the circuit to switch either the field of the motor or the armature voltage. In either case the contacts will have to be carefully rated. Both alternatives are shown in Fig. 15.17 along with a system using a thyristor to perform the high power switching, controlled by the relay but using an a.c. supply— always an advantage. The practical electromagnetic relay introduces another problem into the analysis in the form of hysteresis, the 'drop-out' signal being smaller than the 'pull-in' signal. Further, the time taken for the contacts to change over adds a finite time delay, which, however, can usually be neglected. The R–L time constant of the relay coil must be catered for as conventionally.

15.7.2 Mark–Space Ratio Control

First look at a simple continuous control system as shown in Fig. 15.18, where the motor armature current is controlled by a variable series resistance, in this case a transistor, the effective resistance of which is proportional to the control signal.

The disadvantage of this method of control is the power dissipation in the transistor. However, at the two extremes of zero and maximum current the power dissipated by the transistor is zero, in the first case since the current is zero and in the second since its resistance is zero.

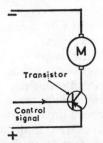

Fig. 15.18 Series 'resistance' control using a transistor.

Thus a method of control which is a cross between relay and continuous control can be devised by switching fully on and off at a fixed repetition rate and varying the ratio of the on to the off period, termed the mark–space ratio, proportional to the control signal; the relay circuit of, say, Fig. 15.17 (*a*) could be used. The useful output torque is now the mean value of the repetitive waveform, as shown in Fig. 15.19.

If the switching rate is high compared to the maximum response

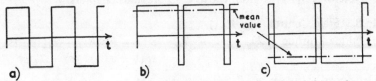

Fig. 15.19 Mark–space ratio control: (*a*) zero mean torque, (*b*) high forward torque and (*c*) high reverse torque.

frequency of the system, then the system acts as a low-pass filter and the torque is proportional to the control signal. This is then effectively a continuous system.

Circuitry for developing a switching signal with mark–space ratio proportional to the d.c. level of the control signal can be found in any good electronics textbook.

15.8 Transfer Functions of Continuous D.C. Motor Systems

15.8.1 Notation

$E(s)$ or $e(t)$ = input voltage, actuating signal, volts

$A(s)$ = voltage amplifier transfer function, volts/volt

i_f = field current, amperes
i_a or I_a = armature current, amperes
e_a or E_a = induced armature e.m.f., volts
T = total motor torque, N m
K_e = e.m.f. constant with constant flux, volts per rad/s
K_T = torque constant with constant flux, N m per ampere i_a
K_2 = torque constant with constant I_a, N m per ampere i_f
R_f = field resistance, ohms
L_f = field inductance, henry
$T_f = L_f/R_f$ = field time constant, seconds
R_a = armature circuit resistance, ohms
L_a = armature circuit inductance, henry
$T_a = L_a/R_a$ = armature circuit time constant, seconds
J = total inertia referred to the output, kg m^2
F = total viscous friction coefficient referred to the output, N m s
$T_m = J/F$ = mechanical time constant, seconds
T_L = applied load torque, N m
n = reduction gear ratio
θ_o = output position, rad
θ_m = motor position, rad

N.B. for a d.c. motor $E_a \propto \Phi\theta_m$ and $T \propto \Phi I_a$.

If Φ is constant, $E_a = K_e\theta_m$ and $T = K_T I_a$.
With the torque in newton metres, $K_e = K_T$.
Note that A/R_f has been called G in previous chapters.

15.8.2 Field Control (Fig. 15.1)

Total load torque $= J\ddot\theta_o + F\dot\theta_o + T_L \equiv s(sJ + F)\Phi_o(s) + T_L(s)$
$$= sF(1 + sT_m)\Phi_o(s) + T_L(s)$$

Developed torque $= n \cdot K_2 \cdot i_f \equiv n \cdot K_2 \cdot A(s) \cdot \dfrac{E(s)}{R(1 + sT_f)}$

$$\therefore s\Phi_o(s) = \frac{nK_2A(s)}{RF(1 + sT_f)(1 + sT_m)} \cdot E(s) - \frac{1}{F(1 + sT_m)} \cdot T_L(s)$$

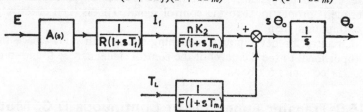

Fig. 15.20 Transfer function of a field-controlled d.c. motor.

Remember that $A(s)$ will involve certain time constants. Fig. 15.20 shows the block diagram of this system, which has been previously derived in Chapter 4.

15.8.3 Armature Control (Fig. 15.2)

$$v = e_a + i_a R_a + L_a \frac{di_a}{dt}$$

$$\therefore \quad V(s) = E_a(s) + (R_a + sL_a)I_a(s)$$

but

$$e_a = nK_e\dot{\theta}_o \quad \text{and} \quad i_a = \frac{T}{K_T}$$

hence

$$V(s) = nK_e s\Phi_o(s) + (R_a + sL_a)\frac{T(s)}{K_T}$$

But

$$nT(s) = sF(1 + sT_m)\Phi_o(s) + T_L(s)$$

$$V(s) = nK_e s\Phi_o(s) + \frac{R_a + sL_a}{nK_T}[(sF(1 + sT_m)\Phi_o(s) + T_L(s)]$$

$$= \left[\frac{FL_aT_m}{nK_T}s^2 + \frac{F(L_a + R_aT_m)}{nK_T}s + \left(\frac{FR_a}{nK_T} + nK_e\right)\right]s\Phi_o(s)$$
$$+ \frac{R_a + sL_a}{nK_T}T_L(s)$$

$$= \frac{FR_a + n^2K_eK_T}{nK_T}(as^2 + bs + 1)s\Phi_o(s) + \frac{R_a + sL_a}{nK_T}T_L(s)$$

where

$$b = \frac{F(L_a + R_aT_m)}{FR_a + n^2K_eK_T} = \frac{T_a + T_m}{1 + \dfrac{n^2K_eK_T}{FR_a}}$$

and

$$a = \frac{FL_aT_m}{FR_a + n^2K_eK_T} = \frac{T_aT_m}{1 + \dfrac{n^2K_eK_T}{FR_a}}$$

Now

$$V(s) = A(s).E(s) \quad \text{and} \quad R_a + sL_a = R_a(1 + sT_a)$$

so

$$s\Phi_o(s) = \frac{nK_T A(s)}{FR_a + n^2K_eK_T} \cdot \frac{1}{as^2 + bs + 1} \cdot E(s)$$
$$- \frac{R_a(1 + sT_a)}{FR_a + n^2K_eK_T} \cdot \frac{1}{as^2 + bs + 1}T_L(s)$$

This function is shown in block diagram form in Fig. 15.21.

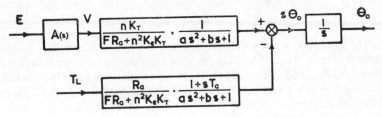

Fig. 15.21 Transfer function of an armature-controlled d.c. motor.

Certain approximations can be made which are adequate for most purposes, viz.:

assume $\quad L_a = 0 \quad$ and $\quad n^2 K_e K_T \gg F R_a$

thus $\quad\quad a = 0 \quad$ and $\quad b = \dfrac{J R_a}{n^2 K_e K_T}$

hence $\quad s\Phi_o(s) = \dfrac{A(s)}{nK_e} \cdot \dfrac{1}{1 + sb} \cdot E(s) - \dfrac{R_a}{n^2 K_e K_T} \cdot \dfrac{1}{1 + sb} \cdot T_L(s)$

This motor time constant, b, is much smaller than the simple time constant, $T_m = J/F$, as applicable for field control. The effective damping is thus much greater, due to the effective velocity feedback of the armature e.m.f.

15.9 Motor Speed–Torque Characteristics

From the previous two sections, expressions relating the steady state torque, speed and control signal can be derived and plotted as in Fig. 15.22.

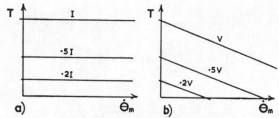

Fig. 15.22 D.C. motor torque-speed curves: (*a*) field, and (*b*) armature controlled.

(*a*) Field control:

$$T = K_2 i_f \quad \text{for all} \quad \dot{\theta}_m$$

(*b*) Armature control:

$$T = K_T i_a$$

and $\quad\quad v = e_a + i_a R_a \quad$ in the steady state

Thus $\quad\quad T = K_T \cdot \dfrac{v - e_a}{R_a} = K_T \cdot \dfrac{v - K_e \dot{\theta}_m}{R_a}$

$$\quad\quad = \dfrac{K_T}{R_a} \cdot v - \dfrac{K_T K_e}{R_a} \cdot \dot{\theta}_m$$

The torque, T, is the total torque produced by the motor. The motor output torque, T_o, is T less the torque losses due to iron, friction, etc., losses in the motor. Since the motor is used with a load and the viscous friction and other motor losses are included as part of J, F and T_L it is most misleading to consider T_o instead of T, although T_o must be measured by necessity in testing the motor.

15.10 D.C. Tachometers

A tachometer is a device for measuring speed, used in servos either as a monitor for speed control, or for providing derivative of output for damping. A d.c. tacho is a precision generator, usually with a permanent magnet field system, so that the output voltage is directly proportional to speed and reverses polarity with reversal of direction. The major problem with tachos is 'brush noise', an electrical voltage generated by the brushes moving from one commutator segment to the next, which is about 2% of the maximum output signal. The noise frequency is equal to the number of commutator segments times speed of rotation. A tacho should therefore be geared, if possible, to run at its highest possible speed, giving a larger output signal and improving the signal to noise ratio. This also helps to keep the noise frequency as high as possible.

Since the tacho is used as an instrument there is little or no electrical loading on it so that armature resistance drop and armature reaction raise no problems. The linearity of the voltage–speed characteristic is inherently good and may need to be particularly so for a speed monitor, but is of little importance when used for damping.

16
A.C. Servomechanisms

16.1 Introduction

A.C. servos, systems using a.c. motors, can be split into two categories:

(i) The demodulator–modulator system, Fig. 16.1, in which the a.c. error signal is demodulated; in this form it resembles the signals in a d.c. servo. This signal can now be compensated by any of the series techniques described in Section 8.4. The compensated

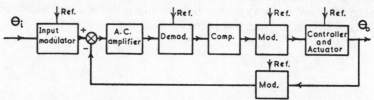

Fig. 16.1 Demodulator–modulator compensated a.c. servo.

signal must then be used to remodulate the carrier into such a form that it can be fed to the actuator.

(ii) The all-a.c. system, Fig. 16.2, in which the compensation is performed directly on the modulated carrier, by using a selective network tuned to the carrier frequency.

Tachometric velocity feedback can be employed in the same way as in d.c. servos to either of the two classes of a.c. servo.

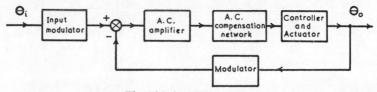

Fig. 16.2 An 'all-a.c.' servo.

Before continuing with this discussion it may be necessary to enlarge upon some of the electrical terms used so far.

(*a*) *Carrier*. The a.c. servomechanism uses devices for monitoring and actuation which require a.c. signals. To vary the 'information content' of these signals, the basic a.c. signal must be modified according to some function such as speed, position, etc. The basic unmodified signal is called the 'carrier' and is of constant frequency, ω_c rad/s. The servo must respond to the modifications of the carrier, but not at all to the carrier itself. For this reason ω_c should be at least ten times the system bandwidth, so that 400 Hz is a more common carrier frequency than the 50 Hz mains.

(b) *Modulation.* The process of modifying the carrier in sympathy with the information signal is called 'modulation'. A modulated carrier is the carrier signal with information impressed on it.

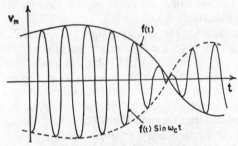

Fig. 16.3 Amplitude modulation.

There are three methods of modulation which stem from the basic equation of the carrier, thus

$$v_c = V_c \sin (\omega_c t + \phi)$$

(i) Amplitude modulation. V_c is modulated, so that if the information signal is $f(t)$ and letting $\phi = 0$:

$$v_m = V_c . f(t) . \sin \omega_c t \quad \text{as shown in Fig. 16.3}$$

Note that the reversal of sign of $f(t)$ is equivalent to 180°-phase shift of the carrier.

(ii) Frequency modulation. ω_c is modulated by $f(t)$

$$\therefore \ v_m = V_c \sin \omega_c (1 + af(t))t$$

(iii) Phase modulation. ϕ is modulated by $f(t)$.

$$\therefore \ v_m = V_c \sin (\omega_c t + bf(t))$$

Amplitude modulation is the only case of consequence in control systems. Note that pulse modulation is a special case of on–off amplitude modulation.

(c) *Demodulation* is the process of extracting the information $f(t)$ from the modulated carrier. For the relevant case of amplitude modulation, $f(t)$ can be extracted by filtering the output from a 'phase sensitive rectifier', a rectifier circuit which retains the essential polarity information of $f(t)$.

In order to sense the polarity of $f(t)$ it is necessary to compare the phase of the modulated carrier with that of the reference signal.

(*d*) *The reference signal.* This is a constant amplitude signal of the same phase and frequency as the carrier. In other applications a similar signal but in phase quadrature with the carrier is required, termed the quadrature reference.

16.2 Modulators and Demodulators

The electronic type of modulator required when re-modulating the compensated, demodulated actuating signal can be found in most standard electronics textbooks.

Devices for producing the actuating signal as a modulated carrier are of specific importance to the a.c. servo; fortunately the Synchro or Resolver, described in Section 14.11, which is one of the cheapest and best position transducers, is ideally suited. Note, however, the use of a.c. servos as position controllers almost exclusively. As explained in Section 14.11 a correctly connected output and input pair of synchros will develop an electrical signal $V \sin (\theta_i - \theta_o) . \sin \omega_c t$. This, then, is a voltage of carrier frequency of amplitude $V \sin (\theta_i - \theta_o)$, whereas the required signal should be modulated by $V(\theta_i - \theta_o)$. If $\theta_i - \theta_o$ is less than $10°$, these two functions are approximately equal, with improving accuracy as desired position is approached.

The simple potentiometer can still be used, employing the reference signal to excite the pots.

The only demodulators of interest are phase-sensitive rectifiers, again best covered in electronics textbooks, but outlined in Section 14.11, particularly Fig. 14.9.

16.3 A.C. Servomotors

The three-phase induction motor is a most desirable machine on many counts, particularly cost and reliability. It is unfortunate from the servo aspect, then, that the starting torque is low and that the torque–speed characteristic is nonlinear and difficult to control. For a servomotor the torque characteristic can be improved by using high resistance rotors, incidentally losing some of the cost advantage, but worse, increasing the power dissipation in the rotor. Work outside the scope of this text is being done on thyristor circuits to utilise the bigger three-phase motors, but the only useful motors in this context are small two-phase motors. Thyristor speed control is explained in Section 16.7.

With an induction motor there is a maximum speed set by the 'synchronous' speed. This is a rotor speed equivalent to the speed of rotation of the magnetic field set up by the applied a.c. signals.

Typical conventional induction motor torque–speed curves are shown in Fig. 16.4, from which it can be deduced that variation of torque with applied voltage is nonlinear as well as being speed dependent.

The two-phase induction motor is wound with two stator coils physically at right angles, which when fed with sinusoidal currents 90° out of phase with each other produce the necessary rotating magnetic field to cause the motor to function.

If stator 1 current is $\qquad i_1 = I_1 \sin \omega_c t$

and the other is $\qquad\qquad i_2 = I_2 \cos \omega_c t$

then the mean torque is proportional to $I_1 . I_2$ provided that the magnitudes I_1 and I_2 are not modulated at a rate higher than about $0.2\omega_c$.

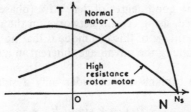

Fig. 16.4 Induction motor torque-speed characteristics.

The method of operation, then, is to make I_2 constant so that the torque is proportional to I_1, which is the amplitude of the control signal; the modulated carrier is used direct. The characteristics are plotted in Fig. 16.5.

These characteristics can be assumed linear over a limited working range, so that a transfer function similar to that of the armature fed d.c.

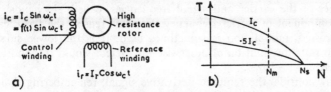

Fig. 16.5 (*a*) An a.c. servomotor, (*b*) torque-speed characteristics. N_m is the maximum 'linear' working speed.

motor can be derived as in Section 15.8.3. The limitation of the speed range can be seen by comparing Fig. 16.5 (*b*) with Fig. 15.22 (*b*).

Thus $\qquad\qquad T = K_I I_c - K_N N$

where K_I is the slope of the torque-control current curve and K_N is the slope of the torque-speed curve. The fact that K_I and K_N are *not* constants is a limiting factor in the analysis and performance of a.c. servos.

Finally, the a.c. servomotor is very limited in size due to the heating caused by the constant reference field. As the motor size is increased the ratio of heat produced to heat dissipation goes up, so that it is not practical to make motors of a rating above about 200 W.

16.4 Modulator–Demodulator Systems

In this type of system the actuating signal is isolated from its carrier by the demodulator, compensated by the standard series techniques, and the corrected signal used to remodulate the carrier before feeding to the motor. The design of such a system is no different to that of a d.c. system, although the signal frequencies must not exceed 20% of the carrier frequency.

16.5 All-A.C. Systems

It is not possible to achieve a wide range of compensation without demodulating, but a good degree of derivative (phase lead) compensation can be achieved, using certain networks on the modulated carrier direct. The big advantage of this technique is that no d.c. amplification is required, maintaining the advantage inherent in a.c. systems of low drift.

To simplify the explanation, consider only a sinusoidal actuating signal,

$$f(t) = V \sin a\omega_c t$$

where a is a constant $<0\cdot 2$ for good servo operation.

The modulated carrier which is to be derivative compensated is thus

$$v_m = V \sin a\omega_c t \,.\, \sin \omega_c t$$

$$= \frac{V}{2} [\cos (1 - a)\omega_c t - \cos (1 + a)\omega_c t] \qquad (16.1)$$

It is thus shown that the sinusoidally modulated carrier comprises two sinusoidal components of frequencies equal to the sum and difference of the carrier and signal frequencies, called 'sidebands', with no component of carrier frequency. For this reason the term 'carrier suppressed' modulation is sometimes used. It is the lower sideband which puts the limit on a, otherwise the system would respond to the frequency $(1 - a)\omega_c$.

Now consider the required derivative signal, remembering that it is $f(t)$ that must be differentiated and not v_m.

Thus $$\frac{df(t)}{dt} = a\omega_c V \cos a\omega_c t$$

The corresponding modulated carrier is

$$v_d = V a\omega_c \cos a\omega_c t \sin \omega_c t$$

$$= V \frac{a\omega_c}{2} [\sin (1 - a)\omega_c t + \sin (1 + a)\omega_c t]$$

$$= \frac{V}{2} a\omega_c \left[\cos \left((1 - a)\omega_c t - \frac{\pi}{2} \right) + \cos \left((1 + a)\omega_c t - \frac{\pi}{2} \right) \right]$$

$$= \frac{V}{2} a\omega_c \left[\cos \left((1 - a)\omega_c t - \frac{\pi}{2} \right) - \cos \left((1 + a)\omega_c t + \frac{\pi}{2} \right) \right] \qquad (16.2)$$

In this form the derivative-modulated carrier can be compared with the direct-modulated carrier (equations 16.2 and 16.1). Hence the requirements for derivative compensation are that the sideband amplitudes must both be proportional to $a\omega_c$ and that the lower sideband should lag the carrier by 90° while the upper sideband leads by 90°. Fig. 16.6

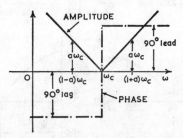

Fig. 16.6 Characteristics of a network which when fed with a modulated carrier will produce an output of the carrier, modulated by the derivative of the input modulation.

shows the phase and amplitude requirements of the derivative action. It must be stressed that straightforward differentiation of the modulated carrier will differentiate the upper and lower sidebands which is *not* differentiating the signal $f(t)$.

The degree of derivative compensation required is a matter of the design specification, particularly the degree of damping, so that in general the compensated actuating signal will be $v = v_c + Av_d$, where A is a constant.

There is no practical network which will achieve this desired result

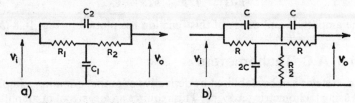

Fig. 16.7 (*a*) Bridged-T network, and (*b*) parallel-T network.

but two types of network give reasonable approximation, particularly for $a < 0.2$:

(i) Bridged T network, Fig. 16.7 (*a*), gives a result approximating to $v_c + Av_d$.

(ii) Parallel T network, Fig. 16.7 (*b*), gives a result approximating to v_d only, so that the proportional term must be scaled and added in separately.

Both these networks have the quite serious problem of being dependent upon the carrier frequency, ω_c.

For the bridged T network, if $R_1C_1 = T_1$ and $R_2C_2 = T_2$, the transfer function is

$$\frac{V_o(s)}{V_i(s)} = \frac{T_1T_2s^2 + (R_1 + R_2)C_2s + 1}{T_1T_2s^2 + (T_1 + (R_1 + R_2)C_2)s + 1}$$

Put $T_1T_2 = \dfrac{1}{\omega_c^2}$ and $\alpha = \dfrac{V_{o(j\omega)}}{V_{i(j\omega)}}\bigg]_{\omega=\omega_c} = \dfrac{(R_1 + R_2)C_2}{T_1 + (R_1 + R_2)C_2}$

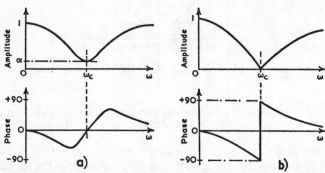

Fig. 16.8 Phase and amplitude versus frequency characteristics of: (*a*) bridged-T (proportional + derivative), and (*b*) parallel-T (derivative only) networks.

The amplitude and phase diagrams of this function are plotted in Fig. 16.8 (*a*).

For the parallel T network, if $CR = T = \dfrac{1}{\omega_c}$, the transfer function is

$$\frac{V_o(s)}{V_i(s)} = \frac{T^2s^2 + 1}{T^2s^2 + 4Ts + 1}$$

The amplitude and phase of this function are plotted in Fig. 16.8 (*b*).

16.6 A.C. Tachometers

A.c. systems can be stabilised by derivative of output damping; simplest by using tachometric feedback. This requires an a.c. signal of amplitude

Reference
$V_r \sin \omega_c t$　　　　　　Aluminium
　　　　　　　　　　　　　cup rotor

　　　　　　　　　　　Output
　　　　　　　　　= K.ω.Sin ω_c t

Fig. 16.9 An a.c. tacho.

proportional to velocity but of the same phase and frequency as the carrier. A conventional a.c. generator will not suffice, since the frequency as well as amplitude are proportional to speed. The tacho used is a 'drag-cup' device in which the rotor is a thin aluminium cup. The stator is wound with two coils physically at right angles, one, the excitation winding, being connected to the reference supply, the other being the output winding. When the rotor is stationary there is no output signal, since the two stator windings are at right angles. When the rotor is moving, however, induced e.m.f.s, due to cutting the reference field, are set up in the rotor which link with the output winding. The magnitude of the signal in the output winding is then proportional to the speed of the rotor over a working range, the frequency and phase being that of the reference signal.

16.7 Thyristor Speed Control of A.C. Motors

The speed of rotation of an induction motor is largely proportional to the frequency of the a.c. supply, provided there is sufficient current to maintain the torque. Thus, over a limited range, about 5% to full, the speed can be controlled by controlling the frequency of the supply. In view of the standard fixed frequency power supplies available, this is a major problem.

Thus a frequency controlled, high power oscillator is required. This is achieved by using thyristors to 'chop' a d.c. supply on and off at a controlled rate, resulting in a square wave of the desired frequency; multi-phase thyristor circuits result in a waveform which is a better approximation to a sine wave. Such a 'power oscillator' is termed a thyristor invertor. The raw d.c. is provided by rectification of the standard a.c. power supplies.

Thus the cascaded rectifier–invertor system acts as a frequency convertor of controlled output frequency.

One further point to be observed is that the magnetic flux should be kept constant. The flux is proportional to the ratio of voltage to speed, so that the invertor must be designed to reduce the voltage at lower speed requirements, i.e. the voltage/frequency ratio is to be kept constant. Commercial systems are now available (ref. 14).

The speed is controlled by adjustment of the frequency (a low power oscillator in the thyristor switching circuitry) so that this is essentially an open loop method of control.

17
Hydraulic Servomechanisms

17.1 Introduction

Electric motors produce torque, due to the electromagnetic forces existing on a current carrying conductor in a magnetic field. The available torque is limited by two factors:

    (i) magnetic saturation;
    (ii) heating due to internal I^2R losses.

Hydraulic motors produce torque, due to pressurised oil. The limit on the available torque is set by the maximum safe value of pressure. Internal heating, which will set a limit to the working range of the motor, is determined by the oil flow rate and therefore is a function of motor speed rather than torque.

In practice, the limit set by magnetic saturation and heating is far more stringent than that set by safe working pressures, so that for a given physical size, the hydraulic motor is capable of producing as much as twenty to thirty times the torque of an electric motor. This is best expressed by saying that the hydraulic motor has a higher torque–inertia ratio, and is therefore capable of higher acceleration and better transient performance.

Another feature of the use of hydraulics is the availability of a linear drive in the form of a ram or jack.

The following analysis of hydraulic servos is considered in three categories:

    (i) the source of hydraulic power;
    (ii) the motor; and
    (iii) methods of controlling the motor.

17.2 Pumps and Motors

As with electric machines, the same unit can be used as either a pump or a motor, although certain types are more suitable as pumps than motors and vice versa. The following table lists most of the common types of rotary hydraulic units, together with a summary of their most common applications.

Type	Range (Approx.)	Comments
Gear	200 rad/s 14 MN/m²	Cheap, general-purpose pump. Rather noisy. Unsuitable as a servomotor, due to short life under transient operating conditions.
Vane	200 rad/s 10 MN/m²	General-purpose pump also unsuitable as a servomotor due to poor low-speed performance.
Screw	300 rad/s 18 MN/m²	More expensive, but quiet-running pump with small pressure fluctuations.
Rolvane	150 rad/s 14 MN/m²	Smooth-running servomotor with very good low-speed performance. Too expensive to use as a pump.
Radial Piston	200 rad/s 14–20 MN/m²	Good pump which can be made variable delivery by varying the eccentricity of the cam track. Good servomotor, particularly so when fitted with modified profile cam track. Available as a cheaper pump, using ball-bearings for pistons.
Axial Piston	200–600 rad/s 12–35 MN/m²	Variable-delivery pump using swash-plate angle control. Small high-speed versions commonly used in aircraft. Low-inertia motor, but cannot be corrected as can radial type.

All the above units are described in detail in Merritt (ref. 16), the only one enlarged upon here being the variable delivery axial piston pump, since it is used as a method of control. A few general comments, however, may be helpful.

(*a*) Pumps. A pump will be driven at constant speed in all practical cases so that the delivery is constant, i.e. the flow Q is nominally constant. The output pressure P_s will be a function of the loading on the pump, i.e. zero pressure on no load. This is of prime importance when considering power packs, i.e. Section 17.3.2.

(*b*) Motors. A motor develops torque proportional to the applied pressure differential across the motor and in the steady state the speed is proportional to the flow of oil.

17.3 Methods of Control

17.3.1 Variable Delivery Pump—High Flow Systems

Since the motor speed is proportional to flow a pump which delivers a variable flow can be used as a means of control, as shown in Fig. 17.1.

(a) *Variable Swash-plate Axial Piston Pump* (Fig. 17.2).

For this pump the slipper pads bear up against the face of the swash plate and therefore drive the pistons in and out once per revolution of the pump. As the piston is driven into the body, oil is pumped out, and as the piston is retracted, oil is drawn in. Thus the valve plates are required to connect the pistons which are delivering oil to the output port and the pistons sucking in oil to the inlet port. (Note the similarity to a commutator in a d.c. machine). The slipper pads are kept in contact

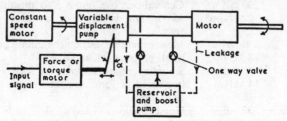

Fig. 17.1 Variable delivery pump control.

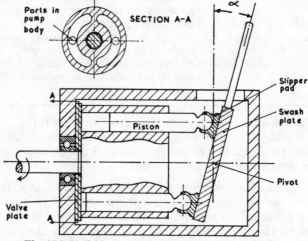

Fig. 17.2 Variable angle swash plate axial piston pump.

with the swash plate on the suction stroke by back pressure, although a mechanical retaining ring or springs are sometimes used to assist low-speed performance.

The delivery is dependent upon the angle of the swash plate, α rads. Thus the flow

$$Q = d_p \sin \alpha \cdot N \text{ m}^3/\text{s}$$
$$\simeq K_p \cdot \alpha$$

where $d_p \sin \alpha \simeq d_p \alpha$ is the pump displacement m³/rad* and N is pump speed, rad/s, which is constant.

* Do not confuse swash-plate angle, α, with pump rotation, i.e. displacement is m³/radian of rotation.

(b) *The Drive Motor.* The prime mover can be any form of cheap, simple motor, and is therefore nearly always a three-phase induction motor. Note the similarity to a Ward–Leonard speed control.

(c) *The Force or Torque Motor.* On the diagram a linear motor is shown as one means of converting the control signal, in an electrical form, to mechanical angle α. A torque-synchro has often been used for this purpose. For a reversible drive α must swing plus or minus. For larger motors the force required to move the control arm is reduced by hydraulic servo assistance, similar to the profile copying unit described later (Fig. 17.6).

(d) *The Reservoir.* The one-way valves in the feeders are included so that oil can only be taken out of the reservoir by the pump in order to replenish any losses in the system. The boost pump is included so as to pressurise slightly (about 1 MN/m²) the low pressure line, thus avoiding 'cavitation' of the oil.

17.3.2 Valve Control—Low and Medium Flow Systems

Valve control systems work from a constant pressure source, the flow to the motor being controlled by the valve, as shown in Fig. 17.3. Note that one power pack can be used for more than one servo.

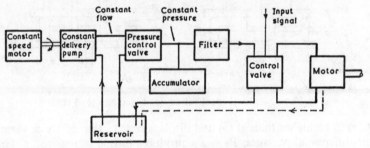

Fig. 17.3 A valve-controlled system.

(a) *Pressure Control Valve.* The output pressure is governed by a spring loaded valve which by-passes oil to the reservoir when not required by the load.

(b) *The Accumulator.* Pressure fluctuations and transient flow demands can be improved by using a pressurised accumulator which has the same effect as a smoothing capacitor in a d.c. supply.

(c) *The Filter.* Due to the very small working clearances of precision control valves and motors filtering of the oil is important. Usually filters of 5 or 10 μm (the maximum size of particle passed) are used. Care must be taken that they do not block.

(d) *The Flow Control Valve* is the subject of the next section.

17.4 Flow Control Valves

17.4.1 Types of Control Valve

There is quite a variety of types of valve by which the flow of oil from a constant pressure source to a load can be controlled. All basically comprise a variable orifice and some means of controlling the orifice opening.

In most cases the control signal will be an electrical one, so that an electric 'motor' of some sort is required to convert the electrical signal into orifice opening. Brief consideration, however, must be given to the 'all-hydraulic' servo, i.e. a position servo with mechanical feedback.

The spool valve is the only valve considered here, since the available commercial units are of this type.

17.4.2 The Simple Spool Valve

The commonly used spool valve is of the cylindrical four port variety, two variants of which are shown in Fig. 17.4.

In each case displacement of the spool to the left causes the pressure

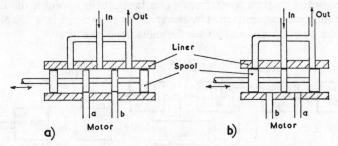

Fig. 17.4 Cylindrical spool valves: (*a*) 4 land and (*b*) 3 land.

at (*a*) to be higher than at (*b*) and displacement to the right vice versa. The differential pressure, $P_a - P_b$, produces torque in the motor. The flow and therefore the motor speed are a function of the spool-valve opening, affected somewhat, as will be shown in deriving the transfer functions in Section 17.5, by the load pressure. The differential pressure between the supply ports is supply pressure, P_s. Note that in each case there are, in fact, two orifices in series. It is the precision required in the machining of this assembly that accounts for the high cost of such components; in fact, the ports are slightly overlapped or underlapped to give differing characteristics in the null position.

Note that leakage of oil across the inlet supply ports is of no consequence, but leakage across the output load ports is of importance.

17.4.3 An All-hydraulic Position Control System

A spool valve is simplest used in a position control system utilising mechanical feedback. Such systems are used as remote position, power amplifiers for such as aircraft elevator controls (Fig. 17.5) and profile-copying systems (Fig. 17.6).

For the elevator control, deflection of the control column opens the spool valve which applies pressure to the actuating jack; the resulting

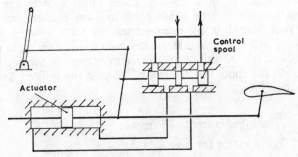

Fig. 17.5 Aircraft elevator control system.

motion of the jack is such as to close the spool valve, so that when it is in its new desired position the valve is closed and motion ceases. Note the inherent regulating action that any external forces on the elevator open the valve slightly and produce reaction forces without effecting the control column.

For the profile-copying system the stylus is maintained in contact

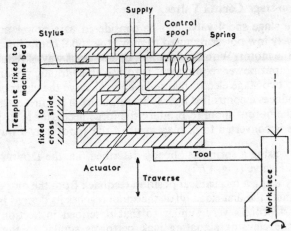

Fig. 17.6 A profile copying unit.

with the template by a spring, the servo action being similar to that described above. The important point here is that the output accurately follows the shape demanded by the template and that cutting reaction forces are produced by the actuator and are not applied to the template.

17.4.4 Electro-hydraulic Control Valves

To make the spool valve suitable for incorporation in a conventional servo using electrical monitoring, data processing and compensating, the spool displacement must be actuated by a linear electric motor, commonly called a 'force' motor. These linear motors are simple permanent magnet devices which produce force proportional to the input current. Since it is displacement and not force which is required proportional to the input signal it is common to use a spring to provide the reaction force. A serious problem encountered is the axial forces produced on the spool due to flow of oil past the orifices, known as 'flow reaction forces'.

Thus:

$$\text{The input current} = i \text{ ampere}$$
$$\text{Force motor force} = K \text{ N/A}$$
$$\text{Spring stiffness} = \lambda \text{ N/m}$$
$$\text{Spool deflection} = x \text{ m}$$
$$\text{Flow reaction force} = P_R \text{ N, which is not constant}$$

then
$$K i = \lambda x + P_R$$

In order to make P_R negligible, $K i$ and λx must be made large, resulting in a high input power requirement which is highly undesirable.

17.4.5 Two-stage Control Valves

The single stage spool valve can be considered as a power amplifier, the relatively low power required to move the spool controlling a large power in the motor. With this in mind, one obvious way of reducing the electrical input power requirement is to use an hydraulic pre-amplifier. Thus, in a two-stage electro-hydraulic control valve the electrical input signal produces a controlled hydraulic output from the first stage which is fed to the output spool to produce the force necessary to displace it. This force is converted to displacement of the spool in two ways:

(a) By using a spring as already described, i.e. the DOWTY–MOOG SERIES 21 valve (Fig. 17.7).

(b) By using a mechanical position feedback from the output spool which shuts the first stage off as the spool reaches its desired position. This is a process very similar to that described in Section 17.4.3, where the elevator actuating jack performs similar to the output spool. Typical valves are the DOWTY–MOOG SERIES 31 and the TELEHOIST–PEGASUS (Fig. 17.8).

Various valves other than spool valves are used for the first stage; the Moog valve of Fig. 17.7, for example, uses a double nozzle and flapper valve, Fig. 17.9. This is a pressure control valve.

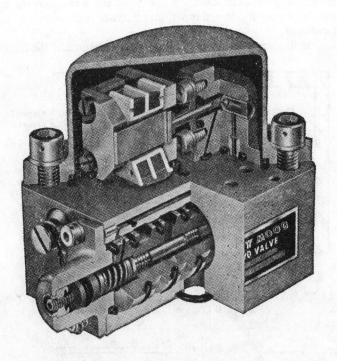

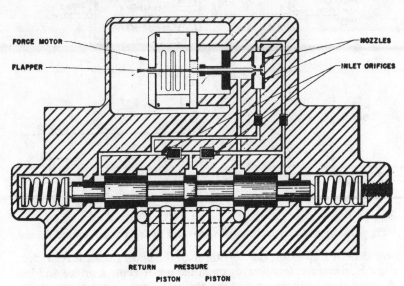

Fig. 17.7 Dowty Moog series 21 two-stage electrohydraulic spool valve.
(Courtesy of Dowty Rotol Ltd.)

The flexural pivot is effectively a free pivot and a reaction spring, so that the *displacement* of the flapper, up or down, is proportional to the input currents.

With the flapper central the system is in equilibrium. If the flapper is

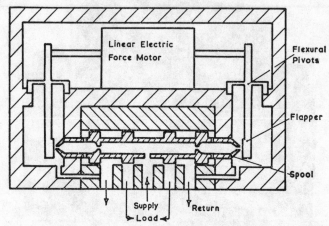

Fig. 17.8 Telehoist Pegasus electrohydraulic spool valve. (Courtesy of Telehoist Ltd. and Pegasus Laboratories Inc.)

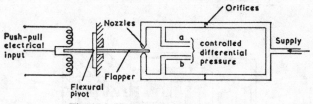

Fig. 17.9 First stage of the Moog valve.

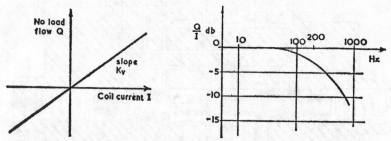

Fig. 17.10 Typical characteristics of a two-stage electrohydraulic control valve.

now deflected upwards, the flow through the upper nozzle is restricted; there is, therefore, less flow through the upper control orifice and less pressure drop across it, so that the pressure down stream of the orifice, at (*a*), increases. By similar reasoning the pressure at (*b*) falls, producing a differential pressure between (*a*) and (*b*) proportional to the input current.* (*a*) and (*b*) are connected to each end of the output spool, so that the spool is deflected by the differential pressure. In this way considerable forces are available which swamp flow reaction forces and produce rapid acceleration of the spool, resulting in very high frequency response. Fig. 17.10 shows typical characteristics of a two-stage electrohydraulic control valve.

17.5 System Transfer Functions†

17.5.1 Spool Valve Controlled System

(i) Motor

Total motor flow

$$Q = Q_m + Q_L + Q_c \text{ m}^3/\text{s}$$

where

Q_m = useful flow to motor
Q_L = leakage flow to motor
Q_c = equivalent compressibility flow

Also let

θ_o = output position, rad
$\omega = \dot\theta_o$ = output velocity, rad/s
d_m = motor displacement, m³/rad
P = differential pressure applied to motor, N/m²
L = leakage coefficient, m³/s per N/m²
J = total inertia, kg m²
F = viscous friction coefficient, N m s
V = volume of oil in motor and pipes, m³
B = bulk modulus of oil, N/m²

Then

$$Q_m = d_m . \omega$$
$$Q_L = P . L$$

and

$$Q_c = \frac{dV}{dt} = \frac{V}{B} . \frac{dP}{dt}$$

since

$\dfrac{\Delta V}{V} = \dfrac{\Delta P}{B}$, where ΔV is the change in V by compression due to the change ΔP in P.‡

---

* This is a hydraulic Wheatstone Bridge.

† Inch-lb-second units are in common use in hydraulic systems. See the conversion table preceding Chapter 1.

‡ This equation serves as a definition of bulk modulus. Note that B must include an allowance for expansion of pipes.

Neglecting pressure drops in the lines between the motor and the control valve, P provides the total motor torque T so that

$$T = d_m \cdot P$$

$$= J\dot{\omega} + F\Omega + T_L$$

$$\therefore \ P(s) = \frac{1}{d_m}[(Js + F)\Omega(s) + T_L(s)]$$

Thus

$$Q(s) = d_m \cdot \Omega(s) + \left(L + \frac{V}{B}s\right)P(s)$$

$$= d_m \cdot \Omega(s) + \frac{1}{d_m}\left(L + \frac{V}{B}s\right)[(Js + F)\Omega(s) + T_L(s)]$$

$$= \left[\frac{VJ}{Bd_m}s^2 + \frac{1}{d_m}\left(LJ + \frac{FV}{B}\right)s + d_m + \frac{LF}{d_m}\right]\Omega(s)$$

$$+ \frac{1}{d_m}\left(L + \frac{V}{B}s\right)T_L(s)$$

(ii) Control valve.

The valve considered here is a two-stage spool valve.

Let:

P_s = supply pressure, N/m²
Q = output flow = total motor flow, m³/s
i = input current to spool valve, amperes
v = input voltage to spool valve, volts
R = resistance of spool valve coil, ohms
T_c = time constant of spool valve coil, seconds
T_v = low frequency approximate time constant of valve, seconds
ζ = damping coefficient of the valve
ω_n = undamped natural frequency of the valve, rad/s
K_v = rated flow, m³/s per A
K_q = slope of flow–pressure characteristic (Fig. 17.11), m³/s per N/m²

For this type of spool valve, with $P = 0$

$$\frac{Q(s)}{I(s)} = K_v \cdot \frac{1}{1 + \frac{2\zeta}{\omega_n}s + \frac{1}{\omega_n{}^2}s^2}$$

and for frequencies up to 50 Hz this can be approximated to

$$\frac{Q(s)}{I(s)} = K_v \frac{1}{1 + sT_v}$$

However, the flow through an orifice is proportional to the square root of the pressure drop across the orifice.

$$\therefore \; Q = \frac{K_v i}{\sqrt{P_s}} \cdot \sqrt{P_s - P}, \quad \text{in the steady state}$$

$$= K_v i \left(1 - \tfrac{1}{2}\frac{P}{P_s} - \tfrac{1}{8}\left(\frac{P}{P_s}\right)^2 - \ldots \right)$$

$$\simeq K_v i \left(1 - \frac{1}{2P_s}P \right) \quad \text{for} \quad P < \tfrac{2}{3}P_s \; (\text{see Fig. 17.11})$$

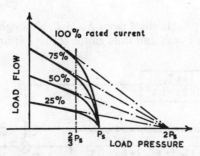

Fig. 17.11 Flow-pressure characteristics of a spool valve.

In the transient condition, for P constant:

$$Q(s) = \frac{K_v I(s)}{1 + sT_v}\left(1 - \frac{1}{2P_s}P \right)$$

and for i constant:

$$Q(s) = K_v i \left(1 - \frac{1}{2P_s}P(s) \right)$$

Applying the principle of superposition:

$$Q = \frac{\partial Q}{\partial i} \cdot i + \frac{\partial Q}{\partial P} \cdot P$$

$$\therefore \; Q(s) = \frac{K_v}{1 + sT_v}I(s) - \frac{K_v i}{2P_s}P(s)$$

The value of i in the second term is an assumed quiescent condition so that provided the variation of i is limited, which was assumed in applying superposition, a good approximation is made by putting $\frac{K_v i}{2P_s} = K_q$, a constant, which must be changed when the steady state value of i is altered. K_q is the slope of the linear part of the flow–pressure curve of Fig. 17.11.

Put $$I(s) = \frac{V(s)}{R}\frac{1}{1 + sT_c}$$

and equate the motor flow to the valve flow, thus:

$$\frac{K_v}{R} \frac{1}{(1 + sT_v)(1 + sT_c)} \cdot V(s) - K_q P(s) = d_m \Omega(s) + \left(L + \frac{V}{B}s\right)P(s)$$

$$\frac{K_v}{R} \frac{1}{(1 + sT_v)(1 + sT_c)} V(s)$$
$$= \left[\left(d_m + \frac{L_1 F}{d_m}\right) + s\left(\frac{L_1 J}{d_m} + \frac{VF}{B d_m}\right) + s^2 \frac{VJ}{B d_m}\right]\Omega(s)$$
$$+ \left(\frac{L_1}{d_m} + s\frac{V}{B d_m}\right)T_L(s)$$

where $L_1 = L + K_q$. In most cases $K_q \gg L$, although for final positioning $K_q = 0$ with $i = 0$, and thus motor leakage is important.

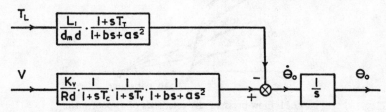

Fig. 17.12 Transfer function of a spool valve controlled motor.

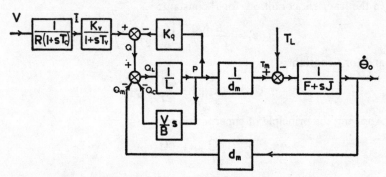

Fig. 17.13 Simulation of the spool valve controlled motor.

$$\therefore \frac{K_v}{R} \frac{1}{(1 + sT_v)(1 + sT_c)} V(s)$$
$$= d(1 + bs + as^2)\Omega(s) + \frac{L_1}{d_m}(1 + sT_T)T_L(s)$$

where
$$d = d_m + \frac{L_1 F}{d_m}$$
$$b = \frac{1}{d d_m}\left(L_1 J + \frac{VF}{B}\right)$$
$$a = \frac{VJ}{d d_m B} \quad \text{and} \quad T_T = \frac{V}{B L_1}$$

Finally

$$\Omega(s) = \frac{K_v}{Rd} \frac{1}{(1 + sT_c)(1 + sT_v)(1 + bs + as^2)} V(s) - \frac{L_1}{dd_m} \frac{1 + sT_T}{1 + bs + as^2} T_L(s)$$

Fig. 17.12 shows the block diagram of this function, and Fig. 17.13 a more complete simulation of the motor and valve.

The following table shows the equivalence of the above calculated hydraulic system and the armature controlled d.c. motor of Section 15.8.3.

D.c. (constant field)	Hydraulic
$i_a{}^*$	P^*
V	Q
Negligible shunt leakage	L
R_a	$K_q \simeq L_1$, since L is small
L_a	V/B
K_e and K_T†	d_m
G	K_v/R
Amplifier time constants	Spool valve time constants
Thus	
$a = \dfrac{JL_a}{K^2 + FR_a}$	$a = \dfrac{VJ}{dd_mB} = \dfrac{J\dfrac{V}{B}}{d_m{}^2 + FL_1}$
$b = \dfrac{JR_a + FL_a}{K^2 + FR_a}$	$b = \dfrac{L_1J + \dfrac{VF}{B}}{dd_m}$
	$= \dfrac{JL_1 + F\dfrac{V}{B}}{d_m{}^2 + FL_1}$

* Electric motor torque $= K_T i_a = \dfrac{p}{a} \dfrac{Z}{2\pi} \Phi i_a = K'\Phi i_a$.
  Hydraulic motor torque $= d_m P$.
  ∴ from a rating viewpoint, Φi_a should be compared to P.
† Note that $K_e = K_T = K$, if SI units are used.

17.5.2 Variable Displacement Pump System

Pump output flow, $Q = K_p \cdot \alpha$ m³/s

where α = swash plate angle, radians

This flow can be equated to the required motor flow as calculated in the previous section, i.e.

Motor flow,

$$Q(s) = \left(\frac{VJ}{Bd_m}s^2 + \frac{1}{d_m}\left(LJ + \frac{FV}{B}\right)s + d'\right)\Omega(s) + \frac{1}{d_m}\left(L + \frac{V}{B}s\right)T_L(s)$$

where $d' = d_m + \dfrac{LF}{d_m}$

$$\therefore \quad \Omega(s) = \frac{K_p}{d'} \cdot \cfrac{1}{1 + \cfrac{LJ + \dfrac{FV}{B}}{d'd_m}s + \cfrac{VJ}{Bd'd_m}s^2} \cdot \alpha(s)$$

$$- \frac{L}{d'd_m} \cfrac{1 + s\dfrac{V}{LB}}{1 + \cfrac{LJ + \dfrac{FV}{B}}{d'd_m}s + \cfrac{VJ}{Bd'd_m}s^2} \cdot T_L(s)$$

Pressure drop in the supply pipes has been neglected, which will often not be possible in practice. To minimise this the connecting pipes must be as short as possible. Pump leakage can be included in L and V must include pump, motor and pipes.

17.5.3 Hydraulic Ram Systems

It is necessary in servomechanisms to use a ram which has equal areas on both sides of the piston, thus necessitating a straight through piston rod, as shown in Fig. 17.14.

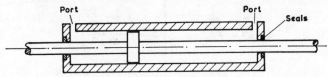

Fig. 17.14 An hydraulic ram or jack.

Let
A = effective piston area, m^2
V = total enclosed volume, m^3
F_L = load force, N
M = load mass, kg
f = load viscous friction coefficient, N per m/s
x = linear output position, m

other symbols as previously defined.

Following the analysis of the previous section, the total flow:

$$Q = Q_m + Q_L + Q_c$$
$$= A\dot{x} + PL + \frac{1}{2}\frac{V}{B}\frac{dP}{dt}$$

The $\frac{1}{2}$ in the Q_c term is a compromise, since if the ram is central the oil in the high pressure half section is pressurised; the volume of oil under compression, however, varies from that of the feeder pipe to V, dependent upon the position of the ram. V in the equation is thus not strictly a constant and the transfer function will be to some extent dependent upon the position, x.

Now
$$A \cdot P = \text{total force}$$
$$= \frac{M}{g} \cdot \ddot{x} + f\dot{x} + F_L$$

Thus

$$Q(s) = AsX(s) + \left(L + \frac{V}{2B}s\right) \cdot \frac{1}{A} \cdot \left[\left(\frac{M}{g}s^2 + fs\right)X(s) + F_L(s)\right]$$

$$= \left[\frac{M}{g} \cdot \frac{V}{2AB}s^2 + \left(\frac{Vf}{2AB} + \frac{ML}{gA}\right)s + A + \frac{Lf}{A}\right]sX(s)$$

$$+ \frac{1}{A}\left(L + \frac{V}{2B}s\right)F_L(s)$$

$Q(s)$ can now be related to the controller output, i.e. the flow from a spool valve as calculated in Section 17.5.1 or from a variable delivery pump as in Section 17.5.2. Note that V/A is approximately the stroke of the ram.

17.6 Oil

The oil used for hydraulic servos must be a particularly good-class oil The following points are important:

    (i) The bulk modulus B must be high.

    (ii) Impurities in the oil must be at a very low level.

    (iii) Lubricating properties should be as good as possible to avoid control valve, motor and pump wear. Hydraulic oils are acceptable but not good lubricants.

    (iv) The oil must be a good anti-corrosive. This is of particular importance when the system is not working and the oil is stationary.

It will be noticed that viscosity has not come directly into any of the transfer functions, but it will effect such values as the spool valve flow constant, K_v. However, the oil should be of low viscosity in order to minimise pressure drops along the feed lines.

Heating of the oil due to losses in the system can be a severe problem, so that oil coolers, usually the cold running water outer jacket type, are normally incorporated in the return line.

Appendix
Laplace and Fourier Transforms

1 Introduction

The Laplace and Fourier transforms are integral transforms used to simplify the solution of linear differential equations by converting the differential equation to an algebraic equation. The time solution can be found by inverse transforming the suitably manipulated direct transform. It is shown in this text, however, that the engineer's desire to simplify design procedure has led to a complete discipline in terms of the direct transform, with no immediate attempt to inverse transform, i.e. the transfer function. The Fourier transform is particularly applicable since the transformed equation has a physical interpretation, i.e. the response in phase and amplitude to a range of sinusoidal frequencies. The Fourier transform and the Fourier series are closely related. For simplicity the Fourier transform can be considered a special case of the Laplace transform with $s = j\omega$, for which reason only the Laplace transform need be enlarged upon.

Cheng (ref. 7) and Aseltine (ref. 8) are excellent mathematical texts for the engineer, and Healey (ref. 10) is a quick reference source.

2 Integral Transforms

(a) Laplace $\qquad \mathscr{L}[f(t)] = \int_0^\infty e^{-st} f(t) dt = F(s)$

(b) Fourier $\qquad \mathscr{F}[f(t)] = \int_{-\infty}^\infty e^{-j\omega t} f(t) dt = F(j\omega)$

s (p is used equally often) is a complex variable of the form $\sigma + j\omega$. σ is introduced so that in multiplying $f(t)$ by $e^{-\sigma t}$ the integral will be finite for a large number of functions. If the integral is not finite, then $f(t)$ is not transformable. Since the Fourier transform does not include the $e^{-\sigma t}$ term a number of functions which are Laplace transformable are not Fourier transformable.

3 Some Laplace Transforms

(i) $\qquad\qquad f(t) = K$, a constant.

then $\qquad F(s) = \displaystyle\int_0^\infty e^{-st} . K\, dt$

$\qquad\qquad\qquad = -\dfrac{K}{s}\,[e^{-st}]_0^\infty = \dfrac{K}{s}$

(ii) $\qquad\qquad f(t) = e^{-at}$

Then $\qquad F(s) = \displaystyle\int_0^\infty e^{-st}e^{-at}\, dt$

$\qquad\qquad\qquad = -\dfrac{1}{s+a}\,[e^{-(s+a)t}]_0^\infty$

$\qquad\qquad\qquad = \dfrac{1}{s+a}$

(iii) The shift theorem:

$\qquad\qquad\qquad f(t) = e^{-at}g(t)$

Then $\qquad F(s) = \displaystyle\int_0^\infty e^{-st}e^{-at}g(t)\, dt$

$\qquad\qquad\qquad = \displaystyle\int_0^\infty e^{-(s+a)t}g(t)\, dt$

Thus treat as $\mathscr{L}[g(t)]$, but with $s + a$ instead of s, i.e. $G(s + a)$.

(iv) $\qquad\qquad g(t) = \dfrac{d}{dt}f(t)$*

Then $\qquad G(s) = \displaystyle\int_0^\infty e^{-st}\dfrac{d}{dt}f(t)\, dt$

$\qquad\qquad\qquad = [e^{-st}f(t)]_0^\infty + s\displaystyle\int_0^\infty e^{-st}f(t)\, dt$ (integration by parts)

$\qquad\qquad\qquad = -f(o) + s\mathscr{L}[f(t)]$

$\qquad\qquad\qquad = sF(s) - f(o)$

where $\qquad f(o) = f(t)$ at $t = 0$

(v) $\qquad g(t) = \displaystyle\int_{-\infty}^t f(t)\, dt$

Then $\qquad G(s) = \displaystyle\int_0^\infty e^{-st}\int_{-\infty}^t f(t)\, dt \,.\, dt$

$\qquad\qquad\qquad = \left[-\dfrac{1}{s}e^{-st}\displaystyle\int_{-\infty}^t f(t)\, dt\right]_0^\infty + \dfrac{1}{s}\displaystyle\int_0^\infty e^{-st}f(t)\, dt$

$\qquad\qquad\qquad = \dfrac{1}{s}\displaystyle\int_{-\infty}^0 f(t)\, dt + \dfrac{1}{s}F(s)$

* This transform is the basis of transfer function theory where the nth derivative is transformed to s^n with all initial conditions zero. Transfer function analysis is thus limited to initially dead circuits.

4 Inverse Transformation

Having manipulated the transform equation into a simpler algebraic form it is necessary to find the inverse transform so as to arrive at an answer in terms of the original variable, t. For most engineering problems it is common to manipulate the transform equation by using partial fractions into a form recognisable from the table of transforms as in the following worked example. For the more difficult problem, however, it is necessary to apply the integral definition of the inverse transform, viz.:

$$\mathcal{L}[f(t)] = F(s) = \int_0^\infty e^{-st}f(t)\,dt$$

$$\text{and} \quad \mathcal{L}^{-1}[F(s)] = f(t) = \frac{1}{2\pi j}\int_{c-j\infty}^{c+j\infty} e^{st}F(s)\,ds \left.\right\} \text{Laplace transform pair}$$

5 Initial and Final Value Theorems

The values of the time function, $f(t)$, at $t = 0$ and ∞ can be found without inverse transformation from limiting values indicated below

Initial value:
$$\lim_{t \to 0} f(t) = \lim_{s \to \infty} sF(s)$$

Final value:
$$\lim_{t \to \infty} f(t) = \lim_{s \to 0} sF(s)$$

6 A Simple Example *

A system is represented by the differential equation

$$\frac{1}{\omega_n^2}\frac{d^2\theta_o}{dt^2} + \frac{2\zeta}{\omega_n}\frac{d\theta_o}{dt} + \theta_o = \theta_i$$

Find the output response as a function of time to a step function input of magnitude, K, applied at time, $t = 0$, with the system initially dead, i.e. $\theta_o = \dot{\theta}_o = 0$, when $t = 0$.

The transform equation is

$$\frac{1}{\omega_n^2}\left(s^2\Theta_o(s) - s\theta_o(o) - \dot{\theta}_o(o)\right) + \frac{2\zeta}{\omega_n}\left(s\Theta_o(s) - \theta_o(o)\right) + \Theta_o(s)$$
$$= \mathcal{L}[K \cdot H(t)] = \frac{K}{s}$$

$$\therefore \left(\frac{1}{\omega_n^2}s^2 + \frac{2\zeta}{\omega_n}s + 1\right)\Theta_o(s) = \frac{K}{s}$$

$$\therefore \Theta_o(s) = \frac{1}{\dfrac{1}{\omega_n^2}s^2 + \dfrac{2\zeta}{\omega_n}s + 1} \cdot \frac{K}{s}$$

* This is an overstatement since this example of a second-order system can be more simply solved by classical means; it is only for more complex problems that this method gives the improvement.

Let this be written

$$\Theta_0(s) = \frac{As + B}{\dfrac{1}{\omega_n{}^2}s^2 + \dfrac{2\zeta}{\omega_n}s + 1} + \frac{C}{s} = \frac{As^2 + Bs + C\left(\dfrac{1}{\omega_n{}^2}s^2 + \dfrac{2\zeta}{\omega_n}s + 1\right)}{\left(\dfrac{1}{\omega_n{}^2}s^2 + \dfrac{2\zeta}{\omega_n}s + 1\right)s}$$

Equating coefficients of the numerator:

$$C = K$$

$$B + \frac{2\zeta}{\omega_n}C = 0 \qquad \therefore \; B = -\frac{2\zeta}{\omega_n}K$$

and
$$A + \frac{1}{\omega_n{}^2}C = 0 \qquad \therefore \; A = -\frac{1}{\omega_n{}^2}K$$

$$\therefore \; \Theta_0(s) = \frac{K}{s} - K\frac{s + 2\zeta\omega_n}{s^2 + 2\zeta\omega_n s + \omega_n{}^2}$$

$$= \frac{K}{s} - K\frac{s + 2\zeta\omega_n}{(s + \zeta\omega_n)^2 + \omega_n{}^2(1 - \zeta^2)}$$

$$= \frac{K}{s} - K\frac{s + \zeta\omega_n}{(s + \zeta\omega_n)^2 + \omega^2} - K\frac{\zeta\omega_n}{(s + \zeta\omega_n)^2 + \omega^2}$$

where $\qquad \omega = \omega_n\sqrt{1 - \zeta^2}$

These three terms are now recognisable from the standard table. Three solutions are possible dependent upon ζ.

(*a*) under-damped $\zeta < 1$, i.e. ω^2 positive.

$$\theta_0(t) = K\left[\underset{\underset{[6]}{\uparrow}}{1} - \underset{\underset{[21]}{\uparrow}}{e^{-\zeta\omega_n t}\cos\omega t} - \underset{\underset{[20]}{\uparrow}}{\frac{\zeta\omega_n}{\omega}e^{-\zeta\omega_n t}\sin\omega t}\right]$$

$$= K\left[\underset{\underset{\text{Steady state}}{\uparrow}}{1} - e^{-\zeta\omega n t}\underset{\underset{\text{Transient}}{\uparrow}}{\left(\cos\omega t + \frac{\zeta\omega_n}{\omega}\sin\omega t\right)}\right]$$

(*b*) Critically damped $\zeta = 1$, i.e. $\omega = 0$

$$\theta_0(t) = K\left[\underset{\underset{[6]}{\uparrow}}{1} - \underset{\underset{[11]}{\uparrow}}{e^{-\omega_n t}} - \underset{\underset{[13]}{\uparrow}}{\omega_n t e^{-\omega_n t}}\right]$$

$$= K[1 - e^{-\omega_n t}(1 + \omega_n t)]$$

(*c*) Over-damped $\zeta > 1$, i.e. ω^2 negative

$$\theta_0(t) = K\left[\underset{\underset{[6]}{\uparrow}}{1} - \underset{\underset{[24] \text{ and } [4]}{\uparrow}}{e^{-\zeta\omega_n t}\cosh\omega t} - \underset{\underset{[23] \text{ and } [4]}{\uparrow}}{\frac{\zeta\omega_n}{\omega}e^{-\zeta\omega_n t}\sinh\omega t}\right]$$

$$= K\left[1 - e^{-\zeta\omega_n t}\left(\cosh\omega t + \frac{\zeta\omega_n}{\omega}\sinh\omega t\right)\right]$$

Table of Common Laplace Transforms

$f(t)$	$\mathscr{L}[f(t)] = F(s)$
1. $\dfrac{d}{dt} f(t)$	$sF(s) - f(o)$
2. $\dfrac{d^n}{dt^n} f(t)$	$s^n F(s) - s^{n-1} f(o) - \ldots - f^{n-1}(o)$
	where $f^r(o) = \dfrac{d^r}{dt^r} f(t)$ at $t = 0$
3. $\displaystyle\int_0^t f(t)\, dt$	$\dfrac{1}{s} F(s) + \dfrac{1}{s} \displaystyle\int_{-\infty}^0 f(t)\, dt$
4. $e^{-at} f(t)$	$F(s + a)$
5. Unit impulse, $\delta(t)$	1
6. Unit step, $H(t)$	$\dfrac{1}{s}$
7. Ramp function, t	$\dfrac{1}{s^2}$
8. Delayed unit step, $H(t - T)$	$\dfrac{e^{-sT}}{s}$
9. Rectangular pulse	$\dfrac{1 - e^{-sT}}{s}$
10. $\dfrac{t^n}{n!}$	$\dfrac{1}{s^{n+1}}$
11. e^{-at}	$\dfrac{1}{s + a}$
12. $1 - e^{-at}$	$\dfrac{a}{s(s + a)}$
13. te^{-at}	$\dfrac{1}{(s + a)^2}$
14. $\dfrac{t^n}{n!} e^{-at}$	$\dfrac{1}{(s + a)^{n+1}}$
15. $e^{-at} - e^{-bt}$	$\dfrac{b - a}{(s + a)(s + b)}$
16. $\sin \omega t$	$\dfrac{\omega}{s^2 + \omega^2}$
17. $\cos \omega t$	$\dfrac{s}{s^2 + \omega^2}$
18. $1 - \cos \omega t$	$\dfrac{\omega^2}{s(s^2 + \omega^2)}$
19. $\omega t \sin \omega t$	$\dfrac{2\omega^2 s}{(s^2 + \omega^2)^2}$
20. $e^{-at} \sin \omega t$	$\dfrac{\omega}{(s + a)^2 + \omega^2}$

Table of Common Laplace Transforms

$f(t)$	$[f(t)] = F(s)$
21. $e^{-at}\cos \omega t$	$\dfrac{s + a}{(s + a)^2 + \omega^2}$
22. $\sin (\omega t \pm \phi)$	$\dfrac{\omega \cos \phi \pm s \sin \phi}{s^2 + \omega^2}$
23. $\sinh \omega t$	$\dfrac{\omega}{s^2 - \omega^2}$
24. $\cosh \omega t$	$\dfrac{s}{s^2 - \omega^2}$

7 Fourier Methods

Any time function can be expressed as the sum of a series of sinusoidal functions, the amplitude and phase of which are defined by the Fourier series (periodic signals) or Fourier integral (transient signals).

Consider a periodic function, $f(t)$, of period T seconds and fundamental frequency $\omega_1 = 2\pi/T$ rad/s. 'Harmonics' occur at multiple frequencies $\omega = n\omega_1$, where n is an integer.

Thus

$$f(t) = \frac{a_0}{2} + a_1 \cos \omega_1 t + a_2 \cos 2\omega_1 t + \dots$$
$$+ b_1 \sin \omega_1 t + b_2 \sin 2\omega_1 t + \dots$$
$$= \frac{a_0}{2} + c_1 \sin (\omega_1 t + \phi_1) + c_2 \sin (2\omega_1 t + \phi_2) + \dots$$

where

$$a_n = \frac{2}{T} \int_0^T f(t) \cos n\omega_1 t \, dt \quad \text{and} \quad b_n = \frac{2}{T} \int_0^T f(t) \sin n\omega_1 t \, dt$$

$$c_n = \sqrt{a_n^2 + b_n^2} \quad \text{and} \quad \phi_n = \tan^{-1} \frac{a_n}{b_n}$$

This can be expressed in exponential form (see Cheng (ref. 7)) as

$$f(t) = \sum_{n=-\infty}^{\infty} \alpha_n e^{\pm jn\omega_1 t}$$

where

$$\alpha_n = \frac{1}{T} \int_{-\frac{T}{2}}^{\frac{T}{2}} e^{-jn\omega_1 t} f(t) \, dt$$

which is a complex number giving the amplitude and phase of the individual harmonics. Note the unpractical aspect of the exponential form in that hypothetical negative frequencies are introduced.

This approach can now be extended to aperiodic functions by treating them as periodic functions with $T \to \infty$ and thus the fundamental frequency, $\Delta\omega \to 0$. 'Harmonics' exist at $\Delta\omega$, $2\Delta\omega$, $3\Delta\omega$, etc., up to $\omega = \infty$, even though $\Delta\omega \to 0$; in other words, at all frequencies instead of discrete frequencies. The Fourier integral is then defined similar to the series with the general frequency ω instead of the discrete values $\pm n\omega_1$, and the summation performed by an integral, thus:

$$f(t) = \frac{1}{2\pi} \int_{-\infty}^{+\infty} F(j\omega)e^{j\omega t}\, d\omega \quad \text{(FOURIER INTEGRAL)}$$

and

$$F(j\omega) = \int_{-\infty}^{\infty} e^{-j\omega t} f(t)\, dt, \text{ which is called the FOURIER TRANSFORM of } f(t)$$

$F(j\omega)$ is a complex function, akin to α_n for the periodic function, which gives the amplitude *density* and phase of the sinusoidal components required for $f(t)$ and is a continuous 'spectrum' having values at all frequencies from $-\infty$ to $+\infty$.

Consider now a system for which it is required to know the output, $f_o(t)$, in response to an input, $f_i(t)$. The system is previously tested with variable frequency signals to determine the output/input amplitude ratio and phase shift at all practically applicable frequencies (theoretically from 0 to ∞). The response to negative frequencies is, of course, a theoretical function and is the conjugate of the response to positive frequencies, i.e. similar in amplitude but lag instead of lead and vice versa in phase. The Fourier transform of the output, $F_o(j\omega)$, can now be found by multiplying the amplitude of the Fourier transform of the input, $|F_i(j\omega)|$, by the amplitude ratio of the system and adding the phase shift of the system to the phase angle of $F_i(j\omega)$; $f_o(t)$ can then be found by applying the Fourier integral.

This technique is intuitively in everyday use, as, for example, in defining a tape-recorder as having a frequency response of, say, 50 Hz to 15 kHz; this implies that this frequency range is adequate to reproduce the Fourier components of any *transient* signals encountered in speech and music.

The similarity between this and the transfer function technique is striking, and is even more so if the equations defining the Fourier and Laplace transforms are compared, thus:

$$F(j\omega) = \int_{-\infty}^{\infty} e^{-j\omega t} f(t)\, dt$$

and

$$F(s) = \int_{0}^{\infty} e^{-st} f(t)\, dt$$

The bottom limit of the integrals are dissimilar, but the transfer function theory is based only on functions which are zero for negative time, the negative half of the Fourier transform integral thus making no contribution. It is shown in Section 4.11 that $s = \sigma + j\omega$ so that the Fourier transform can, for functions that are zero for $t < 0$, be defined as the limiting case of the Laplace transform as $\sigma \to 0$.

References

Basic Closed Loop Control Theory

1. *Automatic Feedback Control System Synthesis*, J. G. Truxal (McGraw-Hill, 1955).
2. *Servomechanisms and Regulating System Design*, Vol. 1, H. Chestnut and R. W. Mayer (Wiley, 1951).
3. As reference 2, Vol. 2.
4. *Analysis and Design of Feedback Control Systems*, G. J. Thaler and R. G. Brown (McGraw-Hill, 1960).
   This book is recommended as one of many alternatives to references 2 and 3.
5. *Transfer Function Techniques for Control Engineers*, D. R. Towill (Illife, 1970).
6. *Handbook of Automation, Computation and Control*, edited by E. M. Grabbe, S. Ramo and D. E. Wooldridge (Wiley, 1958).
   This is an extensive survey of the field in three volumes: Vol. 1, Control fundamentals; Vol. 2, Computers and data processing; Vol. 3, Systems and components.

Mathematical Texts

7. *Analysis of Linear Systems*, D. K. Cheng (Addison-Wesley, 1959).
8. *Transform Method in Linear System Analysis*, J. A. Aseltine (McGraw-Hill, 1958).
9. *Complex Variable Theory and Transform Calculus*, N. W. McLachlan (Cambridge University Press, 1953).
10. *Tables of Laplace, Heaviside, Fourier and Z Transforms*, M. Healey (Chambers, 1967). Another useful handbook is *Mathematical Toolkit For Engineers*, H. A. Webb and D. E. Ashwell (Longmans, 1959).

Texts on Fundamental Principles and Components

11. *Electronics for Engineers*. H. Ahmed and P. J. Spreadbury (Cambridge University Press, 1973).
12. *Theory of Vibrations for Engineers*, E. B. Cole (Crosby-Lockwood, 1957).
13. *Rotating Amplifiers*, edited by M. G. Say (Newnes, 1954).
14. *Thyristor Control*, F. F. Mazda (Butterworth, 1973).
15. *Magnetic Amplifiers and Saturable Reactors*, edited by M. G. Say (Newnes, 1954).
16. *Hydraulic Control Systems*, H. E. Merritt (Wiley, 1967). A more extensive text is *Fluid Power Control*, edited by J. F. Blackburn, G. Reethof and J. L. Shearer (Wiley, 1960).
17. *Electrical Transducers for Industrial Measurement*, P. H. Mansfield (Butterworth, 1973).

The following references are specialised texts, extending the basic theory included here.

18. *Synthesis of Feedback Systems*, I. M. Horowitz (Academic Press, 1963).
19. *Nonlinear Automatic Control*, J. E. Gibson (McGraw-Hill, 1963).
20. *Probability, Random Variables and Stochastic Processes*, A. Papoulis (McGraw-Hill, 1965).
21. *Sampled Data Control Systems*, J. R. Ragazzini and G. Franklin (McGraw-Hill, 1958).
22. *Process Systems Analysis and Control*, D. R. Coughanowr and L. B. Koppel (McGraw-Hill, 1965).
23. *Analogue Computation*, A. S. Jackson (McGraw-Hill, 1960).
24. *Adaptive Control Systems*, edited by E. Mishkin and L. Braun (McGraw-Hill, 1961).
25. *State Variables for Engineers*, P. M. DeRusso, R. J. Roy and C. M. Close (Wiley, 1965).
26. *Systems Engineering Tools*, H. Chestnut (Wiley, 1965).
27. *State Functions and Linear Control Systems*, D. G. Schultz and J. L. Melsa (McGraw-Hill, 1967).
28. *Modern Control Engineering*, A. R. M. Noton (Pergamon, 1973).
29. *Optimal Systems Control*, A. P. Sage (Prentice-Hall, 1968).
30. *Stochastic Optimal Linear Estimation and Control*, J. S. Meditch (McGraw-Hill, 1969).
31. *Computational Techniques for Chemical Engineers*, H. H. Rosenbrock and C. Storey (Pergamon, 1966).
32. *Computer Control of Industrial Processes*, E. S. Savas (McGraw-Hill, 1965).
33. *Modern Foundations of Systems Engineering*, W. A. Porter (Macmillan, 1966).
34. *Adaptive Control Processes—A Guided Tour*, R. Bellman (Princeton 1961).
35. *Modern Concepts in Control Theory*, H. A. Prime (McGraw-Hill, 1969).
36. *Digital Processes for Sampled Data Systems*, A. J. Monroe (Wiley, 1962).
37. *Applied Optimum Control*, A. E. Bryson and Y. C. Ho (Ginn, 1969).

Problems

1. A d.c. generator has $R_a = 1\,\Omega$ and $R_f = 200\,\Omega$. The field is fed by an amplifier with a voltage gain of 50, the input to the amplifier being the difference between a fixed reference voltage, V_R, and the generator terminal voltage, V.

(a) Determine V_R so that $V = 200$ when $I_a = 20$ A.
(b) Find the no-load generator voltage.
(c) Find V when $I_a = 30$ A.

The open-circuit characteristic of the generator is,

Field current (amperes)	0·6	0·8	1·0	1·2
E.M.F. (volts)	180	205	220	230

[204; 201; 198]

2. For a system similar to problem 1 find the amplifier gain (mA/volt) required to keep the line voltage within $\frac{1}{4}\%$ of the full-load value of 200 volts. The generator line voltage rises from 200 to 220 volts with 1 A in the field when full load is thrown off. Neglect saturation. [86.4]

3. For the velocity control system of Fig. P.1.

$$\text{Amplifier gain} = 500 \text{ mA/volt}$$
$$\text{Motor torque} = 10^{-4} \text{ Nm/mA}$$
$$\text{Total inertia} = 5 \times 10^{-6} \text{ kg m}^2$$
$$\text{Viscous friction coefficient} = 3 \times 10^{-6} \text{ Nms}$$
$$\text{Tacho constant} = 0·1 \text{ volt per rad/s}$$

Find the differential equation relating ω to v_i on no load.

Hence calculate (a) the steady speed when $v_i = 10$ volts; (b) the variation in speed in response to a step input from 0 to 18 volts with the system initially at rest.

[100; 180 $(1 - e^{-1000t})$ rad/s]

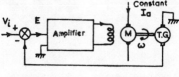

Fig P·1

4. For the system of problem 3 find the differential equation relating ω, v_i and the load torque, T_L. Hence calculate the variation in speed in response to a step change in load from no load to 10^{-2} Nm if the system is initially in the steady state with $v_i = 10$ volts.

[98 + 2e^{-1000t} rad/s]

5. Repeat problems 3 and 4, allowing for an amplifier time constant of

334

0·1 second. What is the undamped natural frequency and the damping ratio of the system?

[N.B. Solve this problem by both classical and Laplace transform methods and compare. Note also that while the extra time constant is almost negligible in the open loop it has a marked effect on the closed loop!]   [97·5; 0·054]

6. Fig. P.2 shows a small speed-control system and its parameters. The generator is driven at constant speed, and the motor is provided with a constant field current. Further data are as follows:

Generator e.m.f. $e_g = 1500$ volts per field ampere; motor e.m.f. $e_m = 1·0$ volt/(rad/s); motor torque $= 1·0$ N m per armature ampere; inertia of motor and load together, $J = 0·48 \times 10^{-4}$ kg m²; friction is negligible; $r_1 = 1000$ ohm, $L_1 = 10$ H, $r = 100$ ohm.

Find an expression relating the instantaneous angular velocity (in radians per second) of the load to the input voltage v_i. Indicate in a sketch the expected form of response to a suddenly applied direct input voltage.

$$[\omega + 1·48 . 10^{-2}\dot{\omega} + 0·48 . 10^{-4}\ddot{\omega} = 1·5v_i]$$

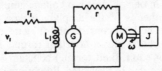

Fig P·2

7. In the speed-regulating system of Fig. P.3 the generator field time constant and that of the motor-generator armature circuit are negligible. The amplifier gain is 4 amperes/input volt; the generator e.m.f. is 50 volt/field ampere; the motor speed is 0·25 rad/s per generator volt with an effective time constant of 1 second; and the tachometer output is 0·2 volt per rad/s. Find the transfer function relating output speed, ω, to input voltage, v_i, on no load. Find the variation in speed in response to a sudden input of 5 volts, applied with the system at rest. $[\omega = 22·7(1 - e^{-11t})$ rad/s]

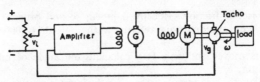

Fig P·3

8. A position control servo has a motor of inertia 10^{-6} kg m² driving a load of inertia 0·01 kg m² through a gearbox of ratio 100 : 1. The viscous friction referred to the load is 0·64 N m s. The amplifier gain is 50 mA/volt and the error detector gives 1 volt/degree misalignment. The motor torque is 10^{-4} N m/mA. Calculate:

(a) ω_n;

(b) the damping ratio;

(c) the value of the tachometer constant, mounted on the motor shaft required for critical damping.   [37·7; 0·422; 0·018]

9. Find for the servo of problem 8 the output response to a step input of

10° and find the peak overshoot as a percentage of the input and the time at which $\theta_o = \theta_i$ initially and at the second instant.   [23%; 0·059; 0·15]

10. For a r.p.c. servo which has $\zeta = 0.25$ and $\omega_n = 8\pi$ rad/s, calculate:

(a) the peak overshoot in response to a step input;

(b) the frequency and decay rate of any oscillatory terms in the output;

(c) the number of oscillations before the output in response to a step input settles within a $\pm 5\%$ band; and

(d) the resonant frequency and the amplitude of the resonant peak.

[44·4%; 24·3, 2π; 23·5, 6·24 db]

11. A remote position system has an inertia J and negligible F. The torque is modified to be proportional to:

$$K_1 e + K_2 \frac{de}{dt} + K_3 \int e \, dt, \quad \text{where} \quad e = \theta_i - \theta_o$$

Find the equation relating θ_i and θ_o.

12. Two simple remote position controllers are each connected as shown in Fig. P.4. The amplifier in one of them is a voltage amplifier with a gain of 10. The amplifier in the other is of the amplidyne type producing x amperes output per volt input. The load consists of 1·0 kg m² of inertia plus viscous damping (0·5 N m per rad/s) plus a constant torque T_L. The motor field is constant and the motor e.m.f. (and torque) constant = 1·5. $R_a = 0.2$ ohms and other inductances etc. can be neglected. Potentiometer constant = 2 volts/rad error.

Find x so that both systems have the same steady state error, due to T_L, and compare ω_n and ζ for the two systems.   [50; 12·25, 0·48; 12·25, 0·024]

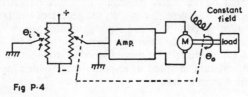

Fig P·4

13. For a viscous friction critically damped, second order, r.p.c. servo, $J = 500$ kg m² and motor torque referred to the output = 2000 N m/rad of misalignment.

(a) If θ_i is moved suddenly 90°, calculate the angle of misalignment 2 seconds later.

(b) Calculate the steady state error when the input is continuously rotated at 1 rev/min.   [8° 15′; π/30]

14. The position of a rotatable mass driven by an electric motor is controlled from a hand-wheel. The damping torques due to viscous friction and velocity feedback are equal. $J = 100$ kg m² and $f_n = 2.5$ Hz. If the system is critically damped, calculate the feedback torque per unit angular velocity and the steady state misalignment when the input is rotated at 1 rad/s.

[1570; 0·127]

15. A water-supply tank with water surface area A m² is filled from the top by a simple float-controlled valve such that the in flow q_i m³/s is proportional to the difference between the full level h_i m and the operating level h_o. The constant of proportionality is k. Water is drawn off from the bottom by a tap-controlled outlet. For any constant outflow demand Q

m³/s what is the steady operating height of the tank, and how long does it take to half refill when the outlet tap is suddenly closed? $\left[h_i - \dfrac{Q}{k}, \dfrac{A}{k} \log_e 2 \right]$

16. A simple two-position discontinuous speed control works as follows: A motor drives a load of constant inertia against a constant resisting torque T_m and the desired speed is ω_m. The control is arranged such that when the speed reaches an upper value ω_1 ($> \omega_m$) a motor torque T_2 ($< T_m$) is applied to the load, and when the speed reaches a lower value ω_2 ($< \omega_m$) a motor torque T_1 ($> T_m$) is applied. Show by plotting motor torque and speed against a time base, the manner in which the motor torque and speed will vary with time under the following conditions:

 (a) $T_m = 40$; $T_1 = 50$; $T_2 = 30$; $\omega_m = 100$; $\omega_1 = 110$; $\omega_2 = 90$
 (b) $T_m = 45$; $T_1 = 50$; $T_2 = 30$; $\omega_m = 100$; $\omega_1 = 110$; $\omega_2 = 90$
 (c) $T_m = 40$; $T_1 = 50$; $T_2 = 30$; $\omega_m = 100$; $\omega_1 = 105$; $\omega_2 = 95$
 (d) $T_m = 40$; $T_1 = 55$; $T_2 = 25$; $\omega_m = 100$; $\omega_1 = 110$; $\omega_2 = 90$

17. A remote temperature indicator has characteristics such that the rate of change of indicated temperature on the dial is proportional to the difference between the actual bulb temperature in the furnace and the indicated temperature. The constant of proportionality is K. At a given instant when the bulb and indicated temperatures are the same, the furnace temperature starts to increase at the constant rate of 5° C per minute.

Find the value of K such that the steady state error of the dial reading is 1° C and derive an expression giving the indicated temperature at any time before the steady state is reached. $[5, \theta_{(o)} + 5t - 1 + e^{-5t}]$

18. A position-control system has zero steady state displacement error. The steady state velocity error coefficient is 0·005. The errors to input signals of higher orders are time-varying quantities. The only component in the 'transient' part of the response is a decaying sinusoid with a decay rate of 16 second⁻¹.

 (a) Assuming the simplest system, determine the open and closed loop transfer functions. Evaluate (i) the undamped natural frequency, (ii) the damping ratio and (iii) the frequency and decay rate of the damped transient. $[80, 0·2, 78·4, 16]$

 (b) Find the tachometer constant required to give critical damping, given that the input and output potentiometer coefficients = 2 volt/rad.
$[0·04]$

 (c) Calculate the new value of velocity error coefficient. $[0·025]$

19. With rated armature current the motor of a velodyne produces 20 N m/ A. (The motor may be assumed to remain unsaturated.) The field circuit resistance is 500 ohms and it is fed from a voltage amplifier with a gain of 100. Calculate the voltage per 1000 rev/min. required from the tachogenerator if the speed is to fall by no more than 25 rev/min as the load torque is increased from zero to a value of 2 N m.

 (a) Neglecting friction and windage torque.
 (b) Taking the friction and windage as 0·02 N m per rev/min. $[20, 15]$

20. Obtain the transfer function which relates the p.d. applied to the armature and the shaft displacement of the armature-fed d.c. motor detailed over page.

With the field current held constant at normal value the machine, driven at 500 rev/min as a generator for test purposes, gave an induced e.m.f. of 52·35 volts. The resistance and inductance of the armature circuit are respectively 1·0 ohm and 30 mH, the moment of inertia is 5×10^{-3} kg m² and the friction and windage torque amounts to 0·05 N m s.

Does the transient response in speed to step input voltage show any over-shoots and undershoots? If so, by how much must the viscous torque be increased to make the damping critical? [0·6]

21. A heat engine has $J = 6$ lb-ft-s² and viscous damping 30 lb-ft per rad/s. The fuel injection corresponds to a time constant of 0·1 second lag, and the governor regulating lever turns 2° for every rad/s change in speed. With the governing system inoperative, 1° change of the regulating lever changes the engine speed by 2 rad/s. Find the frequency and decay rate of the transient following a step change in load. What would be the steady state error for a load increase of 120 lb-ft? [13·9; 7·5; 0·8]

22. A variable delivery pump controls the speed of a similar motor. The motor speed is monitored by a tacho, the output of which is subtracted from a constant voltage and the difference fed to an amplifier and servomotor which controls the pump delivery control spindle through a 10 : 1 gearbox. Pump speed = 1500 rev/min; the pump supplies full delivery for a rotation of 4 rad of its control spindle which has an inertia of 2 lb-in²; the servomotor inertia = 0·1 lb-in² and the motor torque per volt at the amplifier = 35 lb-in. The equivalent damping coefficient at the control spindle is 1 lb-in per rad/s. Neglecting compressibility and leakage oil effects find the open and closed loop transfer functions if the tachometer has a transfer function $K/(1 + 0·05s)$ and thus determine the value of K to just cause instability. [0·38]

23. Two single non-interacting delay elements with time constants T_1 and

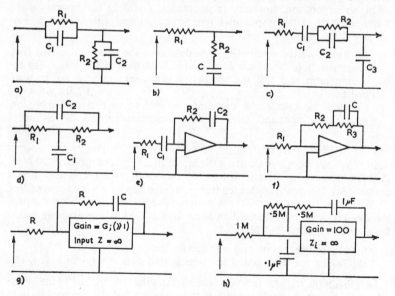

Fig P·5

T_2 are connected in series. If a step input of magnitude H is applied to the system, which is initially dead, find the output as a function of time. If $T_1 = 2T_2$, find the output when $t = T_2$.   [0·156 H]

24. Find the transfer functions of the electrical networks of Fig. P.5.

25. Find the transfer functions of the systems of Figs. P.1 to P.4.

26. Find C/R for the systems of Fig. P.6.

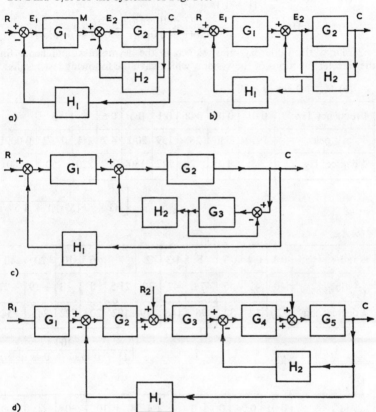

Fig P·6

27. Sketch the Bode diagrams for the following functions. State the type and order of each system represented, and roughly sketch the Nyquist diagrams. Draw in detail the Nyquist diagram for one of the functions so as to prove the greater ease of the Bode method.

(a) $G(s)H(s) = \dfrac{5(1 + s)}{(1 + 2s)(1 + 0·1s)(1 + 0·2s)}$

(b) $G(s)H(s) = \dfrac{3}{s(1 + 0·1s)(1 + 0·5s)}$

(c) $G(s)H(s) = \dfrac{10(1 + 0·1s)}{s^2(1 + 0·2s)(1 + s)(1 + 0·01s)}$

(d) $G(s)H(s) = \dfrac{2(1 + 0.1s)}{s(1 + 0.1s + s^2)}$

(e) $G(s)H(s) = \dfrac{2}{(1 + s^2)(1 + 0.1s)}$

(f) $G(s)H(s) = \dfrac{5(1 + s + s^2)}{(1 + 1.1s + s^2)}$

(g) $G(s)H(s) = \dfrac{3(1 + 0.1s)(1 + 0.5s)}{s^2(1 + 0.05s)(1 + 0.01s)(1 + 0.1s + 0.005s^2)}$

28. Draw the Bode diagrams, sketch in the asymptotes, and hence find the transfer functions of the systems which gave the following test results:

(a) $V_{in} = 2 \sin \omega t$ mV.

Frequency Hz	0·01	0·03	0·06	0·1	0·3	0·6	1	3	6
V_o mV peak	1990	636	285	159	20·2	4·7	1·4	0·071	0·0091
ϕ degrees lag	98	110	126	142	180	202	220	250	260

$$\left[\frac{62.8}{s(1 + 0.159s)(1 + 1.59s)} \right]$$

(b)

ω rad/s	0·1	0·4	0·7	1	1·3	2	3	6	10	20	40
$\dfrac{V_o}{V_{in}}$ db	40	41	45	47·6	42	30	21·5	8	−3	−19	−37
ϕ degrees lag	2	12	32	96	148	179	192	208	222	241	254

$$\left[\frac{100}{(1 + 0.1s)(1 + 0.4s + s^2)} \right]$$

(c)

ω rad/s	0·01	0·04	0·1	0·4	1	4	10	30	60	100
$\dfrac{V_o}{V_{in}}$ db	54	43	37	33·5	31	19	4	−21	−39	−52
ϕ degrees lag	85	70	53	43	68	138	193	240	255	261

$$\left[\frac{5(1 + 10s)}{s(1 + 0.1s)(1 + 0.2s)(1 + s)} \right]$$

29. Find the characteristic equations of the following systems:

(a) $G(s) = \dfrac{25(1 + s)}{s(1 + 0.1s)(1 + 0.5s)}$ $H(s) = 1$

(b) $G(s) = \dfrac{25(1 + s)}{s(1 + 0.1s)(1 + 0.5s)}$ $H(s) = \dfrac{1}{1 + 2s}$

(c) $\quad G(s) = \dfrac{25(1 + s)}{s(1 + 0 \cdot 1s)(1 + 0 \cdot 5s)(1 + 2s)} \qquad H(s) = 1$

(d) $\quad G(s) = \dfrac{2(1 + 0 \cdot 1s)}{s(1 + 0 \cdot 1s)(1 + 0 \cdot 01s + 0 \cdot 005s^2)} \quad H(s) = 4$

(e) $\quad G(s) = \dfrac{3(1 + s)}{s^2(1 + 0 \cdot 1s)^2(1 + 0 \cdot 2s)} \qquad H(s) = 1$

(f) For the system of Fig. P.6 (a):

$$G_1(s) = \frac{4}{1 + s}; \ G_2(s) = \frac{2}{s(1 + 0 \cdot 1s + 0 \cdot 01s^2)}; \ H_1(s) = 1; \ H_2(s) = \frac{1}{1 + 2s}$$

(g) For the system of Fig. P.6 (c):

$$G_1(s) = 5; \quad G_2(s) = \frac{10}{s(1 + 0 \cdot 01s + 0 \cdot 05s^2)}; \quad G_3(s) = \frac{1}{1 + 0 \cdot 2s};$$

$$H_1(s) = \frac{4(1 + 5s)}{1 + 10s} \quad \text{and} \quad H_2(s) = 2$$

30. Plot the poles and zeros for the examples of problems 24, 27 and 29.

31. Plot the root loci for the examples of problems 27 and 29.

32. Use Routh's criterion to determine the stability of the examples of problem 29.

33. Try using Routh's array to find the value of K that just renders the following systems unstable:

(a) $\qquad\qquad G(s)H(s) = \dfrac{K(1 + 0 \cdot 05s)}{s(1 + s)(1 + 0 \cdot 1s)}$

(b) $\qquad\qquad G(s)H(s) = \dfrac{K}{s(1 + 0 \cdot 1s + 0 \cdot 01s^2)(1 + s)}$

(c) $\qquad\qquad G(s)H(s) = \dfrac{K(1 + 0 \cdot 1s)(1 + s)}{s^2(1 + 0 \cdot 5s)(1 + 0 \cdot 05s)}$

$$[24 \cdot 5; \ 9 \cdot 2; \ 0 \cdot 488]$$

34. For the examples of problem 33 plot the Bode diagrams with $K = 1$, and hence find the values of K to render the system just unstable.

35. Draw the Bode diagrams for the examples of problem 29 and hence determine the gain and phase margins. (Note: compare the slope of the Bode amplitude diagram as it crosses the 0 db line with the degree of stability.)

36. Use the M and δ circles, the Nichols Chart and L.F.–H.F. approximations to find C/R as a function of frequency for the system of problem 29 (a), (b) and (c). Express the answer as a Bode amplitude diagram.

37. Examine the Bode diagrams of problem 27 and suggest some form of series compensation (and change of gain factor, K?). Sketch the modified open loop diagram and the Bode diagram of the compensation suggested and devise an electrical network to give this compensation.

38. Show that, if the open loop transfer function of a control system is:

$$\frac{90(1 + 0 \cdot 5s)}{(1 + s)(1 + 2s)(1 + Ts)}$$

there is a certain range of values of T which will render the system unstable and determine the limiting values of T. $\qquad\qquad [4 < T < 8]$

39. The open loop transfer function of a control system is of the form:

$$\frac{K(1 + Ts)}{s^2(1 + aTs)}$$

where K, T and a are positive constants. Show that, with the loop closed, the system would be unstable for any value of K unless the value of a lies in a certain interval.

A system with such an open loop transfer function is stable and gives the following results when tested with a transfer function analyser under open loop conditions:

Excitation signal	Angular frequency (rad/s)	Response
$1{\cdot}0 + j0$	$1{\cdot}0$	$-12 - j4$
$1{\cdot}0 + j0$	$2{\cdot}0$	$-3{\cdot}75 - j1{\cdot}25$

Determine the values of K, T and a.   [a < 1; 10, 1, 0·5]

40. For the open loop transfer function

$$\frac{K}{s(1 + \frac{1}{16}s)(1 + \frac{1}{4}s)}$$

choose K so that gain and phase cross-overs coincide. The system performance is to be modified by the introduction in series of a passive electrical phase lag network. Sketch such a network and, assuming that its time constants are in the ratio of four to one, find the maximum possible phase correction. Show how to choose suitable values of the time constants so that the compensated system shall have a phase margin of at least $20°$ and give an approximate indication of the effect of compensation upon the final logarithmic plots.

[20; 37°]

41. A linear r.p.c. servo is stabilised by velocity feedback and has the following details:

Total moment of inertia referred to output shaft, $J = 10 \text{ N m s}^2$
Motor constant, K_2 $= 1{\cdot}0 \text{ N m/A}$
Amplifier conductance, G $= 0{\cdot}5 \text{ ampere/volt}$
Error detector constant, K_1 $= 10 \text{ volt/rad of error}$
Gearbox ratio $= 100 : 1$
Neglect Friction

Determine the equation of motion of the system and evaluate the tacho-generator constant K_3 for critical damping.

Acceleration feedback is now added by passing the tacho-generator signal through a network having a transfer function,

$$\frac{V_o}{V_i} = \frac{s}{1 + sT}$$

Determine the new equation of motion, and show that velocity lag error is eliminated.   [2·8]

42. A linear type 1 second-order servo is compensated by derivative of error, and is critically damped.

Derive an expression for the closed loop frequency response when the input is driven by a sinusoidal input signal. From this expression determine the maximum value of magnification and the relative frequency at which this occurs.

43. The input voltage to the amplifier of an r.p.c. servomechanism consists of a signal proportional to output shaft velocity which is subtracted from the signal which results when the actuating signal is passed through an operational amplifier having a transfer function

$$\frac{V_o}{V_i} = \frac{(1 + sT_1)}{sT_2}$$

Derive the equation of motion of the system.

Making reference to the equation of motion, illustrate the effect of introducing the integral term upon:

(a) The velocity lag.

(b) The steady state error introduced by a fixed load torque on the output shaft.

(c) The stability of the system.

44. Open loop tests on a system ultimately destined to be a remote position controller gave the following results:

ω rad/s	Magnitude $G(\omega)$	Phase angle $\phi(\omega)$
0·1	10·4	$-96°$
0·2	5·8	$-103°$
0·3	4·7	$-115°$
0·4	4·7	$-138°$
0·5	4·0	$-180°$
0·6	2·0	$-211°$
0·7	1·20	$-234°$
0·8	0·71	$-243°$
0·9	0·46	$-248°$
1·0	0·32	$-251°$

Investigate the possibility of rendering the closed loop stable by the introduction of an active phase advance network, the time constants of which are in the ratio 10 : 1, and the L.F. gain is unity.

If stability is not possible under the above conditions estimate the fractional extent to which the attenuation of the inserted network can be offset to leave suitable gain and phase margins.

45. A machine tool drive has the following open loop transfer function:

$$G(s) = \frac{K}{s(1 + 0 \cdot 001s)(1 + 0 \cdot 025s)(1 + 0 \cdot 1s)}; \quad H(s) = 1$$

It is to have a phase margin of 45°. Find the value of K for the system:

(a) Uncompensated.

(b) With passive phase lead compensation having break frequencies at 10 and 40 rad/s.

(c) With additional passive phase lag compensation having break frequencies at 1 and 4 rad/s.

Compare the resulting closed loop frequency response of the three systems and comment on the choice of break frequencies of the compensation networks.   [8.7, 70, 160]

46. The system of Fig. P.4 is to operate with a stiffness of 10 N m/degree. The system response to a step input must have a peak overshoot not exceeding 30%. Determine the required amplifier gain and design any necessary compensation.

Gear ratio between motor and load = 50 : 1

Load inertia, $J_L = 1$ N m s^2

Motor inertia, $J_m = 8 \times 10^{-4}$ N m s^2; Viscous friction coefficients of the load = 1.43×10^{-3} N m s and of the motor = 0

Motor torque = 0.92 N m/A

Motor e.m.f. constant = 1.25 volt per rad/s

$R_a = 24.4$ ohms

Potentiometer voltage = 100 volts, with a stroke of 6 radians [0.75 A/V]

47. A speed control system (Fig. P.1) is compensated by a series active network of Fig. P.5 (f). Without the compensation the speed drops from 2500 rev/min to 2490 rev/min from no load to full load. What is the L.F. gain of the network required to reduce the 'droop' to 0.005% of no load speed? Derive the complete transfer function for the closed loop system.   [80]

48. A process has a transfer function, $G(s) = \dfrac{5}{s(1 + 0.5s)}$. It is supplied from a controller having a transfer function, $G_c(s) = \dfrac{K}{1 + 0.01s}$ and the loop is closed by direct feedback. Find the limiting value of K. With this value of K, if the 0.01 second controller time constant is replaced by a finite time delay of 0.01 second, investigate the system for stability, calculate the gain and phase margins and design a series compensation to bring the gain and phase margins to acceptable values.

49. Express $[(x_3 - x_1)^2 + (x_3 - x_2)^2]$ in the form $x'Ax$ when A is symmetric and x is a three-vector. Show that $\dfrac{\partial}{\partial x}(x'Ax) = 2Ax$.

50. A control system has the transfer function

$$\frac{C(s)}{U(s)} = \frac{s^2 + 3s + 2}{s(s^2 + 7s + 12)}$$

Draw the 'state variable' diagram for this system. Express in the form of a matrix differential equation. Find the eigenvalues and the transition matrix.

51. A system is represented by

$$\frac{d^2x}{dt^2} + \frac{3dx}{dt} + 2x = 2u(t)$$

Determine the transition matrix and hence the vector difference-equation if $u(t)$ is a sample and hold signal of period 0.1 seconds.

52. For the network of Fig. P.7, treat $v_i(t)$ as the 'control' and $v_o(t)$ as the 'output' and express the system equations in state variable form. (Use the state vector $x = [i_1\ i_2\ v_c]'$.)

53. A chemical process plant has the forward path transfer function $\dfrac{K}{(1 + sT_1)(1 + sT_2)}$ and unity feedback. $T_1 = 0.25$ min, $T_2 = 1$ min and $K = 0.5$.

Obtain a state-variable representation of the system and find the state transition matrix. Hence calculate the system output and rate of change of output at 1, 2 and 3 minutes with zero input and initial conditions: output = 1 rate of change of output = 0.

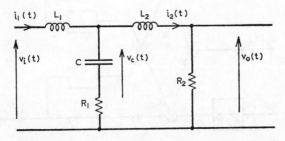

Fig. P.7

54. For the system

$$\dot{x} = \begin{bmatrix} 0 & 1 \\ -3 & -2 \end{bmatrix} x + \begin{bmatrix} 0 \\ 1 \end{bmatrix} u; \quad z = [1 \quad 1]x$$

Use the linear transformation

$$y = \begin{bmatrix} 1 & 1 \\ 0 & 2 \end{bmatrix} x$$

to find the system equations in terms for the variables y and u.

55. Find the transfer function of the system

$$\dot{x} = \begin{bmatrix} 0 & 1 & 1 \\ 0 & 2 & -1 \\ -1 & -2 & 0 \end{bmatrix} x + \begin{bmatrix} 1 \\ 0 \\ 1 \end{bmatrix} u$$

$$y = [1 \quad 0 \quad 1]x$$

56. Find a set of state equations for the system of Fig. P.8.

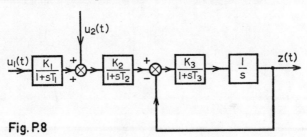

Fig. P.8

57. Transform the following system of equations to canonical form (diagonal):

$$\dot{x} = \begin{bmatrix} 0 & 1 & 0 \\ 0 & 0 & 1 \\ 0 & -2 & -3 \end{bmatrix} x + \begin{bmatrix} 0 \\ 0 \\ 2 \end{bmatrix} u$$

$$z = [1 \quad 0 \quad -1]x$$

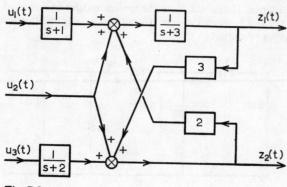

Fig. P.9

58. Find a set of state equations for the system of Fig. P.9.

Hence find the state difference equations if $u(t)$ is sampled and clamped with a period 0·1 second. Thus if $u_1(k) = 1$ and $u_2(k) = -1$ for all k and $u_3(k) = k$ for $k \geqslant 0$, find z_1 and z_2 at $k = 0$, 1 and 2, assuming zero initial conditions.

59. Find the matrix transfer function for the system of Fig. P.9.

Index

a.c. servos, 267, 300
  motors, 302, 307
  tachos, 306
acceleration error coefficient, 122
  feedback, 152
  gain constant, 123
  input, 121
active networks, 147, 262
actuating signal, 6, 11, 45
actuator, 265
adaptive control, 195
A.I.E.E. terminology, 10
aliasing, 223
amplidyne generator, 285
amplitude conditional stability, 182
  criterion, 100
  spectrum, 211, 330
analogue computers, 262
analogous systems, 23
analytical design, 195
angle of departure, 100
armature control of d.c. motors, 285, 297
assisted braking, 277
asymptotes for Bode diagrams, 77
  for root loci, 99
attenuation, 77
attenuation diagrams—*see* Bode diagrams
autocorrelation functions, 205, 211
auxiliary equation, 35
average values, 199

backlash, 169, 268
bandwidth, 124
bang-bang servo—*see* relay servo
block diagrams, 43
Bode diagrams, 77
  theorems, 109
braking, 277
branches, 57

break frequencies, 80, 83
bridged T network, 161, 305
buffer amplifier, 44
bulk modulus, 317

cancellation compensation, 161
canonical form, 246
carrier frequency, 300
cascade compensation—*see* series compensation
chain codes, 208, 216
characteristic equation, 89, 240
clamp circuit, 219
classical solution of linear differential equations, 33
closed loop control, 6
  frequency response, 111
coarse–fine systems, 275
coefficient matrix, 236
column matrix or vector, 232
command, 10
compensation, 131
  acceleration feedback, 152
  derivative, 135
  input, 154
  integral, 137, 149, 160
  parallel, 149
  phase lag, 144, 159
  phase lead, 139, 158
  series, 135
  velocity feedback, 152, 161
complementary function, 35, 89
complex frequency plane—*see* s-plane
  conjugate, 65
  variable, s, 40, 42, 324
  roots, 35, 55, 80
compressibility, 317
conditional stability, 106, 108
contactor servo—*see* relay servo
controllability, 247

controlled rectifiers, 286
controlled variable, 10, 45
convolution, 200
  integral, 201
cooling of oil, 323
corner frequency—*see* break frequency
correlation, 204
  autocorrelation, 205, 211
  crosscorrelation, 205, 209
coulomb friction, 31
critical damping, 35
cross-field generator—*see* amplidyne
cross-modulation, 171
cross over frequency, 110
current limiting, 289

damping, 32
  critical, over- and under-damping, 35
  ratio, 35, 81
db-phase diagrams—*see* Nichols diagram
dead zone, 168
decay rate, 35, 126
decade, 78
decibel (db), 75
decrement—*see* decay rate
definitions of symbols, 10
delay, 168
delta circles, 111
demodulators, 302
derivative of error compensation, 135
  of output feedback, 32, 48, 134, 152, 161
describing functions, 171
  frequency invariant, 171
  frequency variant, 186
  of hysteresis, 177
  of relay servo, 178
  of saturation, 174
  of saturation and dead zone, 176
  validity of, 172
desired value, 11, 202
deterministic signal, 119
digital filtering, 229
Dirac function, 220

direct digital control, 218, 229
direct feedback, 47
discrete time systems, 217
disturbance, 11

eddy current couplings, 290
  brakes, 292
eigenvalues, 240, 248
eigenvectors, 240, 248
ensembles, 196, 199
equivalent inertia, 268
equalisation—*see* compensation
ergodicity, 196
errors, 17, 38
  acceleration lag, 38, 121
  coefficients, 121, 126
  criteria, 202
  integral square, 200, 202
  mean square, 199, 202, 213
  position, 38, 120
  signal—*see* actuating signal
  velocity lag, 38, 120
expected values, 199
external load torques, 38, 51, 129, 132

feedback amplifier, 16
  compensation, 149
  controller, 5
  direct, 47
  negative, 6, 10, 45
  positive, 6, 45, 88
feedforward compensation, 154
field control of d.c. motor, 266, 278
  time constant, 39
figure of merit—*see* error criterion
final value theorem, 136, 326
finite time delay, 168
firing angle of controlled rectifiers, 288
first-order system, 24
first probability density function, 197
flow control valve, 312
flywheel diode, 290
forced response, 239
forcing function, 235
Fourier series, 173, 329
  transforms, 324, 330

frequency response, 64
  response functions, 65, 69
  normalised, 67
  of oscillation, 35, 125
friction, 31, 268

$G(s)$—*see* transfer function
$G(j\omega)$—*see* frequency response function
gain constants, 122
  margin, 105
  open loop, 69, 91
Gaussian distribution, 198
gearing, 29, 269
Green's function, 201
grey noise—*see* pink noise
Guillemin's synthesis method, 156

harmonics, 329
harmonic response—*see* frequency response
Heaviside's partial fraction expansion, 65
high–low frequency approximation, 116
high–mid–low frequency approximation, 150
hold circuit, 219
hydraulic systems, 308, 317
  amplifier, 314
  electrohydraulic valves, 314
  pumps and motors, 308
  valves, 312
hysteresis, 169

I.T.A.E., 203
ideal value, 202
impedance, 66
impulse response—*see* weighting function
index of performance—*see* error criteria
induction motors, 267, 285, 302, 307
inertia, 26, 268
initial conditions, 25, 40, 238, 325
  condition response, 239
  value theorem, 326

input compensation, 132, 154
  parabolic, 121, 146
  polynomial, 244
  ramp, 37, 120, 245
  sinusoidal, 64, 245
  step, 25, 36, 119, 245, 326
insertion loss, 141
integral control, 137, 146
  of error compensation, 137, 146
  square error, 200, 202
integrated circuit amplifiers, 147
integrating amplifier, 263
interaction, 43
instability, 14, 88
inverse Nyquist diagram, 108, 180
  Fourier transform, 330
  Laplace transform, 326
isoclines, 190

jump resonance, 170, 185

lag network, 144
Laplace transforms, 324
lead network, 139
lead–lag network, 148
leakage flow, 317
limit cycle oscillation, 170, 182, 192
linear equations, 23, 167, 324
load disturbances—*see* external load torques
log magnitude and phase diagrams—*see* Bode and Nichols diagrams

M circles, 111
machine tool control, 13, 271, 313
magnetic amplifiers, 290
Magslips, 279
major and minor loops, 46
manipulated variable, 11
manual control, 14
mark–space ratio control, 294
Mason's flow formula, 60
mathematical models, 8
matrices, 232
matrix representation of transfer functions, 53, 249

mean modulus of error, 203
  square error, 199, 213
  value, 199
measurement of transfer functions, 255
memory in describing functions, 177
mercury-arc rectifiers, 287
minimum-phase systems, 57, 72
modal matrix, 246, 248
modulation, 301
motors
  a.c.—*see* induction motor
  d.c., 24, 267, 283, 295
  force, 314
  hydraulic, 267, 308
  linear, 268
  servo, 267, 283, 302
multiloop systems, 46
multivariable systems, 52, 236

natural frequency, 30
neper, 108
network synthesis, 156
Nichols chart, 114, 143
  diagram, 69, 76
nodes, 57
noise, 195
nonlinear elements, 168
non-minimum phase systems, 57, 72
normal distribution, 198
Nyquist diagram, 69
  stability criterion, 102, 104

observability, 247
octave, 78
on–off control—*see* relay servo
open loop frequency response locus —*see* Nyquist diagram
  gain constant, 91
  transfer function, 45
operational amplifier, 147, 262
operational form, 40
order of a system, 71
output derivative feedback—*see* velocity feedback
over damping, 36
overshoot, 30, 119

P + D, 135
P + I, 137
P + D + I, 138
parallel compensation, 132, 149, 161
  T network, 305
Parseval's theorem, 200
particular integral, 36
performance index—*see* error criteria
piecewise linearity, 170
phase angle criterion, 100
  canonical, 246
  lag compensation, 144, 159
  lead compensation, 139, 158
  margin, 105
  plane, 186
  portrait, 188
  sensitive rectifier, 280
  variable—*see* state variable
pink noise, 208
polar plot—*see* Nyquist diagram
pole factor, 55
poles, 55
position control—*see* servo
position error coefficient, 122
  gain constant, 123
potentiometer error detector, 29
power density spectrum, 210
  measurement of, 215
probability density functions, 196
profile copying unit, 313
proportional error control, 132
pseudo random binary sequence (p.r.b.s.), 216
pulsed transfer function, 224

R.P.C. servo—*see* servo
ramp input, 37, 120, 245
random signal, 196
reference input, 10, 45
regeneration in d.c. motors, 277, 289
relay servos, 170, 178, 293
residues, 215
resolvers, 279
resonance, 81, 123
resonant frequency, 81, 123
response time—*see* settling time
rise time, 119

root loci, 94
  construction rules, 98
  of conditionally stable system, 108
  use in design, 156
rotating amplifiers, 285
Routh's array, 92
row matrix or vector, 233

s, complex variable, 324
s-plane, 57, 74, 89, 156
S.C.R., 287
sampled data, 217
sampling theorem, 223
second-order system, 33, 123, 326
second probability density function, 197
Selsyn, 279
sensitivity functions, 127
series compensation, 132, 135
servo (servomechanism), 12, 265
  continuous, 9
  motor, 267, 283, 302
  on–off, 9, 293
  relay, 170, 178, 293
  remote position control (r.p.c.), 29, 48, 132
set point, 229
settling time, 119
Shannon's theorem, 223
signal flow, 45
  diagrams, 57
simulation, 264
singular point, 187
sinusoidal input, 64, 245
small perturbations, 167
spectral density, 210
spool valve, 312
spring–mass–damper, 33
springy couplings, 274
stabilisation—*see* compensation
stability, 88
standard deviation, 198
state space, 237
  plane—*see* phase plane
  variable, 235
  vector differential equation, 236, 238
static friction, 31
stationary signal, 195

steady state error, 37
step input, 25, 36, 119, 245, 326
stiffness, 28, 129, 133
stochastic signals, 196
straight-line approximation—*see* asymptotes
sub-harmonic signal, 171
summing amplifier, 262
  point, 11
superposition integral, 201
  theorem, 52, 167
synchro, 279
synthesis, 156

tables of correction for Bode diagram, 79
  db-amplitude, 78
  Laplace transforms, 328
tachometer, 32, 299, 306
  feedback, 32, 48
terminology, 10
three-term controller, 138
thyratron, 287
thyristor, 287
  invertor, 307
time constant, 26, 39, 49
  delay, 168
  scaling, 264
torque constant, 26, 296
torsion, of shafts, 274
transconductance, 26, 39
transducers, 278
transfer functions, 40, 42
  closed loop, 45
  general form, 54
  matrices, 249
  of d.c. motor system, 48, 295
  of hydraulic system, 317
  of mechanical components, 275
  of networks, 54, 139, 141
  open loop, 45
  sampled, 224
transform impedance, 66
transformation of state vector, 246
transients, 37
transition matrix, 238
transmittance, 57
twin T network, 305
two-sided Laplace transform, 211
type number, 70

ultimately controlled variable, 10
uncontrollable mode, 71, 248
undamped natural frequency, 30
under-damping, 36
unit impulse response, 200
unstable system, 88

valve, spool, 312
   electrohydraulic, 314
Van der Pol's equation, 171
variable brakes, 292
   couplings, 290
variance, 198
velocity control system, 26
   error coefficient, 122
   feedback, 32, 48, 134, 152, 161
   gain constant, 123
   input, 37, 120, 245
   lag, 38, 120

Velodyne, 284
virtual earth amplifier, 263
viscous friction, 31
voltage regulator, 13, 18

Ward–Leonard system, 285
weighting function, 200
white noise, 208, 212

z-plane, 227
Z transfer functions, 224
Z transforms, 224
zeros, 55
zero factors, 55
zero order hold, 219
zeta—*see* damping ratio